Key Biodiversity Areas of Iraq

Priority Sites for Conservation & Protection

Library of Congress Cataloging-in-Publication Data:
     Key biodiversity areas of Iraq: priority sites for
              conservation & protection / Nature Iraq.
              pages cm
              LCCN 2015952207
              ISBN 978-0-9886514-6-3

              1. Biodiversity conservation—Iraq. 2. Natural areas
              —Iraq. 3. Natural resources surveys—Iraq.
              I. Nature Iraq (Nonprofit organization)

    QH77.I645K49 2017   333.95'1609567
         QBI15-1640

Recommended citation: Nature Iraq (2017). Key Biodiversity Areas of Iraq. Sulaimaniyah, Iraq: Tablet House Publishing, 327pp.

First published in 2017

Tablet House Publishing, 16835 Algonquin St #257, Huntington Beach CA 92649
www.tablethousepublishing.com

Printed in China through i Book Printing Ltd.

Key Biodiversity Areas of Iraq

Priority Sites for Conservation & Protection

NATURE IRAQ

TABLET HOUSE PUBLISHING

2017

This report has been prepared to summarize information on Key Biodiversity Areas throughout Iraq, which have been surveyed through the biodiversity project of the Iraqi Ministry of Health and Environment and Nature Iraq with support from the Italian Ministry of Environment, Land & Sea and the Canadian International Development Agency. For more information please contact:

Iraqi Ministry of Health & Environment
Department of Environment
Baghdad, Iraq
E-mail: moen_iraq@yahoo.com
Website: www.moen.gov.iq

Nature Iraq
Baghdad, Sulaymaniyah, Chibaish
Mailing address: PO 249, Sulaymaniyah, Iraq
Email: info@natureiraq.org
Website: www.natureiraq.org

Compilers of the Inventory and KBA Accounts: : N. Abdulhasan, S. Abdulrahman (S.A. Ahmad), O.F. Al-Sheikhly, K. Ararat, A. Bachmann, A.H. Kadhim, L.A. Al-Obeidi, M. Rahim, H.A. Raza, & M. A. Salim.

Edited by: A Bachmann, Azzam Alwash, & Ali Lami

Maps: A. Atwani

Design and Layout by: Rafique Suchwani and Susan Bard

Images: © To photographers listed & Nature Iraq

Images on front cover: Sakran Mountain (K. Ararat), Zubeidaat (M. Salim), & Mesopotamian Marshlands (M. Salim)

Images on rear cover: Desert Monitor (Salwan A. Abed), Wild Goats (Claude Balcaen), Marbled Teal (M. Salim), Basra Reed Warbler (M. Salim), *Fritillaria crassifolia* sp *hakkarensis* (S. Ahmed), Gazelle (M. Salim), *Tulipa kurdica* (S. Ahmed), Egypian Vulture (L.A. Al-Obeidi), & Spur-thighed Tortoise (K. Ararat)

TABLE OF CONTENTS

TABLE OF TABLES

TABLE OF FIGURES

MAPS

PREFACE

This book was compiled over a period of three years after several years of field work and at the final stages of publication, the New Key Biodiversity Area (KBA) Criteria was made public during the IUCN World Conservation Congress in Hawaii in early September 2016. Further in early July of 2016, UNESCO voted to include the marshes as a World Heritage Site ("The Ahwar of Southern Iraq: Refuge of Biodiversity and the Relict Landscape of the Mesopotamian Cities").

The Iraqi Ministry of Health and Environment and Nature Iraq are proud to have helped in making the marshes a cause for Iraq to celebrate and now, with their inclusion as a World Heritage Site, it is no doubt a symbol of the rebirth of this nation. We will treat this as a living document and will issue modified editions in the future that will re-evaluate the sites based on the new KBA criteria as well as additional field data to be collected by Nature Iraq and other like-minded groups.

Meanwhile we will continue to work, with like-minded NGOs and governmental agencies as appropriate, for the creation of protected areas as identified in this book and will work with conservation groups and the government agencies in Iraq to accomplish this goal.

ACKNOWLEDGMENTS

This publication would not have been possible without the efforts of many people throughout Iraq and the world, all united by a desire to see a network of protected sites throughout Iraq. It is impossible to fully list all the contributors to this effort but clearly the field team members who participated in the survey work from 2005 to 2012, often under difficult security conditions, deserve mention. The Key Biodiversity Areas (KBA) team was made up of staff from the former Iraqi Ministry of Environment (IMoE[1]) (including its regional offices in Southern Iraq, the Middle Euphrates, Central Iraq and the Kurdistan Environmental Protection & Improvement Commission (KEPIC) in northern Iraq) and staff from Nature Iraq (NI) as well as occasional members from the Universities of Baghdad (UoB), Basrah, Karbala (UoK) and Sulaymaniyah (UoS), and Blue Horizon Laboratories (BH):

Dilzar (KEPIC), Bestoon (Trainee), A.N. Abass (IMOE), A.S. Abed (IMoE), I.M. Abd (IMoE), H.S. Abd, M.E. Abd (IMoE), S.A. Abed (Trainee), N.J. Abdel Zahra, S. Abdulrahman (S.A. Ahmad) (NI), Q.N. Abid (IMOE), N. Abood (NI), H. Ahmad (IMOE), B. Ahmed (Uof Karbala), Z.Ch. Al-Asadi (NI), Dh.M. Al-Mandalawi (IMOE), L. A. Al-Obeidi (NI), S.A. Al-Razak (IMOE), M.A.T. Al-Saffar (NI), O.F. Al-Sheikhly (UoB), G.B. Al-Waili (IMoE), A.S. Al-Zubaidi, M.J. Al-Asadi (NI), R.A. Alewy (BH), B. Ali (IMoE), M. Ali (IMOE), S. Ali (NI), A.A. Alwan (UoBasrah), K. Ararat (NI), H.N. Asad (IMOE), O. Ashraf (NI), A.N. Awla (KEPIC), S. Babarasul (NI), A. Bachmann (NI), H. Barwary (Duhok Environmental Police), Dr. M.Kh. Haba (UofB), A. Haloob (IMoE), A. Hamad, A.M. Hassan (KEPIC), A. Hillawi (NI), S.H. Hussen (Trainee), H. Ibrahem (IMoE), H. Ismail (Uof Salahadin), M.H. Jalal (KEPIC), M. T. Jabbar (Trainee), K. Jawad (Trainee), S. Kadhim, A.Kh. Ala (BH), A. Kamel (IMOE), K. Faik (UoS), L.S. Khidr (KMoE), H. Mahdi (NI), A.M. Mahir (BH), A. Mohammad (IMOE), K.H. Mohammed (KEPIC), D.A. Mohammed (KEPIC), S.I. Mohammed (Uof Salahadin), N. Abdulhasan (NI), M. Qadir (KEPIC & NI), A. Qais, A.G. Radhi (NI), M.A. Rasul (KEPIC), H.A. Raza (NI), M.Sh. Minjal, Gh. Sabah (NI), S. Sadiq (Uof Duhok), M. Saeed (KEPIC), Z.H. Salih, B.A. Salih (KEPIC), M.A. Salim (NI), M. Salim (Trainee), A.N. Salman (IMoE), A.O. Salman (IMoE), Z. Salman (Trainee), H.A. Sfeyh (Trainee), H. Shibil, M.B. Slewa (KEPIC), and M. Talib (Trainee).

Core members of the team who conducted the data analysis and wrote the assessments in this Inventory included: Ahmed Atwani (NI GIS Expert), Dr. Saman Abdulrahman (UofS Plant Taxonomist & NI), Hana Ahmed Raza (NI Mammal expert, Kurdistan Region Team Leader from 2012-2013), Korsh Ararat (NI Ornithologist, Kurdistan Region Team Leader from 2008 to 2011), Laith Ali Al-Obeidi (NI Ornithologist), Mariwan Qadir Rahim (KEPIC Ornithologist), Mudhafar A. Salim (NI Ornithologist, KBA Program Manager (2005-2006), South Team Leader (2005-2012), Central Iraq Team Leader (Winter 2009)), Nabeel Abdulhasan (NI Botanist), and Omar Fadhil Al-Sheikhly (UoB Ornithologist, Central Iraq Team Leader from Summer 2009 to 2012). Anna Bachmann (NI) was the KBA Program Manager (2007 to present) and along with Richard Porter (Nature Iraq Ornithology Advisor) contributed text to the KBA Inventory of Iraq.

Other contributors to the data collection and analysis effort include Elnaz Najafi Majd (Ph.D candidate, Ege University, Turkey), Randle Rogers, Chiara Teatini (Studio Galli Ingegneria S.p.A. (SGI) and Jörg Freyhof (Leibniz-Institute of Freshwater Ecology and Inland Fisheries).

Expert review of and key advisors on the accounts covered in the inventory was done for Bird Assessments by Richard Porter (advisor for the NI Bird Team and the entire KBA Process, formerly with the Royal Society for the Protection of Birds); for Mammal Assessments by David Mallon (Manchester Metropolitan University & IUCN Species Survival Commission Member); for Herpetology assessments by Dr. Hanyeh Ghaffari (Director, Pars Herpetologists Institute, Tehran, Member, IUCN/SSC Tortoise and Freshwater Turtle Specialist Group) and Barbod Safaei Mahroo (Chairman, Pars Herpetologists Institute, Tehran); for fish assessments by Brian Coad (Canadian Museum of Nature, Ottawa) with additional assistance from Jörg Freyhof

---

1 In 2015, the Ministry was combined with the Health Ministry to form the Iraqi Ministry of Health & Environment. In the main chapters of this book, the Ministry will hereafter be refered to as the Ministry of Health & Environment (IMoHE).

(Leibniz-Institute of Freshwater Ecology and Inland Fisheries, Berlin), and for botanical assessments by Tony Miller and Sophie Neale (Centre for Middle Eastern Plants/Royal Botanic Garden Edinburgh). Additional advisors include Amirhossein Khaleghi (Plan for the Land, Tehran), Mia Fant (Studio Galli Ingegneria Spa (SGI), Alessandra Rossi (SGI), and Dr. Zuhair Amr (Jordan University of Science & Technology). Richard Porter deserves particular thanks for his dedication, perseverance and persistence throughout the many years of the program, all virtues that were tested to the limit. Nadheer Abood and Haider Ahmed Falih of Nature Iraq also worked tirelessly to provide coordination with the former IMoE (now the Ministry of Health and Environment).

Michael Evans, Lincoln Fishpool, Sharif Jbour and particularly Richard Porter of BirdLife International gave extensive guidance on the methodology and procedures for identifying Important Bird & Biodiversity Areas and Key Biodiversity Areas. Richard Porter, Simon Aspinall, Clayton Rubec, Michael Evans, Sharif Jbour, Tony Miller, Sophie Neale, Sabina Knees, Ihsan Al-Shehbaz, and Derek Scott helped in the training programmes for the survey teams coordinated by Richard Porter from 2005 to 2011. For the trainings undertaken in Syria and Jordan (2004-2008), we are grateful for the organization and support given by the Syrian Society for the Conservation of Wildlife and the Ministry of the Environment. We would also like to acknowledge the authors of the Kazakhstan Important Bird Areas book, which served as a model for the KBA Inventory of Iraq.

Funding for the training and fieldwork was generously provided by the Canadian International Development Agency from 2004-2006. From 2006 to 2010, support was provided by the Italian Ministry of Environment, Land and Sea (IMELS) and Nature Iraq. Additional support for the publication of the Inventory was provided by the Ford Motor Company Conservation Fund, the United Nations Development Program, the Ornithological Society of the Middle East, and the Royal Society for the Protection of Birds.

Partners in the program included the former Iraqi Ministry of Environment (now the IMoHE) and Nature Iraq (NI). Key thanks go to Minister Dr. Adela Hammoud and to former Ministers Sargon Lazar Slewa and Narmin Othman for their commitment to the program. Thanks also go to Dr. Ali Al-Lami for his support of the fieldwork and review of the KBA Inventory. The NI founder, Dr. Azzam Alwash, is also responsible for maintaining a long-term commitment to the KBA effort, which has been pivotal in its successful completion. Additional partners in the program include the Kurdistan Environmental Protection & Improvement Commission (KEPIC). Our thanks go to Samad Mouhammed, the current head of the Board, and to the former head Dr. Rezan Hasan Mawlood and former Minister Dara Ameen. Additionally, a key partner and source of critical technical and moral support has been BirdLife International in the United Kingdom.

Finally additional editors, advisors and translators have been important in making this book organized, clear and readable. We extend our gratitude to Dr. Sinan Abood, Nadheer Aboud, Hana Ahmed, Samad Ali, Dr. Azzam Alwash, Anna Bittman, Penny Butler, Mike Evans, Dr. Lincoln Fishpool, Adrian Long, Dr. David Mallon, Richard Porter, Virginia Tice, and James Wudel.

Additional local partners have assisted the program logistically and with information about the survey sites. These have included the NI Administration and Logistics Team as well as the guides, herders, fisherman and hunters who have helped the team. For those who provided critical support and assistance who are not mentioned here we express our sincerest apologies but they are include in our thanks as well.

ACRONYMS

CBD	Convention on Biological Diversity
CIDA	Canadian International Development Agency
CIMI	Canada-Iraq Marshlands Initiative
GEF	Global Environment Facility
IBA	Important Bird & Biodiversity Areas
IMELS	Italian Ministry of Environment, Land & Sea
IMoE	Iraq Ministry of Environment (now Iraq Ministry of Health & Environment)
IMoHE	Iraq Ministry of Health & Environment (formerly the Iraq Ministry of Environment)
IPA	Important Plant Areas
IOCN	Iraqi Organization for Conservation of Nature
IUCN	International Union for the Conservation of Nature
ITPGRFA	International Treaty on Plant Genetic Resources for Food and Agriculture
IMoWR	Iraq Ministry of Water Resources
KBA	Key Biodiversity Areas
KEPI	Kurdistan Region Environmental Protection and Improvement Commission (or Board)
MoHE	Ministry of Higher Education
NBSAP	National Biodiversity Strategy and Action Plan
NESAP	National Environmental Strategy and Action Plan
NI	Nature Iraq
PoWPA	Programme of Work for Protected Areas
CMEP/RBGE	Centre for Middle Eastern Plants/Royal Botanic Gardens Edinburgh
ROWA	Regional Office for West Asia
UNEP	United National Environmental Programme
USAID	United States Agency for International Development
WWF	World Wide Fund for Nature

© K ARARAT/NI

برنامج مناطق التنوع الأحيائي الرئيسة

هو مشروع قامت به منظمة طبيعة العراق بالتعاون مع وزارة البيئة العراقية وجهات أخرى بهدف تحديد وتوثيق وحماية مجموعة من المواقع المهمة بيئياً في العراق على أساس تنوعها البيولوجي. بدأ المشروع عام ٢٠٠٤ وشمل أبحاث موقعية ورصد وحملات توعية ومناصرة ومبادرات وخطط إدارة وحماية قانونية على الصعيدين الوطني والعالمي.

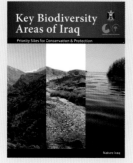

معطيات لترتيب الاولويات وصنع القرار

يلعب جمع وتحليل المعطيات العلمية دوراً مهما للغاية في إعداد حماية وإدارة مناطق التنوع الأحيائي. إن الهدف من المطيات المقدمة هنا إرشاد الإدارة التطبيقية لهذه المناطق واستهداف الآليات السياسية والقانونية لتحقيق حماية مناسبة لهذه المناطق المهمة محلياً ووطنياً وعالمياً حيث أن تلك المعطيات مفيدة جداً للعلماء والجهات الحكومية والمنظمات ومتخصصي الحماية لتحقيق هذه الغايات.

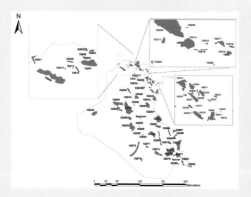

شبكة مواقع مهمة عالمياً

إن الأهمية العالمية لمجموعة المواقع المعينة هنا أمر مفروغ منه نظراً للتطبيق الدقيق للمعايير العالمية الخاصة بالحياة الفطرية. إن كل موقع من هذه المواقع يطابق المعايير العالمية الكمية في الغالب مما يسهل المقارنة بين المواقع على الأصعدة المحلية الوطنية والاقليمية. مثال على ذلك تتطابق تلك المعايير مع تلك المستخدمة لتعيين مسطح مائي ذو أهمية عالمية بموجب إتفاقية رامسار.

© 2010 A BACHMANN/NI

إسهامات شبكة من الافراد على الصعيد الاقليمي

أتت هذه الإحصائية من إسهامات العديد من خبراء الطيور والنبات واللبائن والأسماك والحماية وغيرهم من العراق والبلدان المجاورة وأوروبا وكندا. فقد إشترك أكثر من ٥٠ شخصاً بشكل مباشر في جمع المعطيات وإجراء الإستطلاعات الميدانية الحيوانية والنباتية منذ ٢٠٠٥ بمساعدة كوادر المجلس العالمي لحماية الطيور ومركز الشرق الأوسط للحديقة الملكية البريطانية والوكالة الكندية للتنمية الدولية ووزارة البيئة الايطالية.

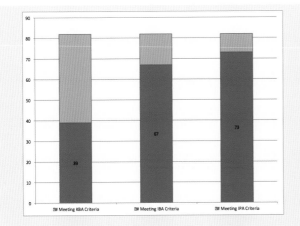

تحديد ما مجمله ٩٣ منطقة تنوع احيائي رئيسة (وفق المعايير التي لا تشمل الطيور) و٧٦ منطقة مهمة للطيور والتنوع البيولوجي، و٣٧ منطقة مهمة للنباتات؛ تغطي كلها ٣٫٤٪ و٣٫٦٪ و٦٫٥٪ من مساحة العراق على التوالي

لقد تم تعيين شبكة تتكون من ٢٨ موقعاً إستنادا الى مختلف المعايير العالمية لتغطي مساحة ٨٨٣٫٨٢ كم٢ (٥٫٦٪) من مساحة العراق. ان الحماية والإدارة الكفوءة لهذه المواقع التي تشكل نسبة قليلة من مساحة البلد هو هدف معقول سيسهم جدياً في حماية الكثير من الأنواع الحيوانية والنباتية.

©L.A. AL-OBEIDI

المواقع المحددة حسب المعايير المختلفة

تم إنتخاب 28 موقعا في العراق إستنادا الى ثلاث مجموعات مختلفة من المعايير:

المواقع المحددة وفق معايير مناطق التنوع الحيوانية من غير الطيور

وقد أختير ٩٣ موقعا إستناداً الى تواجد أنواع مهمة صونياً، منها ١٢ موقعا تطابق معايير اللاتعويضية لتواجد نوع محصور نطاقيا: sucitarhpuesutefaR.

© RF PORTER

المواقع المحددة وفق معايير المناطق المهمة للطيور

تم تحديد ٢٨٪ من المناطق المهمة للطيور والتنوع وفق وجود ٩ أنواع طيور مهددة بالإنقراض (١ حرج الوضع، ٤ مهددة، ٤ معرضة للتهديد)، وأختير ٢٤٪ من المناطق المهمة للطيور لتميزها بتجمعات لأنواع محصورة النطاق. وان ٦٢ من هذه المواقع (٩٣٪) ملائمة لطيور الماء واليابسة المتجمعة فيها بأعداد كبيرة.

© S. ABDULRAHMAN

المناطق المحددة وفق معايير المناطق والموائل المهمة للنبات

تم إنتخاب ٥٥٪ من المناطق المهمة للنبات وفق وجود أنواع نباتية متوطنة وشبه متوطنة ونادرة وطنياً. يملك العراق ٢٨١ نوع نباتي مهم صونياً (٢٦ متوطن، ٤٦ شبه متوطن، ٦٥ نادر وطنياً)، منها تحددت وفق معلومات تاريخية، وهناك ٧٦ (٢٩٪) موقعا محددة وفق الموائل الغنية بالأنواع حيث ١١ موقعاً منها (٥١٪) تعد مواقع بقاء أساسية و٧١ (٣٢٪) محددة وفق وجود ٣ موائل مهددة بالزوال (غابة صنوبر، غابة نهرية، والأهوار الوسطى).

© RF PORTER

مواقع تشكل شبكة لحماية أنواع متجمعة

يعرف المجلس العالمي لحماية الطيور ٥ مناطق حياتية جغرافية رئيسة في العراق وقد أختير ٢٨ موقعا لإسنادها معظم الأنواع المسجلة وطنياً مرتبطة بهذه المواقع وهذا كان أسهل بالنسبة للطيور. توفر مناطق التنوع العراقية موائل لحوالي ٦٦ نوع من الطيور (٣٧٪ من مجموع الأنواع المسجلة في الشرق الاوسط)، منها ٣٤ نوعا متواجدة دوريا ٥ منها شاردة أو متواجدة لسبب غير معلوم: *(atallicipsnoc aivlyS, agypocuelehtnaneO, iinietsnethcilselcoretP, alahpeconalem aivlyS specinurbazirebmE)*

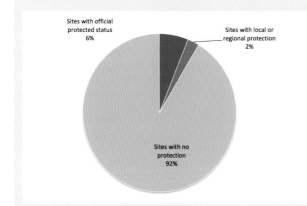

تتطلب مواقع التنوع حماية بموجب القوانين الوطنية والعالمية

لـدى بلـدان كثـيرة في العـالم قوانـين ومؤسسـات معنيـة بالمحميات. لا بـد مـن أن تكـون منطقـة التنوع محميـة وفق القانـون الوطني المناسب. هناك موقعـان مـن أصل ٢٨ موقعـا للتنوع في العـراق تتمتـع منهـا خمسـة (١,٦٪) بنـوع مـن الحمايـة القانونيـة وموقعـين (٤,٢٪) محميـان بقانـون محـلي ليبقـى ٥٧ موقعـا (٥,١٩٪) غـير محمـي حاليـاً. ورغم الأهميـة العالميـة لهـذه المناطـق فـان خمسـة مواقـع فقط (هـور الحويـزة، بحـيرة سـاوة، الأهـوار الوسـطى، هـور الحـمار الشرقي والغربي) إسـتفادت مـن الاتفاقيـات الدوليـة (إتفاقيـة رامسـار لـلأراضي الرطبـة المهمـة عالميـاً وإتفاقيـة الـتراث العالمـي).

© M SALIM/NI

مناطق التنوع في سياق البيئة الأوسع

ان الكثـير مـن المشـاكل والمهـددات البيئيـة هـي إقليميـة أو عالميـة مـن حيـث النطـاق أو المصـدر ولا يمكـن معالجتهـا بإنفـراد بمجـرد حمايـة وإدارة مناطـق التنوع فقط. تعيـش أنـواع كثـيرة وتعشـش باعـداد قليلـة بشـكل متفـرق وإسـلوب إنفـرادي وبالتـالي لا يمكـن حمايـة أعدادهـا باسـلوب حمايـة الموقـع وحـده. مـن المهـم السـعي لحمايـة البيئـة ككل بالتـوازي مـع حمايـة مواقـع التنوع مـن خـلال غايـات بيئيـة توضـع في السياسـات القطاعية.

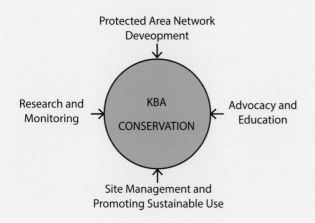

الإجراءات المطلوبة

بالنظـر للضغـوط الهائلـة التـي تتعـرض لهـا البيئـة في العـراق والعـالم ككل، تشـكل مناطـق التنوع العنصـر الأسـاس لتقويـة إسـتراتيجية الحمايـة. لقـد تـم تحديـد مجموعـة مـن الغايـات والنشـاطات الإسـتراتيجية المستحسنة وفق ٤ مبـادرات أو مواضيـع: البحـث والرصـد، إنشـاء المحميات، إدارة المواقـع والترويـج للإسـتخدام المسـؤول للمـوارد الطبيعيـة، المناصـرة والتثقيـف. تعـبر الغايـات المستحسـنة السـبع نتائـج الجهـود المضنيـة لتحقيـق حمايـة التنوع البيولوجـي في العـراق وفـق هـذه المبـادرات. وإن النشـاطات الإسـتراتيجية التسـع عشرة المحـددة في التقريـر للوصـول لتلـك الغايـات وضعـت لتكـون مباشرة وتحقـق أهدافهـا وقابلـة للتطبيـق.

برنامج مناطق التنوع الاحيائي الرئيسة هو مشروع لوزارة البيئة العراقية ومنظمة طبيعة العراق.

مەبەست لە پرۆگرامی ناوچە گرنگەکانی جۆراوجۆری زیندەوەری چیە؟

پرۆگرامی ناوچه گرنگهکانی جۆراوجۆری زیندهوهری (KBA)، دهستپێشخهرییهک بوو لهلایهن وهزارهتی تهندروستی و ژینگهی عیّراق و ریکخراوی سروشتی عیّراقهوه، ئامانج لیّی دیاریکردن، به دۆکیومێنت کردن، وه پاریّزگاری کردن لهو ناوچانهی که گرنگن بۆ پاریّزگاریکردن له جۆراوجۆری زیندهوهری. ئهم پرۆژهیه له سالّی ٢٠٠٤ دهستی پیکرد، که پیکهاتبوو له توّیژینهوهی تایبهت به ناوچهکان، چاودیّریکردن، پهروهردهو فیّرکردن، بانگهشه، چالاکی، بهرپوهبردن، وه پاریّزگاری یاسایی لهسهر ئاستی ناوخۆیی و جیهانی.

کۆکردنهوهی داتا بۆ دیاریکردن و بریاردان له گرنگی ناوچهکان

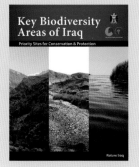

کۆکردنهوهی داتا زانستییهکان و شیکردنهوهیان، رۆلیکی گرنگ و کاریگهر دهبینن له دانانی بناغهیهکی پتهو بۆ پاریّزگاریکردن و بهرپوهبردنی ناوچه گرنگهکانی جۆراوجۆری زیندهوهری. ئهم داتایانهی که لهم پهرتووکهدا خراونهته بهردهست، سوودهبهخشه بۆ شارهزایانی بواری زانستی بالّندهناسی، رووهکناسی، شیرهدهرهکان، خشۆک و وشکاوهکییهکان، دامهزراوه حکومییهکان، ریکخراوهکانی کۆمهلّگهی مهدهنی، وه شارهزایانی بواری پاریّزگاریکردنی ژینگه. ئهم زانیاریانهته خراونهته بهردهست گرنگن له رابهریکردنی بهرپوهبردن و کارکردنی پراکتیکی له ناوچه گرنگهکانی جۆراوجۆری زیندهوهری، که کاریان پیّدهکریّت وهک میکانیزمیکی سیاسی و یاسایی لهپیّناوی هیّناوهی پاراستنی ئهم ناوچانهی که گرنگن لهسهر ئاستی ناوخۆیی و نهتهوهیی و جیهانی.

پیّکهیّنانی تۆریّک لهو ناوچانهی که گرنگن لهسهر ئاستی جیهانی

به بهکارهیّنانی چهند پیّوهریّکی وردی جیهانی له دیاریکردن و ناسینهوهی بالّنده و رووهک، که لهم پهرتووکهدا پشتیان پیّبهستراوه، پیّکهیّنانی تۆریّک لهو ناوچانهی که گرنگن لهسهر ئاستی جیهانی سنوّرکراوه.

ههر ناوچهیهک لهو ناوچانهی که لهم پهرتووکهدا دیاریکراون، یهکیک یان زیاتر لهو پیّوهرانه دهیانگریّتهوه که گرنگن بۆ بریاردان له دیاریکردنی ئهم ناوچانه، وه بهکارهیّنانی ستانداردی جیهانی له پیّوهرهکاندا هاوکاره له بهراوردکردن و دیاریکردنی ناوچهکان لهسهر ئاستی ناوخۆیی و نهتهوهیی و ناوچهیی.

بۆ نمونه، ئهم پیّوهرانه یهکدهگرنهوه لهگهلّ ئهو پیّوهرانهی که بهکاردههیّنریّن له دیاریکردنی ناوچه شیداره گرنگهکان لهسهر ئاستی جیهانی بهپیّی یاساکانی پهیماننامهی راسار.

به بهشداریکردنی تۆریّک له خهلّک له سهرانسهری ناوچهکه

ئهم پهرتووکه له ئهنجامی بهشداریکردنی ژمارهیهک له شارهزایانی بواره جیاوازهکانی بالّنده، رووهک، شیرهدهرهکان، وشکاوهکی و خشۆکهکان، ماسی، شارهزایانی بواری پاریّزگاریکردنی ژینگه، وه شارهزایانی تر له عیّراق و ولّاتانی دهوروبهری و ئهوروپا و کهنهدا، هاتوّته بهرههم، زیاتر له ٥٠ کهس بهشیّوهیهکی راستهوخۆ بهشداربوون له کۆکردنهوهی داتا و جیّبهجیّکردنی کاری مهیدانی بۆ رووهک و گیانهوهرهکان له سالّی ٢٠٠٥ و به هاوکاری ستاف له ریّکخراوی جیهانی ژیانی بالّنده (BirdLife International) و سهنتهری رووهکی رۆژههلّاتی ناوهراست له باخچهی شاههانهی ئیدنبره Botanic Garden Edinburgh (CMEP/RBGE). دامهزراوهی نیّودهولّهتی گهشهپیّدانی کهنهدی The Canadian International Development Agency، وه وهزارهتی ژینگهی ئیتالی زهوی و دهریا .(The Italian Ministry of Environment Land & Sea (IMELS. ئهم کتیبه هاوکاری لهلایهن چهند سپۆنسهریکهوه پیّگهشتهوه، لهسهر ئاستی بچووکتهرو پیوهست به پرینت کردن و بلّاوکردنهوهی.

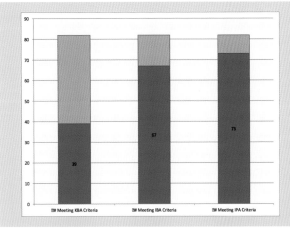

لەکۆی گشتی؛ ٣٩ ناوچە گرنگن بەپێی پێوەر بۆ گیانەوەرە کێوییەکان جگە لە باڵندە، ٦٧ ناوچە گرنگن بۆ باڵندە، وە ٧٣ ناوچە گرنگن بۆ رووەک، کە ٤.٣٪، ٦.٣٪، وە ٥.٦٪ ی هەموو عێراق دادەپۆشن، بەدوای یەکدا.

بەگوێرەی ئەم پێوەرە جیهانیانە، تۆزێک لە ٨٢ ناوچە دیاریکراون، کە نزیکەی (٦.٥٪) ٢٨٬٣٨٨ کیلۆمەتر دووجای عێراق دەگرێتەوە.

پارێزگاریکردن و بەرێوەبردنی کاریگەری ئەم ناوچانە، لەسەر روبەرێکی کەمی وڵاتەکە، دەکرێت بکرێتە ئامانجێکی راستەقینە کە بایەخێکی گرنگیان دەبێت لە پارێزگاریکردنی ژمارەیەکی زۆر لە باڵندەو گیانەوەر و رووەک لە عێراقدا.

©L.A. AL-OBEIDI

ئەو ناوچانەی لەسەر چەند پێوەرێکی جیاواز دیاریکراون

هەشتاو دوو ناوچە لە سەرتاسەری عێراقدا دیاریکراون، بەپێی سی سیستمی پێوەری جیاواز.

ئەو ناوچانە کە لەسەر پێوەری تایبەت بە گیانەوەرە کێوییەکان، جگە لە باڵندە، دیاریکراون

سی و نۆ ناوچە هەڵبژێردراون بەپێی پێوەری دەگمەنێتی لەسەر هەبوونی گیانەوەرێکی کێوی لە ناوچەکدا کە مەترسی لەناوچوونی لەسەرە. لە کۆی ئەم ٣٩ ناوچە دیاریکراون بەپێی پێوەری هاوشێوەنەبوون (irreplaceability)، لەسەر هەبوونی جۆرێک لە کیسەڵ (Rafetus euphraticus) کە لە وێنەکەدا دیارە، تایبەتمەندە بە ناوچەکە.

© RF PORTER

ئەو ناوچانەی دیاریکراون لەسەر پێوەری ناوچە گرنگەکان بۆ باڵندە

لە سەدا هەشتاو دووی ناوچە گرنگەکان بۆ باڵندەو جۆراوجۆری زیندەوەری، لەسەر بنچینەی هەبوونی ٩ جۆر باڵندە کە مەترسی لەناوچوونیان لەسەرە لە عێراقدا (١ جۆر مەترسی زۆری لەسەرە بۆ لەناوچوون، ٤ جۆر مەترسی لەسەرە بۆ لەناوچوون، ٤ جۆر بەرەورووی مەترسی لەناوچوون دەبنەوە لە داهاتوودا)؛ ٤٢٪ ی ناوچە گرنگەکان بۆ باڵندەو جۆراوجۆری زیندەوەری (IBA) لەسەر بنچینەی هەبوونی کۆمەڵێک لەو جۆرە هەڵبژێردراون کە تایبەتمەندن بە ناوچەیەکی ژینگەیی لەو ناوچەیەدا دەژین.

بیست و شەش ناوچەی گرنگ بۆ باڵندەو جۆراوجۆری زیندەوەری (IBA) (٣٩٪)، شیاون بۆ باڵندەی ئاوی و وشکانی کە بە ژمارەیەکی بەرچاو لەو ناوچاندا کۆدەبنەوە.

© S. ABDULRAHMAN

ئەو ناوچانەی دیاریکراون لەسەر پێوەری ناوچە گرنگەکان بۆ رووەک و نشینگە ژینگەییەکان (habitat)

لە سەدا پەنجا و پێنجی ناوچە گرنگەکان بۆ رووەک، هەڵبژێردراون لەسەر بنچینەی لەخۆگرتنی جۆری رووەکی نیشتیمانی و رووەکی دەگمەن لەسەر ئاستی نەتەوەیی.

عێراق ١٨٠ جۆر رووەکی تێدایە کە مەترسی لەناوچوونیان لەسەرە [٥٨ رووەکی نیشتیمانی (تایبەت بە عێراق)، ٦٦ تایبەت بە عێراق (بەڵام بوونیشی هەیە لە وڵاتانی دراوسی بە رێژەیەکی دیاریکراو)، وە ٥٦ دەگمەن لەسەر ئاستی نەتەوەیی].

لە ناوچە گرنگەکان بۆ رووەک، زۆریکیان دیاریکراون لەسەر بنچینەی زانیاری مێژووییی. وە هەروەها ٦٧ ناوچە (٩٢٪) دیاریکراون لەسەر بنچینەی نشینگەی ژینگەی دەوڵەمەند بە جۆری جیاوازی رووەک، کە ١١ ناوچیان (١٥٪) وەک پەناگابەک گرنگ دادەنرێن بۆ پارێزگاری کردن لە رووەک، وە ١٧ ناوچەش (٢٣٪) دیاریکراون لەسەر بنچینەی سی نشینگەی ژینگەیی کە مەترسی لەناوچوونیان لەسەرە، ئەوانیش بریتین لە (دارستانی سنەوبەر، دارستانی قەراغ چەم و روبارەکان، وە هۆرەکانی مێزۆپۆتامیا).

پێکهێنانی تۆڕێک له ناوچهی گرنگ بۆ پاریزگاریکردن له کۆبوونهوهی جۆره جیاوازهکانی بائنده

© RF PORTER

ڕێکخراوی جیهانی ژیانی بائنده، به پێی توانا ڕاڤهی پێنج نشینگهی ژینگهیی (یان ناوچهی جوگرافیی زیندهیی) کردووه بۆ عێراق، ئهو ٨٢ ناوچهیه دیاریکراون بۆ عێراق، لهسهر ئهو بنهمایه هەڵبژێردراون که هەمووییان به یهکهوه، پاڵپشتی زۆرترینی ئهو جۆرانه دهکهن که لهسهر ئاستی نهتهوهیی تۆمارکراون و تایبهتمهندن بهم نشینگه ژینگهییانهن.

دیاریکردنی ئهم نشینگه ژینگهییانهن زیاتر بۆ بائندهکان ئاسانتره وهک له گیانهوهر و رووهک. ناوچه گرنگهکانی جۆراوجۆری زیندهوهری له عێراق، نشینگهی ژینگهیی بۆ ٦٦ جۆر له بائندهی تایبهتمهند دابین دهکهن (که دهکاته ٧٣٪ ی ئهو بائندانه تایبهتمهندن به نشینگهیهکی ژینگهیهوه به گشتی له رۆژهەڵاتی ناومراستدا تۆمارکراون). له کۆی ئهم ٦٦ جۆره، ٤٣ جۆریان به هەمیشهیی له عێراقدا روودهدهن و پێنج جۆریان

(Sylvia conspicillata, Sylvia melanocephala, Pterocles lichtensteinii, Oenanthe leucopyga, & Emberiza bruniceps) یان پهرتهوازهن یاخود بوونیان ئاشکرا نیه.

ناوچه گرنگهکانی جۆراوجۆری زیندهوهری پێویستیان به پاریزگاری لێکردنه لهرێگای یاسا نهتهوهیی و جیهانییهکانهوه.

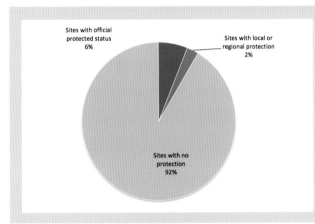

Sites with official protected status 6%

Sites with local or regional protection 2%

Sites with no protection 92%

لهسهر ئاستی جیهانی، زۆربهی وڵاتان چوارچێوهیهکی دهستووری و یاساییان هەیه بۆ دیاریکردن و پاریزگاریکردنی ناوچه پاریزراوهکان. ناوچه گرنگهکانی جۆراوجۆری زیندهوهری پێویسته وهک ناوچهی پاریزراو دیباری بکرێن له چوارچێوهی یاسای نهتهوهیی و بهگوێرهی پێویست. لهو ٢٨ ناوچهیه دیاریکراون بۆ عێراق، له ئێستادا تهنیا ٥ ناوچه (٦٫١٪) به تهواوی یاخود بهشێکی کهمی لهژێر یاسای نهتهوهییدا پاریزراون، وه ٢ ناوچهی دیکه (٢٫٤٪) لهژێر یاسای ناوخۆیی و هەریمی پاریزراون، که بهمش ٥٧ ناوچه (٩١٫٥٪) له ئێستادا نهپاریزراون. سەرهڕای ئهم ناوچانهی لهم پهرتووکهدا دیاریکراون لهسهر ئاستی جیهانی، له ئێستادا تهنیا پێنج ناوچه (هەزرهکانی حەوێزه، دهریاچهی ساوا، هەزرهکانی ناومراست، وه هەزرهکانی حەمار (رۆژهەڵات و رۆژئاوا)) سوودمەند بوون لهوهی که لهژێر پهیماننامهی ژینگهیی نێودهوڵهتیدا (پهیماننامهی رامسار بۆ ناوچه شیدارهکان و پهیماننامهی شوێنهواره جیهانییهکان) دانیان پێدانراوه وهک ناوچهی پاریزراو.

ناوچه گرنگهکانی جۆراوجۆری زیندهوهری له چوارچێوهی ژینگهیی فراواندا

© M SALIM/NI

زۆرێک له کێشه ژینگهییهکان و مهترسییهکان بۆ سهر جۆراوجۆری زیندهوهری که بهردهوام له زیادبوودان، کێشه و مهترسی ناوخۆیی و جیهانین له مەوداو سەرهەڵدانیاندا، وه ناتوانرێت به تهنیا لهرێگهی بهرپوبردنی کاراکی ناوچه گرنگهکانی جۆراوجۆری زیندهوهری چارهسهر بکرێن. زۆربهی جۆری زیندهوهران بهشێوویهکی پهرش و بڵاو به چری کهم دهژین و زۆردهبن، وه کۆمەڵگا نشینگهییهکانیان بهشێوویهکی سهرکهوتوو ناتوانرێت تهنیا لهرێگهی پاراستنی ناوچهکانی نیشتهجێبوونهوه بپاریزرێت، بۆیه زۆر گرنگه که شوێنکهوتهی پاریزگاریکردنی ژینگه بهرفراوانهکهیان بین له چوارچێوهی پاراستنی ناوچه گرنگهکانی جۆراوجۆری زیندهوهری و لهرێگهی یهکخستنی ئامانجه ژینگهییهکان له هەمرو جوگهکانی بریاریداندا.

ئهو رێکارانهی که پێویسته بگیرێنه بهر

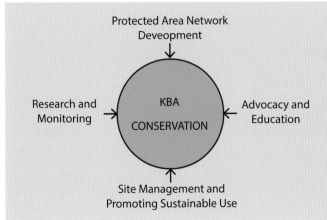

Protected Area Network Deveopment

Research and Monitoring

KBA CONSERVATION

Advocacy and Education

Site Management and Promoting Sustainable Use

بههۆی ئهو پهستانه گهورهیه که خراوته سهر ژینگه له سهرتاسهری عێراق و له جیهاندا به گشتی، ناوچه گرنگهکانی جۆراوجۆری زیندهوهری توخمێکی سەرهکیبه له دانانی ستراتیجێکی بههێز بۆ پاریزگاریکردن له ژینگ. زنجیرهیهک له راسپاردهی ئامانج و رێکاری ستراتیجی لهژێر چوار دهستپێشخهری پان بابهتدا دیاریکراون: توێژینهوهی زانستی و چاودێری کردن؛ دروستکردنی ناوچهی پاریزراو؛ بهرپوبردنی ناوچه و هاندانی بهکارهێنانی هۆشیارانهی سهرچاوه سروشتییهکان؛ وه بانگەشه و پهروهرده کردن. ئهو حهوت ئامانجهی که له چوارچێوهی ئهم بابهتاندا پێشنیار کراون وینای ئهو هەوڵه دووڵایهنیه دهکهن که لهپیناو پاریزگاریکردن له ژینگهی عێراق کیراونەبهر. ئهو ١٩ پلانه ستراتیجییهی که پێشنیارکراون به مەبهستی گهشتن بهم ئامانجانه، زۆر به شێوویهکی روون و ئاشکرا و کرداری نهخشهیان بۆ داریژراوه.

SUMMARY

WHAT IS THE KEY BIODIVERSITY AREAS (KBA) PROGRAM?

The Key Biodiversity Areas (KBA) Program is an initiative of the Iraqi Ministry of Health and Environment and Nature Iraq aimed at identifying, documenting and protecting a network of sites critical for the conservation of Iraq's biodiversity. The KBA Program, started in 2004, includes site-orientated research and monitoring, education, advocacy, action, management, and national and international legal protection.

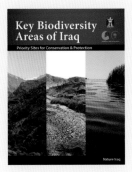

DATA FOR PRIORITY-SETTING AND DECISION-MAKING

Scientific data collection and analysis play a highly influential role in underpinning the conservation and management of Key Biodiversity Areas (KBAs). Useful for ornithologists, botanists, mammalogists, herpetologists, governmental agencies, civil society organizations, and a host of other conservation professionals, the data presented here are intended to guide practical management and action at KBAs and to target political and legal mechanisms to achieve the adequate protection of these locally, nationally, and globally important areas.

A NETWORK OF GLOBALLY IMPORTANT SITES

Because of a rigorous application of internationally agreed ornithological, non-avian fauna and botanical criteria, the global importance of the network of sites identified in this document is assured. Each site meets or exceeds the required, often quantitative, selection thresholds and the global standardisation of the criteria facilitates comparison between sites at local, national and regional levels. For example, these criteria are compatible with those used to designate wetlands of international importance under the Ramsar Convention.

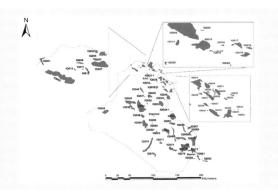

CONTRIBUTIONS BY A NETWORK OF PEOPLE ACROSS THE REGION

This inventory results from the contributions of several ornithologists, botanists, mammal experts, herpetologists, fish experts, conservation experts and other specialists from Iraq, neighbouring countries, Europe and Canada. More than 50 people have been directly involved in the collection of data and carrying out field surveys of flora and fauna since 2005, assisted by staff of BirdLife International, the Centre for Middle Eastern Plants/ Royal Botanic Garden Edinburgh (CMEP/RBGE), the Canadian International Development Agency and the Italian Ministry of Environment, Land & Sea (IMELS). Smaller contributions were made by foundations to sponsor various aspects related to the printing of the book.

A TOTAL OF 39 KBAS (BASED ON NON-AVIAN CRITERIA), 67 IMPORTANT BIRD & BIODIVERSITY AREAS (IBAS), AND 73 IMPORTANT PLANT AREAS (IPAS) WERE IDENTIFIED COVERING 4.3%, 6.3%, AND 5.6% OF IRAQ RESPECTIVELY

A network of 82 sites based on these different international criteria has been identified covering 28,388 km² (6.5%) of Iraq. The effective protection and management of these sites over a relatively low percentage of the area of the country is a realistic goal that would make a significant contribution to the conservation of many bird species and other fauna and flora in Iraq.

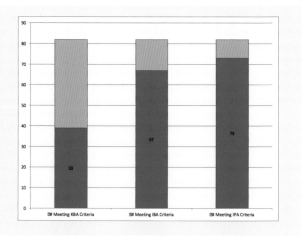

SITES IDENTIFIED UNDER DIFFERENT CRITERIA

Eighty-two sites have been selected in Iraq based on three different criteria systems.

SITES IDENTIFIED ON KBA CRITERIA FOR NON-AVIAN FAUNA

Thirty-nine KBAs (selected for non-avian fauna) have been chosen for vulnerability criterion based on the presence of certain conservation concern species. Out of these 39 sites, 21 meet an irreplaceability criterion for the presence of the restricted range species *Rafetus euphraticus*.

©L.A. AL-OBEIDI

SITES IDENTIFIED ON IBA CRITERIA FOR BIRDS

Eight-two percent of the Important Bird & Biodiversity Areas (IBAs) were identifed based on the presence of 9 of Iraq's threatened birds species (1 Critically Endangered, 4 Endangered, 4 Vulnerable); 42% of the IBAs in Iraq have been chosen for the assemblages of biome-restricted species that occur within them. Twenty-six of the IBA sites (39%) qualify for the waterbirds and/or terrestrial species that congregate in them in significantly large numbers.

© RF PORTER

SITES IDENTIFIED ON IPA CRITERIA FOR PLANTS AND HABITATS

Fifty-five percent of the Important Plant Areas (IPAs) have been chosen based on the presence of national endemic, near endemic, or nationally rare species they hold. Iraq has 180 conservation concern plants (58 endemics, 66 near endemics, & 56 nationally rare species). Of these IPAs, many were identified based on historical information. In addition, 67 sites (92%) were identified based on species rich habitats with 11 (15%) of these sites also considered key refuge sites, and 17 sites (23%) identified based on three threatened habitats (Pine Forest, Riverine Forest of the Plains (Al-Ahrash) and the Mesopotamian Marshlands).

© S. ABDULRAHMAN

SITES FORMING NETWORKS TO PROTECT SPECIES ASSEMBLAGES

BirdLife International defines five key biomes (or biogeographic areas) in Iraq and as far as possible, the 82 sites chosen in Iraq have been selected so that, taken together, they support the majority of the nationally recorded species associated with these biomes. This has been easiest to determine for bird species. Iraq KBA sites provide habitat for 66 characteristic bird species (73% of the total number of biome-restricted species recorded in the Middle East region as a whole). A total of 43 of these species occur regularly in Iraq with five species (*Sylvia conspicillata, Sylvia melanocephala, Pterocles lichtensteinii, Oenanthe leucopyga,* & *Emberiza bruniceps*) being either vagrants or of uncertain occurrence.

© RF PORTER

KBAS REQUIRE PROTECTION UNDER NATIONAL AND INTERNATIONAL LAW

Internationally many countries have legal and institutional frameworks for the designation and conservation of protected areas. KBAs should be designated as protected areas under national law where appropriate. Of the 82 KBAs identified in Iraq to date, currently five (6.1%) are wholly or partly under some form of protection by national law, an addition two (2.4%) have local or regional protection but this leaves 75 sites (91.5%) currently unprotected. Despite the global importance of the KBAs identified in this publication, at this time only five sites (Hawizeh Marshes, Sawa Lake, Central Marshes, & Hammar Marshes (East and West)) currently benefit from recognition under ratified global environmental agreements (the Ramsar Wetland and World Heritage Conventions).

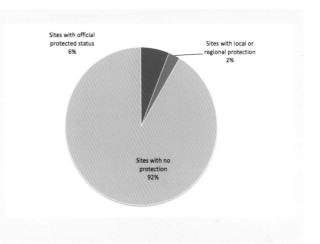

Sites with official protected status 6%

Sites with local or regional protection 2%

Sites with no protection 92%

KBAS IN THE CONTEXT OF THE WIDER ENVIRONMENT

Increasingly, many environmental problems and threats to biodiversity are regional or global in scope or origin, and cannot be addressed solely through the effective protection and management of KBAs alone. Many species live and breed at low densities in a dispersed, non-congregatory manner, and their populations therefore cannot be conserved successfully by a site-based approach alone. It is vital, therefore, that conservation of the wider environment is also pursued in addition to and in the context of, the protection of KBAs, through the integration of environmental objectives into all policy sectors.

© M SALIM/NI

WHAT ACTIONS NEED TO BE TAKEN?

Given the immense pressures that are placed on the environment throughout Iraq and in the world as a whole, KBAs form a key element in a strong conservation strategy. A series of recommended objectives and strategic actions have been developed under four initiatives or themes: Research & Monitoring; Protected Area Development; Site Management & Promoting Sustainable Use of Natural Resources, and Advocacy & Education. The seven recommended objectives envision the results of concerted efforts towards achieving conservation of Iraq's biodiversity under these key themes. The nineteen strategic actions proposed to reach these objectives are designed to be straightforward, pragmatic and ultimately achievable.

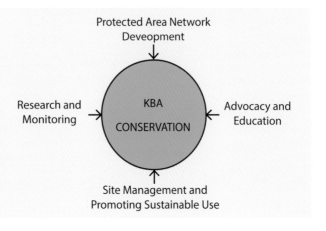

Protected Area Network Deveopment

Research and Monitoring

KBA CONSERVATION

Advocacy and Education

Site Management and Promoting Sustainable Use

This inventory documents the Key Biodiversity Areas (KBAs) of Iraq, the result of widespread and comprehensive surveys from 2005 to 2010 (with some additional survey work in 2011-2012). This ambitious program was a joint effort by the Ministry of Water Resources, the former Iraq Ministry of Environment (IMoE), and Nature Iraq (NI). In 2015, the Ministry of Environment was combined with the Ministry of Health to form the Ministry of Health and Environment. Even though this change did not take place until 2015, hereafter in the main chapters of this book the Ministry will be referred to under its new name and the acronym IMoHE. The program was originally sponsored by the Canadian International Development Agency (CIDA), which supported the initial training of Iraqi scientists in modern observation techniques and monitoring of biological and physical indicators. It was later expanded with support from the Italian Ministry of Environment, Land & Sea (IMELS) and Nature Iraq.

The goal of the KBA Program was to identify areas of outstanding global or regional importance for their biodiversity and to provide a foundation for developing a protected area network in Iraq.

The KBA surveys, conducted primarily during summer and winter, began in southern Iraq in the winter of 2005 and were extended to Kurdistan, northern Iraq in 2007 and to Central and Western Iraq in 2009. Over 220 individual survey sites throughout 17 of Iraq's 18 governorates (excluding Nineva) were visited, often over several years. The surveys covered birds and other fauna, plants, habitats and threats to these sites.

1. A WHAT ARE KEY BIODIVERSITY SITES?

The International Union for the Conservation of Nature (IUCN) defines Key Biodiversity Areas or KBAs as important sites for worldwide conservation and protection.

They are identified nationally using simple, standard criteria, based on their importance in maintaining species populations. As the building blocks for designing the ecosystem approach and maintaining effective ecological networks, key biodiversity areas are the starting point for conservation planning at the landscape level. Governments, intergovernmental organizations, NGOs, the private sector, and other stakeholders can use key biodiversity areas as a tool for identifying national networks of internationally important sites for conservation. (IUCN, 2014a).

KBAs need to be large enough, or sufficiently interconnected, to support viable populations of the conservation concern species that are found at these sites. Here, a 'site' means an area of any size that can be delineated and managed for conservation. The challenge of this approach lies in how best to determine which areas qualify for conservation attention. One effective method is to select sites based on criteria that evaluate the distribution of (apparently viable) populations of key species (species that require conservation actions).

Such species fall into two main, non-exclusive classes: threatened[2] or geographically concentrated. These correspond to the two fundamental variables for planning networks of conservation sites: vulnerability and irreplaceability (Langhammer et al., 2007). To assess the biodiversity at a site these two criteria (sometimes with several sub-criteria) are evaluated to identify KBA sites.

When a site meets one or more of these criteria, the site would be considered an area that is globally important for its biological diversity (note that countries can also develop criteria at the national scale but this has not yet been attempted in Iraq). Within the overall KBA framework a variety of different criteria-based systems focused on specific fauna and flora groups can also be applied. The two most important, which were utilized for site selection in the Iraq Program, are Important Bird and Biodiversity Areas (IBAs) and Important Plant Areas (IPAs). Ultimately if a site meets one or more of these criteria systems (general KBA for non-avian fauna, IBA for birds, or IPA for plants & habitats) the site is considered a Key Biodiversity Area in Iraq.

It should be noted that as this book was being published, IUCN released the new KBA criteria in early September 2016. The new criteria may affect the assessments presented in this book, however, the general outcome for the protection of unique habitats will not change noticeably though priorities may differ. We have endeavored to publish this book using the criteria previously adopted and, in the near future, will collect new data by re-visiting the sites to observe the changes that have occurred since the original site vists. We will also visit areas that were not possible to visit before for security reasons. We will treat this document as a living document

---

2 An important tool in the determination of species status is the International Union for the Conservation of Nature (IUCN) Red List of Threatened Species, which provides information on the conservation status for plants and animals that have been globally evaluated using the IUCN Red List Categories and Criteria (IUCN, 2012). The Red List identifies those plants and animals that are facing a risk of global extinction (i.e. those listed as Critically Endangered, Endangered and Vulnerable). See Web Resources for further information.

and we will update these assessments in accordance with the newly published and revised IUCN criteria.

KEY BIODIVERSITY AREA CRITERIA USED FOR NON-AVIAN FAUNA

For non-avian fauna including mammals, reptiles and amphibians, surveys were limited in Iraq, as these groups generally require more intensive and focused field methodologies than could be applied in the Iraq Program. Data collection on such species was obtained primarily through opportunistic sightings and interviews with locals at the sites. However, important species of conservation concern were identified at some of the sites though clearly more targeted work is needed for these taxa.

As stated above, Vulnerability and Irreplaceability are the two main criteria used to identify KBAs overall. Irreplaceability is further defined into five sub-criteria with thresholds used to determine KBA Status. The full criteria used for non-avian fauna are provided in Appendix 1 but only two criteria could be assessed for certain species in the Iraq Program. These were Vulnerability and Irreplaceability sub-criterion Ia (Restricted-range species). In regards to Irreplaceability, several species may have qualified under the other sub-criteria but quantitative supporting data were generally lacking, so they have not been evaluated at this stage. Collecting detailed information on these species and others to provide insights for updates to the KBA Inventory will be a priority for the future.

Fish surveys formed only a limited part of the KBA Program. Much of the information obtained came from interviews with local fishermen and venders at fish markets, as comprehensive surveys were not possible at the sites. In addition, little information was available on conservation concern fish species in Iraq. Later, at the conclusion of 2014, under a Mediterranean-wide Freshwater Key Biodiversity Area project[3] the assessments for several freshwater taxa for the Levant and Turkey (including Iraq) became available. These assessments were done for fishes, molluscs, odonates (dragonflies and damselflies), and aquatic plants. See the Eastern Mediterranean Freshwater Biodiversity Red List report (Smith, Barrios, Darwall, & Numa, 2014) as well as the Mediterranean KBA report (Darwall et al., 2014). With the completion of this work, information for Iraqi fish species became available on the IUCN Red List, the comprehensive, on-line inventory of the conservation status of the world's species. While other taxa could not be evaluated due to lack of expertise, incomplete species checklists and insufficient information on species conservation status, the KBA Team was able to use this information to improve the site assessments

based on the presence of at least critically endangered or endangered fish species. Still much work remains to be done on these and other taxa for a future edition of this book.

IMPORTANT BIRD AND BIODIVERSITY AREAS (IBAS)

BirdLife International, an organization devoted to bird and habitat conservation throughout the globe, has developed criteria for the designation of IBAs and in 1994 BirdLife published a book entitled *Important Bird Areas in the Middle East*, which listed 42 IBAs in Iraq (Evans, 1994). The IBA criteria are based on the presence of globally threatened species (Criterion A1), Restricted-range species (Criterion A2), Biome-restricted species (Criterion A3) and species that congregate in large flocks (Criterion A4). The full description of the IBA criteria used by the Iraq Program is found in Appendix 1.

Important Bird Areas were renamed Important Bird and Biodiversity Areas in 2013, without a change of acronym, to reflect more accurately the fact that the IBA network also captures many species of conservation significance in other taxonomic groups.

As the Iraq KBA Program has developed a strong ornithological team and all surveys were focused on birds as major indicator species, the application of the IBA criteria to the data from the survey sites has been the most straightforward and comprehensive way for determining if a site is of global importance.

IMPORTANT PLANT AREAS (IPAS)

In addition, Plantlife International, an organization involved in international plant conservation measures, has developed criteria for the designation of Important Plant Areas (IPAs) worldwide. Plantlife International states that the identification of IPAs is based on three broad criteria (Plantlife, 2008). A site qualifies as an IPA if it fulfills one or more of these criteria:

A. Sites with threatened species (sites that hold significant populations of species of global or regional concern);

B. Sites of botanical richness (sites with exceptionally rich flora in a regional context in relation to its biogeographic zone);

C. Sites with threatened habitats (sites that are outstanding examples of a habitat type of global or regional importance).

Since being developed for Europe, the IPA criteria have been adopted and adapted by regional initiatives to take into account differences in both regional biodiversity and data quality (Al-Abbasi et al., 2010). During a meeting of the Arabian Plant Specialist Group (APSG) in 2005 a refinement

---

3 This project was funded by the Critical Ecosystem Partnership Fund (CEPF) and the MAVA Foundation

of the criteria was instigated for the Arabian region and these APSG IPA criteria were adopted for use in Iraq (See Appendix 1 for the full IPA criteria list).

In terms of plants and habitats, Iraq is only in the initial stages of assessing sites based on these three criteria. Comprehensive lists for plant species in Iraq do not yet exist, as they do for birds, and information on threatened plant species (IPA Criterion A) is incomplete. The Flora of Iraq (Guest, 1966, Guest and Townsend, 1966, 1968, 1974, 1980a, 1980b, 1985, & Ghazanfar & Edmonson, 2013) is in the final stages of completion only now and the taxonomy needs updating. Additionally only a provisional list of Iraqi endemics and near-endemics is available (Knees, Zantout, Gardner, Neale & Miller, 2009). But the KBA Program has collected extensive botanical information and the botany work has also helped to develop a broader understanding of species/habitat relationships.

1.B PROGRAM RATIONALE

When the KBA Program was initiated, Iraq had no network of protected areas designed to achieve species and/or habitat conservation. Internationally and within the Middle East region, many countries have developed such networks to ensure the protection of these resources for future generations and preserve the host of ecosystem services that their biodiversity provides. Unfortunately, it was not known what areas of the country were important for such conservation efforts or even what the current status of many species within the country were. Thus the KBA Program was initiated in Iraq to provide this information. KBAs have a number of important attributes and functions that make them suitable as foundation sites for a protected area network.

*One hundred years ago an Inventory of KBAs would have almost certainly included sites that were important for globally threatened species that are now extinct in Iraq — and in one case, **Slender-billed Curlew** Numenius tenuirostris, a species that is probably extinct in the world. **Asiatic Lion** Panthera leo (last seen 1918), **Cheetah** Acinonyx jubatus (1952) and **Persian Fallow Deer** Dama mesopotamica would all have featured amongst the mammals, though there is still a chance this last species occurs in one remote site within Iraq. Globally threatened birds would have included the **Critically Endangered Northern Bald Ibis** Geronticus eremita (last recorded in the 1920s) and the previously mentioned **Slender-billed Curlew** (last record in 1979), whilst the **Critically Endangered Sociable Lapwing** Vanellus gregarius would have been found in many more sites on migration than it does now. This demonstrates the importance of this Inventory as a guide for conservation actions at the national level. No more species should become extinct in Iraq and protection of these KBA sites could help prevent this from happening.*

PROVIDE SITE-SCALE CONSERVATION

Sites are an effective unit for conservation action. They are discrete areas of habitat that can be delineated, protected and managed effectively for conservation. Such protection can address habitat loss and over-exploitation, two of the major causes of biodiversity loss. Protection of a network of sites is a cost-effective conservation approach because, if carefully selected, a relatively small network of sites can support a large proportion of the species, communities and habitats within any given region. In addition, protection of a network of sites is consistent with sustainable development and poverty alleviation agendas because it allows a significant degree of human use of landscapes. For these reasons, sites are a major focus of conservation investment by governments, donors and civil society. They form the basis of most protected area networks.

Plate 1: Marbled Duck © M Salim/NI

For example, Map 1 shows important sites for the globally threatened Marbled Duck *Marmaronetta angustirostris* (Vulnerable) found in the Iraq Program. A protected area network that includes these sites could play an important role in the conservation of this species. In Iraq, Marbled Duck is found primarily in the marshlands in Central and Southern Iraq. It was recorded at 22 sites (67%) out of a total of 33 KBA wetland sites but seven of these sites (21%) did not qualify under IBA Criteria (shown by the small blue squares in Map 1). Of the 15 qualifying sites (sites that meet Criterion A1), six held 1% or more of the biogeographical population of Marbled Duck (shown by the largest blue squares in the map). These six sites include North Ibn Najm (IQ062), Dalmaj Marsh (IQ064), Hawizeh

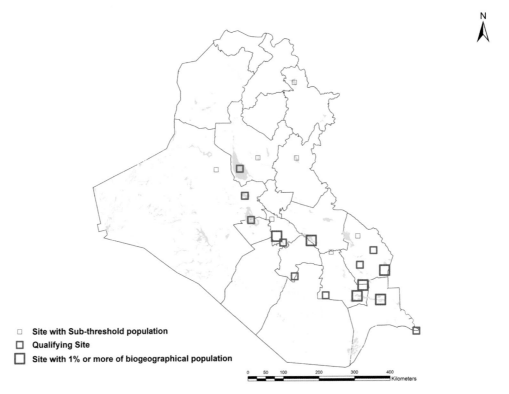

N

□ **Site with Sub-threshold population**
☐ **Qualifying Site**
◻ **Site with 1% or more of biogeographical population**

Map 1: Presence of Vulnerable Marbled Duck at KBA Sites

Marsh (IQ073), Central Marshes (IQ075), West and East Hammar (IQ076 & IQ077). It is possible that the remaining qualifying sites also support large populations of Marbled Duck at different times of the year, thus a protected area system that protects a network of all of these qualifying sites would best ensure the conservation of this species.

CONTRIBUTE TO SOCIO-ECONOMIC DEVELOPMENT

KBAs are important not only for biodiversity but also for socio-economic development at local and national levels. The natural resources provided by KBAs can contribute significantly to human livelihoods, for example, fish, fuel, land for grazing and as the basis for ecotourism. In addition, conservation of the Iraq KBA network would bring significant benefits to national economies because many KBAs provide high-value benefits such as water storage, catchment protection and flood control. Consequently, provided that the socio-economic benefits of KBAs can be equitably shared, and their biological values simultaneously maintained, KBA conservation should be an objective shared by conservationists, local communities and governments alike.

ADDRESS GAPS IN PROTECTED AREAS SYSTEMS ON THE REGIONAL AND NATIONAL LEVELS

In the Middle East, the majority of investment in site-based conservation by national governments and donor agencies has been in the development of protected area systems. However, protected area systems are rarely developed coherently and globally there are major gaps with regard to critical habitat types, ecological regions and threatened species. To ensure that globally important biodiversity is conserved, systematic ecological networks are needed. These should consist of key areas of the highest biodiversity value that are interconnected within a managed landscape. Because KBAs are identified according to objective, scientific criteria, irrespective of current protection status, the establishment of a list of proposed sites to use by the Iraqi government in developing a protected area network is founded on sound conservation science and principles.

Safeguarding these key areas requires a variety of governance approaches, including national parks, community and indigenous conservation areas, and private reserves – the best approach will vary from place to place. However, they will need to be managed with the entire network in mind so as to effectively conserve the important biodiversity that they shelter.

ENCOURAGE NON-FORMAL APPROACHES TO SUPPORT SITE PROTECTION

It may not be feasible to designate every KBA as a formal protected area due to resource limitations, conflicting land ownership and high site management and protection costs on relatively large sites. Indeed, in some circumstances, formal protected area designation could be counterproductive to conservation objectives, particularly where regulations might restrict traditional land and natural resources uses.

Traditional land management practices, which have often developed sustainably over hundreds of years, typically support and conserve the biological diversity of a site quite well. Therefore, there is a need to develop alternative approaches to site-based protection of KBAs. One regional example is the "Hima" system, which encourages traditional land management by local communities (Gari, 2006). In many cases, such an approach may be more cost effective as it engages support from non-traditional sources. Moreover, this approach may provide greater opportunities for sustainable human use of natural resources, and therefore make a greater contribution to poverty alleviation among people for whom natural resources form a critical component of their livelihood strategies.

BUILDING BLOCKS FOR INTERCONNECTED NETWORKS

While protecting individual sites can be an effective approach to conservation for many species, at least in the medium term, the long-term conservation of all species requires the protection of interconnected networks of sites.

This is particularly important for species with large territories, low natural densities and/or migratory behavior, for which individual sites cannot support long-term viable populations. In addition, protection of interconnected networks is essential for the maintenance of broad-scale ecological and evolutionary processes (Schwartz, 1999). Furthermore, interconnected networks may be less susceptible to the impacts of global climate change, as species are better able to "track" changes

in habitat distribution. Therefore, site-level approaches to conservation must be complemented by regional-level approaches, which maintain or establish habitat connectivity among individual sites. Such approaches help to integrate into broader socio-political agendas the goal of interconnecting networks of core areas, linked by habitat corridors, protected by buffer zones and, in some cases, the restoration of degraded areas.

In Iraq the current proposed KBA network represents the most comprehensive assessment of nationally important sites for conservation so far. Map 2 demonstrates that the current list of KBAs could greatly extend the core area of Iraq's protected area network, with further key areas being identified when more comprehensive data for important taxonomic groups become available through future surveys.

It is of note that as this book was being published, IUCN published the new KBA criteria in early September 2016. The new criteria will more than likely affect the classifications presented in this book, however, the general outcome for the protection of unique habitats will not change noticeably, but the priorities may change. We have endeavored to publish this book using the criteria previously adopted and work on collecting newer data in the near future and revise the classification in accordance with newer data and new criteria.

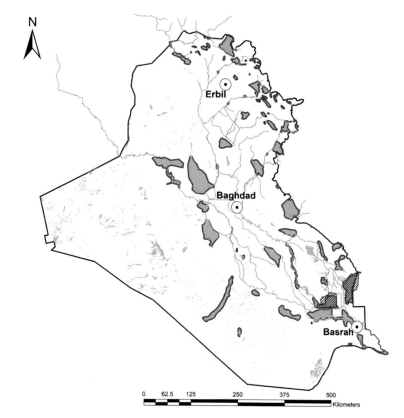

Map 2: The potential for KBAs to extend and develop the existing network in Iraq.
(blue polygons = KBAs and hatched areas = Current Protected Areas)

MAINSTREAM BIODIVERSITY INTO OTHER POLICY SECTORS

Incompatible land-use, development schemes, uncontrolled exploitation and under-resourcing of the infrastructure of protected areas are among the major threats to biodiversity in Iraq. Additionally, the economy of Iraq is expanding, which is likely to result in increased pressure on currently remote and undisturbed regions, especially to exploit resources such as oil, minerals or through increased recreational pressure.

These threats typically arise from insufficient integration of biodiversity conservation objectives into the plans and policies of other sectors, leading to site-based conservation efforts being undermined by incompatible development projects and patterns of land use. Consequently, there is a need to "mainstream" biodiversity into other policy sectors, particularly agriculture, forestry, fisheries, mining, transport, energy and tourism. In order to do this it is essential that accurate, up-to-date information on the conservation importance and status of sites is made available to decision-makers in various government and donor agencies. It is also essential that such information is based on clear, objective and universally accepted criteria. Consequently, KBAs represent a valuable tool for integrating biodiversity into policy and planning.

SUPPORT NATIONAL COMMITMENTS UNDER MULTILATERAL AGREEMENTS

Conservation of the Iraq KBA network will aid the Iraqi government and donor agencies in meeting their commitments under multilateral environmental agreements. These agreements include the Convention on Biological Diversity (CBD), the Ramsar Convention on Wetlands of International Importance (Ramsar Convention), the Convention on Migratory Species (CMS), the African-Eurasian Waterbird Agreement (AEWA), and the International Treaty on Plant Genetic Resources for Food and Agriculture (ITPGRFA) to which Iraq is a signatory nation. Because of the strong similarities between the criteria for identifying important sites for conservation under these multilateral agreements and the criteria used to identify KBAs, such areas fulfill the obligations for site designations under these agreements.

Article 8(a) of the Convention on Biological Diversity (CBD) enjoins parties to establish "a system of protected areas or areas where special measures need to be taken to conserve biological diversity." These obligations for site conservation have been reinforced by the targets of the Millennium Development Goals, and by decisions at the World Summit for Sustainable Development (WSSD). Part of the WSSD Plan of Implementation, for example, is to "promote and support initiatives for hotspot areas and other areas essential for biodiversity, and promote the development of national and regional ecological networks and corridors." Building on this, the March 2003 meeting of CBD Parties (held to decide the Convention's Multi-year Programme of Work) recommended that the protected areas agenda should include the issues of identifying and conserving ecological networks of key sites for biodiversity. The World Parks Congress in September 2003 also emphasized that the international system of protected areas must safeguard all sites that have been empirically determined to be important to conserve biodiversity.

INFORM MULTIPLE STAKEHOLDERS

This inventory provides essential information on currently known sites of global importance for the conservation of wildlife and habitats in Iraq. It will be of essential reading for:

- Decision and policy-makers
- Donors
- Land-use planners and regulators
- Land owners and managers
- Biologists, ecologists and conservationists
- Environmental consultants
- Academic and research bodies
- Interested locals
- Local, national, regional, and international NGOs

The Inventory of KBAs will help these stakeholders to:

- Identify, document and promote awareness of the most important sites in Iraq for the conservation of wildlife and habitats.
- Direct conservation activity and available funding towards these sites.
- Understand the wildlife value of each site in a standardized, reliable way.
- Provide a tool for planning and management, at practical and political levels, through the presentation of key information on wildlife, habitats, threats, legal protection and conservation status.
- Develop networks of local experts, fieldworkers and volunteers and motivate them to monitor and protect KBAs.
- Stimulate national and international cooperation and coordination in conserving the most important sites for birds and other species and habitats in Iraq.
- Establish a more rigorous baseline for measuring success (or otherwise) in conserving Iraq's most important sites for wildlife.

• Promote the value of a site-based approach to the conservation of biodiversity.

KEY BIODIVERSITY AREAS

Key biodiversity areas are places of international importance for their biological diversity. Their conservation is often achieved through protected area networks and other mechanisms. They are identified nationally using simple, standard criteria, based on their importance in maintaining habitats and populations of species. As the building blocks for designing the ecosystem approach and maintaining effective ecological networks, key biodiversity areas are the starting point for regional-level conservation planning. Governments, intergovernmental organizations, NGOs, the private sector and other stakeholders can use key biodiversity areas as a tool to identify and conserve a national networks of internationally important sites.

A LOCALLY-LED PROCESS

While identifying Key Biodiversity Areas is based upon internationally defined criteria that have been consistently applied, to guarantee the success of conservation plans and the sustainability of implementation, the process of identifying these sites must occur at a local or national level. Thus, this process aims not only to identify important conservation sites but also to:

• Develop technical and conservation capacity in a country;

• Build partnerships among key organizations (both governmental and non-governmental) concerned with biodiversity conservation;

• Build widespread understanding of the process and broad ownership of the final site list, and

• Focus new survey work on the most important gaps in knowledge in a coordinated fashion.

For governments, the Key Biodiversity Area criteria provide a tool to identify national networks of globally important sites that will be priorities both for national investment and for channeling resources from international instruments such as the Global Environment Facility. Meanwhile, local, national and international NGOs can use the Key Biodiversity Areas approach to target conservation investment within their priority areas.

1.C KBA PROGRAM PARTNERS

Iraq Ministry of Health & Environment. As the primary partner in the KBA Program, the ministry participated in all surveys and facilitated coordination with other Iraqi ministries and authorities. Formed in 2004 as the Ministry of Environment, in 2015 it was combined with the Ministry of Health and became the Ministry of Health & Environment. It has directorate offices in all governorates of Iraq except the northern Kurdish governorates of Sulaymaniyah, Erbil & Dohuk. These three governorates contain directorate offices of the Kurdistan Environmental Protection and Improvement Commission (KEPIC), whose staff has also participated in the program and helped in logistical support. For more information: IMoHE, Department of the Environment Website: www.moen.gov.iq; KEPIC Website: zhenga.net.

Iraq Ministry of Water Resources, Center for the Restoration of Iraqi Marshes and Wetlands (CRIM-W). The Ministry of Water Resources has been an instrumental partner in the program of the restoration of the marshes of southern Iraq and the management of the restored areas. Iraqi law gives this ministry the lead in enforcing the requirements of the Ramsar agreement and the management of the water resources needed for the survival of the wetlands as well as agricultural activities in Iraq not to mention municipal water needs. The initial trainings for the use of field equipment to collect data and samples for physical and chemical properties of the water in the marshes were held in the Ministry of Water Resources headquarters and included personnel from academia as well as Ministry of Environment and Ministry of Water Resources. MOWR-W personnel participated in most expeditions that included water sampling tasks. www.mowr.gov.iq/en

Nature Iraq. This Iraqi non-governmental, conservation organization has coordinated the KBA surveys together with the Iraqi Ministry of Environment. Founded 2004, it now maintains offices in Baghdad, Chibaish in southern Iraq, and Sulaymaniyah (Sulaimani), Kurdistan, northern Iraq and works in the areas of biodiversity conservation, water resource protection and sustainable development throughout the country. For more information: www.natureiraq.org.

ORGANIZATIONS SUPPORTING THE KBA PROGRAM
The Italian Ministry of Environment, Land & Sea (IMELS). The Italian Ministry support for the KBA Program began in 2006 and supported the continuation and expansion of the survey efforts as well as the continued training program for the field teams and the development of the KBA Inventory. www.minambiente.it

BirdLife International. This global partnership strives to conserve birds, their habitats and biodiversity worldwide, working with people towards sustainability in the use of natural resources. It is one of the World's largest partnership of conservation organizations, with over 100 partner groups, of which Nature Iraq is an affiliate. BirdLife provided training and advisory support throughout the surveys. www.birdlife.org

Centre for Middle Eastern Plants (CMEP)/Royal Botanic Gardens Edinburgh (RBGE) Established by the Royal Botanic Garden Edinburgh in 2009, the CMEP is an authority on the Middle Eastern plants and environment. CMEP provided essential support for the botanical work of the KBA Program. www.cmep.org.uk

Missouri Botanical Gardan. MBG provided assistance through the contribution to training programs for KBA botanical staff. www.missouribotanicalgarden.org

Old Dominion University. Old Dominion provided training support for some of the KBA botanical staff. www.odu.edu

Canadian International Development Agency (CIDA). Initial training and survey support came from the Canada-Iraq Marshlands Initiative (CIMI) funded by CIDA that supported the program from 2004 to 2006.

Iraqi Universities. Various universities of Iraq contributed support by providing information and occasionally staff to the KBA Program including the Universities of Baghdad, Basrah, Dohuk, Karbala, Qadisiyah, Sulaymaniyah (Sulaimani), Salahadin, & Thi Qar.

OTHER SPONSORS

Ford Motor Company. Ford launched their Conservation and Environmental Grants Program in 1983. Since launching the global grants program in 1983, Ford has awarded more than US$2 million in grants to hundreds of environmental projects in 60 countries, spanning Asia Pacific, the Caribbean, Central America, the Middle East and Puerto Rico.

United Nations Development Program (UNDP). The UNDP assistance to the KBA Program began in 2015 with their support for the publication and distribution of this KBA Inventory. www.undp.org

Ornithological Society of the Middle East, the Caucasus and Central Asia (OSME). Promotes ornithology and conservation throughout the region and publishes the journal Sandgrouse.

Royal Society for the Protection of Birds (RSPB). The BirdLife Partner in the UK and one of the largest conservation organizations in the world with over one million members. It has 220 nature reserves covering over 145,000 hectares as well as many education centers.

British Birds (BB). A monthly journal, established 1907, for professional and amateur ornithologists, publishing papers on conservation, identification, status and taxonomy. BB is regarded as the journal of record in Britain and provided funds to the Iraq program.

South Oil Company (SOC): One of Iraq's national oil companies, SOC printed additional copies of this book for distribution in Iraq.

CHAPTER 2: OVERVIEW OF THE COUNTRY

2.A DESCRIPTION OF IRAQ

Map 3 shows Iraq, its six neighbor states, major features and the Tigris and Euphrates Rivers. The country covers an area of 434,320 km², while the Gulf coastline is just 58 km in length. The land upon which the modern state of Iraq has been founded came to be called 'Mesopotamia' (from the Greek meaning 'land between the rivers'). This region is known as the cradle of civilization and from the 6th Millennium has been the home of a series of important empires (Sumerian, Assyrian, Babylonian, Akkadian). Other civilizations that later controlled Mesopotamia includes the Median, Achaemenid, Hellenistic, Parthian, Sassanid, Roman, Rashidun, Umayyad, Abbasid, Mongol, Safavid, Afsharid, and Ottoman empires. The modern boundaries of Iraq were drawn in 1920 to 1925 by the League of Nations in the Treaty of Sèvres upon the collapse of the Ottoman Empire after the First World War. The country became the British Mandate of Mesopotamia under the control of the United Kingdom and starting in 1921 was a monarchy under British administration until its independence in 1932. The Republic of Iraq was created with the overthrow of the King in a coup d'état on the 14th of July 1958 but this led to a period of instability and a series of military coups until the Ba'athist Party took control from 1968 to 2003. Saddam Hussein, its 5th president, ruled the country from July 1979 to April 2003. During his presidency Iraq fought three major and very destructive wars (the Iran-Iraq War in 1980-88, first Gulf War in 1990-91, and the second Gulf War in 2013); conducted a genocidal campaign against its Kurdish population (Anfal in 1986-89); drained the Mesopotamian Marshlands displacing and repressing people of southern Iraq (early 1990s), and suffered through 13 years of economic sanctions (1990-2003). The Ba'ath Party was removed from power in 2003 after a United States-led invasion of the country followed by an occupation that lasted until 2011. Iraq is now governed as a federated Republic with a president, a Prime Minister heading the Council of Ministers and an elected Parliament. There is a federal regional government in the northern part of the country governing the Kurdish moutnians and plains. Unfortunately, there remains considerable unrest with the remnant of the Islamic State of the Levant being removed from Mosul, the country's second largest city.

As of 2012, the World Bank estimates Iraq's population as 32.6 million people, living in 18 provinces in Iraq and since 1970 three of the northern governorates (Dohuk, Erbil, and Sulaymaniyah have been officially designated as a Kurdish autonomous region, with a separate elected legislature, the Kurdish Regional Government (KRG). In 2016 the district of Halabja was designated as a separate governorate, making the KRG into four governorates.

Map 3: Iraq and its governorates

GEOGRAPHICAL AREA

The *Desk Study on the Environment of Iraq* (UNEP, 2003) described the four major geographical zones of the country as follows (see also Map 4):

- **Desert plateau:** Approximately 40% of Iraqi territory consists of a broad, stony plain with scattered stretches of sand, lying west and southwest of the Euphrates River and sparsely inhabited by pastoral nomads. A network of seasonal watercourses — or wadis — runs from the country's western borders towards the Euphrates River.

- **Northeastern highlands:** Covering approximately 20% of the country, this region extends north of a line between Mosul to Kirkuk towards the borders with Turkey and Iran, where mountain ranges reach up to 3,600 m in altitude.

- **Uplands region:** About 10% of Iraq comprises a transitional area between the highlands and the desert plateau, located between the Tigris north of Samarra and the Euphrates north of Hit, and forming part of a larger natural area that extends into Syria and Turkey. Much of this zone may be classified as desert because watercourses flow in deeply cut valleys, making irrigation far more difficult than in the alluvial plain.

- **Alluvial plain:** Approximately 30% of Iraq is composed of the alluvial plain formed by the combined deltas of the Tigris and Euphrates Rivers. This region begins north of Baghdad and extends south to the Gulf coast bordering Iran. Dams, water diversion projects, and large-scale drainage works carried out in Turkey, Syria, Iran and/or by the Iraqi regime have decimated the once extensive wetlands of the region.

BIOMES & ECOREGIONS OF IRAQ

Iraq is part of the Palearctic Realm, the largest of the eight terrestrial biogeographic areas or ecozones that have been defined for the Earth. The World Wide Fund for Nature (WWF) developed the classification system used in the Inventory of KBAs of Iraq. Under the WWF system

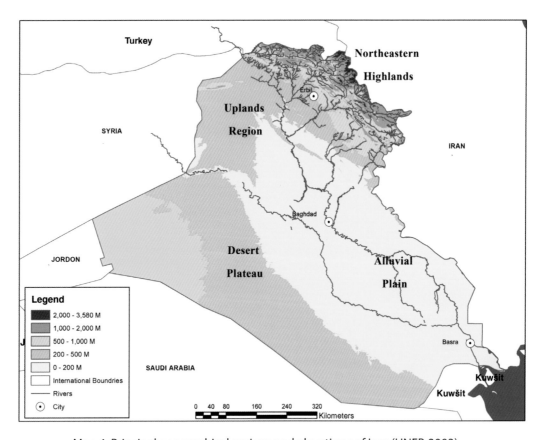

Map 4: Principal geographical regions and elevations of Iraq (UNEP, 2003)

(2014), ecozones are divided into biomes or major habitat types and these biomes are further divided into over 800 ecoregions, which are large areas of land or water that contain geographically distinct assemblages of natural communities that:

• share a large majority of their species and ecological dynamics;

• share similar environmental conditions, and;

• interact ecologically in ways that are critical for their long-term persistence.

The World Wide Fund for Nature (2014) defines five terrestrial biomes and nine ecoregions in Iraq. These are listed in Table 1 with their conservation status, area and percentage within Iraq.

Table 1: WWF Biomes and Terrestrial Ecoregions of Iraq

WWF Biome	Terrestrial Ecoregion*,**	Conservation Status	Total Area (km²)	Area in Iraq (km²)	% in Iraq
Temperate Broadleaf and Mixed Forests	Zagros Mountains Forest Steppe (PA0446)	Critical	397,800	29,376	7%
Mediterranean Forest, Woodland and shrubs	Eastern Mediterranean conifer-sclerophyllous-broadleaf forest (PA1207)	Critical	143,800	1,475	1%
Temperate Grasslands. Savannas and Shrublands	Middle East Steppe (PA0812)	Vulnerable	132,300	37,598	28%
Deserts and Xeric Shrublands	Arabian Desert and East Sahero-Arabian Xeric Shrublands (PA1303)	Critical	1,851,300	192,853	10%
	Mesopotamian Shrub Desert (PA1320)	Vulnerable	211,000	129,995	62%
	Red Sea Nubo-Sindian Tropical Desert and Semi-Desert (PA1325)	Critical	651,300	5,189	1%
	South Iran Nubo-Sindian Desert and Semi-Desert (PA1328)	Critical	351,500	7,993	2%
	Gulf Desert and Semi-Desert (PA1323)	Critical	72,600	1,480	2%
Flooded Grasslands and Savannas	Tigris-Euphrates alluvial salt marsh (PA0906)	Critical	35,600	28,795	81%
		Total	3,847,200	434,753	

*Eastern Anatolian Montane Steppe (PA0805) is an ecoregion with only 3 hectares considered part of Iraq

**Each terrestrial ecoregion has a specific ID code in the format XXnnNN (where XX is the ecozone, nn is the biome number, and NN is the individual ecoregion number).

Of these nine main ecoregions, five account for 96% of the total area of Iraq and these are discussed in more detail below. Note, that much of the information in the list below comes directly from the WWF's (2014) descriptions of these specific ecoregions but it should also be noted that in the Iraq Inventory 'biome-restricted bird species' were evaluated based on a slightly different list of biomes used by BirdLife International. For Iraq, the BirdLife list of biomes includes the Mediterranean (ME12), Eurasian High-Montane (ME05), Irano-Turanian (ME06), Eurasian Steppe and Desert (ME04) and Sahara-Sindian Desert (ME13) Biomes. Except for the Mediterranean biome (ME12), which does not constitute a major part of Iraq, these biomes are identified in brackets with their key, corresponding ecoregion in the following list.

Plate 2: Zagros Mountains Forest-Steppe - Shanidar Valley in Barzan (IQ004)

1. Zagros Mountains Forest-steppe (Western Asia: Northern Iraq and Western Iran)

Location: This region extends northwest to southeast and roughly paralleling much of the Iraq and Iran's border along the Zagros Mountain chain and includes parts of southern Azerbaijan province, Iranian Kurdistan and Faristan, to Makran in southeastern Iran. In Iraq this area falls mainly within the Kurdistan region of northern Iraq. [BirdLife International includes this region in the Irano-Turanian Biome (ME06), with higher elevations part of the Eurasian High-Montane Biome (ME05)].

Key Habitat Features: The climate is semi-arid, with warm summers and cold winter and annual precipitations (falling mainly in winter and spring) averaging from 400 to 800 mm. It features scrub and woodland in the mountains and foothills; inland cliffs and rocky slopes; screes and boulders. Vegetation consists mostly of deciduous broad-leaved trees and shrubs with a wide array of forest layers where more light and heat can penetrate to lower canopy plants. Oak and pistachio trees are a major component of these forests but with variations in species at higher altitudes.

Plate 3: Middle East Steppe Ecoregion - Chamchamal in Kirkuk Governorate (IQ035)

2. Middle East Steppe (Southwestern Asia: Northeastern Syria and Northern Iraq)

Location: Located to the south of the Zagros Mountains Forest-Steppe, this ecoregion forms a broad band of dry grass and shrub lands extending from Syria and western Jordan into northern Iraq. [BirdLife International includes this area within the Eurasian Steppe and Desert Biome (ME04)].

Key Habitat Features: Contains vegetation suited to drier, warmer summers with lower moisture content particularly progressing to the south as the region enters the Mesopotamian

Shrub Desert Ecoregion. According to the WWF, "Herbaceous and dwarf shrub sagebrush (*Artemisia herba-alba*) communities tend to dominate in deeper, non-saline soils and often occur in association with grasses (*Poa bulbosa*) where disturbed by grazing. *Hammada scoparia* characterizes stony soils. Islands in the Euphrates River continue to support remnants of the native riverine woodland. *Tamarix* spp. and *Populus euphratica* are encountered near water as are *Phragmites* spp. reeds."

Plate 4: Mesopotamian Shrub Desert Ecoregion – Western Edge of Tharthar Lake (IQ051)

3. Mesopotamian Shrub Desert (Western Asia: Northern Iraq into Syria and Jordan)

Location: This forms a wider band across most of the center of Iraq. [BirdLife International incorporates this area within the Sahara-Sindian Desert Biome (ME13)].

Key Habitat Features: This transitional area between the deserts of the south and the steppes of the north includes the Syrian Desert. Umbrella-thorn acacia (*Acacia tortillis*), shrubby rock-rose species (*Cistus* spp.), and many dwarf shrubs are characteristic flora. Wetland areas feature reeds and rushes with poplar (*Populus euphratica*) and tamarisk (*Tamarix*) found along river channels (UNEP, 2006). It provides important wintering grounds for migrating Eurasian birds and sparse populations of wolves, hyenas, gazelles, and wild boars. Historically leopard and oryx would have been found here as well.

Plate 5: Tigris-Euphrates Alluvial Salt Marsh Ecoregion - Baghdadia Area in the Central Marshes (IQ075)

4. Tigris-Euphrates alluvial salt marsh (Southern Iraq)

Location: The Mesopotamian Marshlands of southern Iraq, the largest wetland ecosystem in western Eurasia, are located in this ecoregion. [BirdLife International incorporates this area within the Sahara-Sindian Desert Biome (ME13) and considers the area an important Endemic Bird Area (BirdLife International, 2014)].

Key Habitat Features: The main feature here are shallow wetlands with vast areas of aquatic vegetation (*Phragmites* (reeds), *Typha* (rushes), and submerged vegetation) with species adapted to both perennial and seasonally flooded lands. In Iraq, these areas historically fluctuated in size due to seasonal conditions with spring flooding followed by evaporation and drying under hot, dry conditions in summer. These flood pulses have declined in recent decades due to the construction of upstream dams. This area represents some of the most critical wintering areas for migrating birds and is highly susceptible to increased soil salinization.

Plate 6: Arabian Desert and East Sahero-Arabian Xeric Shrublands Ecoregion found near Dalmaj (IQ064)

5. Arabian Desert and East Sahero-Arabian Xeric Shrublands (Southwestern Asia: Most of Saudi Arabia, extending into Oman, United Arab Emirates, Yemen, Egypt, Iraq, Jordan, and Syria)

Location: This is the largest ecoregion of the Arabian Peninsula, stretching from the Yemeni border to the Arabian Gulf and from Oman to Jordan and Iraq. [BirdLife International includes this within the Sahara-Sindian Desert Biome (ME13)].

Key Habitat Features: Characterized by semi-arid to arid conditions, with hot dry summers and cold winters and species adapted to these conditions. The main features of this ecoregion are the sandy terrain of the central Saudia Arabian Peninsula known as the ad-Dahna Desert, the large An-Nafud Desert to its north (which also includes the stony Harrat al-Harrah Desert) and the Rub'al-Khali in the south. These desert regions extend into western and southern Iraq and the plant associations classified by Zohary (as cited in WWF, 2014) include Saharo-Arabian desert vegetation of *Anabasetea articulatae* and Sahara-Arabian interior sand desert vegetation of *Haloxylo-Retametalia raetami* (including *Haloxylion persici arabicum*).

Map 5 provides an overview of these areas and shows the location of Iraq Key Biodiversity Areas within these ecoregions.

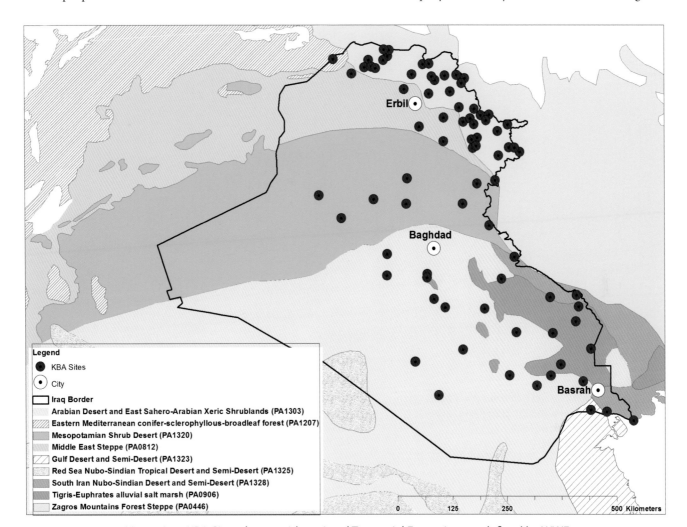

Map 5: Iraq KBA Sites shown with regional Terrestrial Ecoregions as defined by WWF

Aquatic ecoregions of the world are also defined and in Iraq including three different freshwater ecoregions: Arabian Interior (440), Lower Tigris and Euphrates (441), Upper Tigris and Euphrates (442) and one marine ecoregion (Arabian Gulf (90)). (See Table 2 and Map 6).

Table 2: Freshwater and Marine Ecoregions of Iraq

Aquatic Ecoregions	Total Area (km²)	Area in Iraq (km²)	% in Iraq
441 Lower Tigris and Euphrates River Basin	340,633	227,497	67%
442 Upper Tigris and Euphrates River Basin	507,236	64,745	13%
440 Arabian Interior	2,334,454	142,494	6%
90 The Arabian Gulf	251,000	Territorial sea 4,910	2%

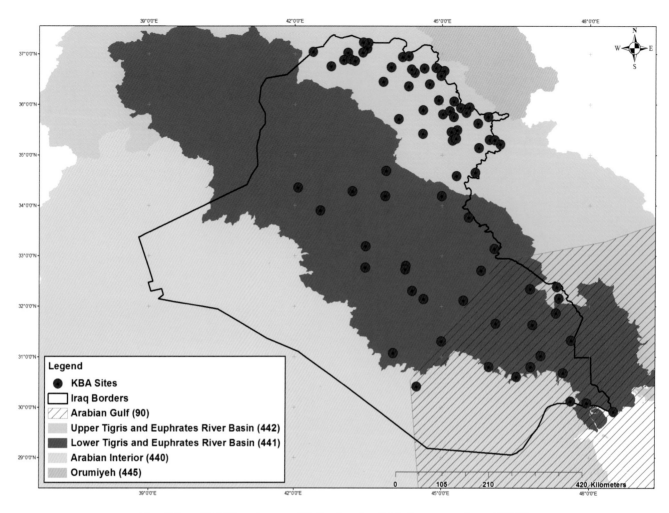

Map 6: Iraq KBA Sites shown with Freshwater & Marine ecoregions (WWF)

Please note these ecoregion maps are not highly accurate and should be viewed at a very rough, global scale (that additionally do not take into account expected climate changes effects). Nevertheless in the KBA assessments, sites are placed spatially within their ecoregion and these are listed in each site account. It is important to remember though that these areas are highly complex and they may contain habitat attributes and species that are commonly associated with different biomes & ecoregions.

LANDSCAPES & SPECIES INFORMATION

Due to the variety of habitats in Iraq from the snow-capped mountains to the hot deserts, the region is biologically diverse. In terms of birds, more than 400 species have been recorded – among them several that are Critically Endangered, such as Northern Bald Ibis *Geronticus eremita*, Slender-billed Curlew *Numenius tenuirostris* and Sociable Lapwing *Vanellus gregarius*. There are a total of 32 globally threatened bird species (3 Critically Endangered, 5 Endangered, 8 Vulnerable and 13 Near Threatened species). Several migration routes cross Iraq, including the African-Eurasian flyways. Millions of

birds depend on the wide variety of habitats that Iraq has to offer as well as vital stopover sites in their annual migrations.

This diversity extends to plants and other fauna as well, with over 2500 plant species (the largest percentage of which are found in the mountain-forest region in the north) and 90 mammals, 98 reptiles, 10 amphibians, and 100 freshwater and marine fish species. It is important to note that most species checklists for Iraq are still incomplete. The botanical diversity of the Zagros Mountains Forest-steppe is quite rich due to the variety of landscapes and terrains within the Zagros range that extends through northern Iraq. The dominant habitat in this region is oak woodlands and important genera representative of the Irano-Turanian Steppe including Tragacanth dominated by species of *Astragalus, Cousinia* and many endemic species. Low slopes are clothed in open, oak forests again home to a good assemblage of endemic and rare species. Key threatened fauna species of this region include Wild Goat *Capra aegagrus*, Persian Leopard *Panthera pardus saxicolor*, Persian Fallow Deer *Dama mesopotamica* (reported by locals at only one site within Iraq), Spur-thighed Tortoise *Testudo graeca*, Azerbaijan Mountain Newt *Neurergus crocatus*,

Kurdistan Mountain Newt *Neurergus derjugini,* and Spotted Belly Salamander *Salamandra infraimmaculata.* Recent fish surveys in the Kurdistan region of northern Iraq found a number of new, un-described species (Freyhof, 2012) and there are likely other new species to be discovered in this region.

Desert landscapes in Iraq, which comprise much of Iraq's west and southwest are extensions of the deserts of Syria and Saudi Arabia in the west and south. Additional desert landscapes lie between the Tigris and Euphrates extending from lower Babil to Nasiria, and areas east of the Tigris from Wasit southward. Average annual rainfall in this area is less than 100 mm but increases towards the north and east to between 100 and 240 mm (Batanouny, 2001). The majority of the deserts in Iraq lie within the Sahara-Sindian Desert biome/Arabian Desert and East Sahero-Arabian Xeric Shrublands Ecoregion. The genera *Artemisia* and *Haloxylon* are important here and there are relic areas of tropical species that date back to a time of climatic optimum e.g. *Acacia gerrardii* and *Razya stricta.* In addition to being relatively rich with unique plant life, these areas are also important with considerable fauna taxa (mammals, reptiles, and birds) of which many are threatened or vulnerable.

Major river ecosystems of Iraq include the Tigris and Euphrates Rivers but a number of other important tributaries to these two main systems such as the Khaboor, Greater Zab, Lesser Zab, Diyala, Kharun Rivers, Gharraf, Shatt Al-Arab, Shatt Al-Hilla and others. Whereas the majority of the Euphrates River is sourced in the upstream, riparian countries of Turkey and Syria, approximately 40% of the waters of the Tigris River arise from within the borders of Iraq (the remainder coming from Turkey & Iran). Lake ecosystems in Iraq are largely formed by dam/reservoir systems such as Mosul, Dukan, Darbandikhan, Habbaniyah, Himreen, Samara, and Qadassiya Lakes as well as depressions that have been used for water storage such as Tharthar and Razzaza Lakes (in fact, these lakes have their origins in the flood control era of the 1950's through 1980's. However since that time major dams have been built upstream making the flood control function secondary to water storage). There are some natural lakes within Iraq including Sawa Lake in southern Iraq, small mineral ponds such as Genaw Lake near Rania in the north and many desert oases. Often waterbodies have been created by water management systems in Iraq that subsequently became biologically rich and important natural resources such as Dalmaj, Suweibaat, Al-Rashed Lake, and the marshes of the southern part of East Hammar.

Many of these systems now contain introduced, sometime invasive species, but the important native fish species include: *Acanthopagrus cf. arabicus, Alburnus mossulensis, Arabibarbus grypus* (Vulnerable)*, Luciobarbus esocinus* (Vulnerable) and *Mesopotamichthys sharpeyi* (Vulnerable). There are also two important cave-dwelling fish reported in Iraq *Caecocypris basimi* and *Garra widdowsoni* (both Critically Endangered). Much fieldwork remains to be done on fish in Iraq. The same is true for herpetology though some important species are known such as the endangered Soft-shell Turtle *Rafetus euphraticus* associated with the Tigris-Euphrates River Basins.

One of the most notable landscapes in Iraq is the Lower Mesopotamian Marshlands, once the 3rd largest wetlands in the world (the largest in western Asia). This extensive network of marshes formed by the lower Tigris and Euphrates River formerly fluctuated seasonally up to 12,000-15,000 km^2. As mentioned in the insert, these marshlands were subject to a major drainage campaign in the 1990s. Alwan (2006) made an inventory and comparison for the aquatic plants historically recorded in the Iraqi marshlands before desiccation (between the 1970s and 1990s) with what were recorded after the reflooding of these marshlands (after 2003). He mentioned that more than one hundred aquatic plants species have historically been recorded in the marshes and 23 of them were lost after their desiccation and later reflooding after 2003. However, during the KBA survey of 2007, the team recorded two of these lost aquatic species (*Utricularia australis* and

Plate 7: Basra Reed Warbler © M Salim/NI

Nymphoides indica) at several sites in the Iraqi Marshlands. Though not botanically rich, the thick beds of *Phragmites australis* and *Typha domingensis* occasionally shelter water lilies Nemphoides indica and other locally unique species and this area remains an important habitat for mammals, amphibians, fish, and of particular importance for migrating birds making the long journeys between Eurasia and Africa.

Other key species found in the Mesopotamian Marshlands include the Vulnerable Smooth-coated Otter (*Lutrogale perspicillata*), which has a sub-species *L. p. maxwelli* and the Endangered Bunn's Short-tailed Bandicoot Rat *Nesokia bunnii* (both associated with this area); the endangered Euphrates Soft-Shell Turtle, and many important fish species such as those previously mentioned. Many threatened bird species such as the vulnerable Marbled Teal (*Marmaronetta angustirostris*) and Iraq's two endemic bird species the Basra Reed Warbler (*Acrocephalus griseldis*) and Iraq Babbler (*Turdoides altirostris*) are also found in the marshlands and this area is the main breeding ground for the latter species.

The Lower Mesopotamian Marshlands are also of global cultural significance. It was in this region that humans first developed writing and agriculture. The marshlands are the legendary location of the biblical Garden of Eden and the indigenous Ma'dan Marsh Arabs can trace their lineage directly to the ancient Sumerians whose lands were centered in the vast wetlands formed around the confluence of the Tigris and Euphrates River.

HISTORICAL LEGACY

The conservation of important sites and species in Iraq has been undermined by extensive political and social upheaval within the country since the 1970s. Iraq has fought or been involved in three major wars, 13 years of economic sanctions, an occupation and a number of internal conflicts during this time.

In addition to these problems there was an almost complete cessation of most research and monitoring activities since the late 1970s and many highly experienced conservationists and researchers moved to other jobs or

DESTRUCTION OF THE MESOPOTAMIAN MARSHLANDS

The destruction of the Mesopotamian Marshlands began with the drainage of some areas for oil development in southern Iraq. Later the Iran-Iraq war, fought between 1980 and 1988, caused extensive damage to Hawizeh (a transboundary marsh) and the Central Marshes (located between the Tigris and Euphrates Rivers). But the Ba'athist Regime pursued the most extensive drainage campaign against the marshes after the 1991 Gulf War. As documented in a Human Rights Watch Briefing Paper (HRW, 2003), despite claims made by the regime at the time, the drainage of the marshes was conducted as "a retaliation and to deprive enemies of the state their hiding place." By the mid-1990s, these marshes had shrunk to less than 10% of their former area and the drainage was called the worst human-caused environmental disaster of the last century by the United Nations Environmental Programme.

It was not until the 2003 Gulf War that parts of the marshlands were re-flooded, largely through local action. The process of restoring the Iraqi Marshlands has occurred slowly since this time. By 2007 approximately 70% of the area had been re-flooded and birds, fish as well as the Ma'dan, the indigenous Marsh Arabs with direct links to ancient Sumer, were returning to the area. Unfortunately declining water levels due to upstream dams and diversions both within Iraq and in upstream countries as well as poor management of water resources overall, declining water quality and climate change, has resulted in a lowering of the bar for marshland restoration in southern Iraq.

Fortunately, in July of 2013, the government of Iraq decided that the marshes will be the site of Iraq's first national park and the Ministries of Water Resources and Environment are actively engaged in activating the decision and dedicating the water resources needed as part of the management plan for the water resources of the country over the next 25 years.

More recently, the Iraqi government presented the proposal that the marshes as well as some archeological sites located around the former marshes be considered as World Heritage Sites. That was approved in early July 2016 conditional on Iraq presenting proper management plans given the limited water supply in the region. This more or less assures the survival of the restored marshes for future generations as Iraq's government has proposed listing voluntarily. It is hoped that Iraq and Turkey (mainly) and Iran and Syria will agree at some point in the near future on the common management of the water basins of the Tigris and Euphrates though that requires many months of difficult negotiations when the political will is present.

THE ANFAL CAMPAIGN

The genocide known as Anfal was a campaign against the Kurds in northern Iraq by Saddam Hussein's Baathist Regime in the last years of the Iran-Iraq War. It lasted from 1986 to 1989 and included both conventional and chemical bombardments, destruction of settlements and deportations, firing squads and mass graves. The chemical bombing of Halabja, in which between 3200 and 5000 (BBC, 2013) people were killed and thousands more were injured by an unknown combination of blister and nerve agents, was the only event during the Anfal campaign to gain international attention, but in fact up to 180,000 people were killed during the entire campaign (Sinan, 2007).

The main human health effects from chemical weapon attacks (CWA) in Halabja (and likely other places across the Kurdistan region) were tearing, shortness of breath, blindness, blistering, burns, and psychological impacts. Many of these chemicals break down in the environment into both toxic and non-toxic components that then combine and interact with the environment around them in likely complex ways. These biochemical interactions are poorly understood but various studies conducted within Iraq indicate long-term, serious human health issues in areas where CWA occurred. Studies focused on environment effects of CWA overall, for example on soil and water interactions of the breakdown chemicals as well as the long-term impacts on biodiversity and species survival are even more limited and not generally available to Iraq stakeholder and decision-makers.

emigrated from the country in search of better security and opportunities. Economic sanctions during the 1990s and early 2000s resulted in a lack of basic field equipment (binoculars, telescopes and other scientific equipment), reduced access to the latest information and resources (human and transport). Sanctions also cut off or at least greatly curtailed direct communication between Iraqi conservationists and their peers outside the country. As a result of these factors, no modern, standardized conservation research methodology was applied in the region for 20 to 25 years.

In areas of Iraq that achieved greater stability such as the Kurdistan region of northern Iraq rapid economic growth from 1991 onwards enjoys a much higher priority in political decision-making than environmental and nature conservation concerns. Consequently, a rapid loss of wildlife and habitat in terms of numbers and diversity has been observed in this region over the last two decades (and in Iraq overall since 2003) due to a range of activities including uncontrolled and/or illegal fishing, hunting, logging and unsustainable uses of habitats.

A vivid example of how conservation policy has to react to (man-made) changes in ecosystems is the drainage and now current attempts at restoration of the Mesopotamian Marshlands. Destruction of the area, as described in the insert, led to massive human and wildlife displacements with ensuing declines to the economic, health and social conditions of the region. Successful restoration of the marshes will require the involvement of multiple stakeholders and decision-makers who have their own diversity of goals and agendas.

The situation has changed significantly in recent years. Iraq has signed onto important international conservation conventions and agreements; GEF co-funded projects to support the implementation of convention goals are now underway in Iraq, and a protected area network is currently under development.

3.A IRAQI INSTITUTIONAL FRAMEWORK

Iraq has a fairly robust environmental regulatory regime (see Table 3). Although the Ministry of Environment was officially established in 2003 (it has since been merged with the Ministry of Health to become the Iraqi Ministry of Health & Environment), several laws have been promulgated, which, as discussed below, outline a comprehensive framework of principles for the protection of Iraq's environment and biological diversity. Although the Ministry has yet to issue the regulatory details needed for rigorous implementation of new statutes, the laws nonetheless set the stage for the country's future conservation.

Table 3: Primary Environmental Laws in Iraq*

Title	#	Year	Current Status (April 2015)
Iraq's accession to the to the Cartagena Protocol on Biosafety	47	2013	Effective
Iraq's accession to the Convention of International Trade in Endangered Species (CITES) (Iraq is party No. 180)	29	2012	Effective
Wild animals protection law	17	2010	Effective
Environment Protection and Improvement Board of Iraqi Kurdistan Region (KEPIC, 2014a)	3	2010	Effective
Wild Animals Protection Act to replace the law on the Protection of Wild Animals and Birds (formerly Law #21 of 1979)	17	2010	Effective
Prohibition on the importation of plants to Iraq — attributes to the Ministry of Environment and Agriculture the sole authority to import plants and prohibits the import of all plants in support to Instruction No. 11	74	2010	Effective
Consumer protection law	1	2010	Effective
Iraqi products protection law	11	2010	Effective
Law of Environmental protection and improvement (formerly Law #3 of 1997)	27	2009	Effective
Forests and orchards Law (formerly Law #75 of 1955)	30	2009	Effective
Republic of Iraq's accession to the United Nations Framework Convention to Combat Desertification (UNFCCD)	7	2009	Effective
Protection and Improvement of the Environment in Iraqi Kurdistan Region (KEPIC, 2014b)	8	2008	Effective
Law for the establishment of the Ministry of Environment to replace the former Council on the Protection and Improvement of the Environment (formerly Law #44 of 2003 of the provisional authority)	37	2008	Effective
Republic of Iraq's accession to the United Nations Framework Convention on Climate Change (UNFCCC) and Kyoto protocol	7	2008	Effective
Iraq's accession to the Convention on Biological Diversity (CBD)	31	2008	Effective
Republic of Iraq's accession to the Ramsar Convention on Wetlands	7	2007	Effective
The first amendment of the Iraqi Environmental Protection and Improvement Law #3 of 1997	73	2001	Effective
Iraqi Environmental Protection and Improvement Law	3	1997	Amended
Law of Exploitation of Beaches	59	1987	Effective
Amendment on wild animals diseases law #68 of 1936	61	1985	Effective
Natural pastures act	2	1983	Effective
Amendment to Regulations on Fishing, Use and Protection of Aquatic Life Forms - Law #48 of 1971	10	1981	Effective

Title	#	Year	Current Status (April 2015)
Law of Protection from Ionic Radiation	99	1980	Effective
Implementation Law	45	1980	Effective
Protection of Wild Animals and Birds	21	1979	Effective
Regulating Law of the Fishing and the Exploitation and Protection of Aquatic Organisms	48	1976	Amended
First amendment the Wild Animal Diseases Law #68 of 1936	78	1972	Effective
Fishing, Use and Protection of Aquatic Life Forms Law	48	1971	Amended
Law of Execution of Irrigation Projects	138	1971	Amended
Penal Code - A general law describing penalties to violators	111	1969	Effective
Pasture Protection Act	106	1965	Effective
Irrigation Act	6	1962	Effective
Amendment to Hunting Law #57 of 1938	4	1948	Effective
First amendment to Hunting Law #57 of 1938	19	1939	Effective
Hunting Law	57	1938	Amended
Wild Animal Diseases Law	68	1936	Amended
Animal Conservation Law	No #	1924	Effective

*Full texts for some laws are available from the FAOLEX Database at faolex.fao.org/faolex/index.htm
Texts for additional federal laws may be available in Arabic from the Iraqi Local Governance Law Library: www.iraq-lg-law.org

In addition, at this time the Iraqi Ministry of Health and Environment and other ministries are actively working on the development of new laws, regulations and instructions to address additional environmental concerns. Several draft environmental laws that have been under consideration include, but are not limited to:

1. National Law of Biosafety regarding the GOMs (Genetically Modified Organisms)
2. Water Protection Act of Kurdistan Region
3. Regulation on Kurdistan Region Environmental Award
4. Sustainable Management of Water Resources in Kurdistan Region

GOVERNANCE STRUCTURES IN IRAQ

Before the creation of the Iraqi Ministry of Health and Environment, Iraq had a well-developed system of environmental governance and monitoring. A Human Environment Directorate was created under the Ministry of Health in 1972, after the United Nations Conference on the Human Environment. Then, in 1986, the Environment Protection Centre (EPC) was established within the Health Ministry. With the introduction of the Environment Protection and Improvement Law in 1997, the EPC was transformed into the Environment Protection and Improvement Directorate

(EPID). In 2001, the EPID was designated as an independent body and formally dissociated from the Ministry of Health.

An Environment Protection and Improvement Council (EPIC), consisting of various governmental and non-governmental representatives and experts was set up within this new institutional framework. Its principal duties were to:

• Review EPID's work plan, at both national and governorate levels;
• Approve environmental quality standards;
• Serve as an intra-governmental coordination body;
• Deliver decisions, including sanctions for environmental offences;
• Create an overall environmental policy framework, and
• Formulate Iraq's position in regional and international environmental negotiations and consider accession to international agreements.

In late 2003, the Iraqi Governing Council and Coalition Provisional Authority approved a resolution for the creation of the then new Environment Ministry, within which the existing Environment Protection and Improvement Directorate would form the core. The resolution, known as CPA Order #44, was eventually abolished and replaced by Law 37 of 2008, which officially established the Iraqi Ministry of Environment (which in 2015 became the Ministry of Health

& Environment) and outlined its mandate to protect and conserve Iraq's environment, as well as protect the residents of Iraq from environmental pollutants and environmental risks to human health. Other duties include the development of environmental policies and programs, as well as the creation and enforcement of environmental standards. One year later, in 2009, the Environment Protection and Improvement Law of 1997 was again updated and amended by Law 27, which continues to be the country's main environmental statute.

LAW NO. 27 OF 2009: PROTECTION AND IMPROVEMENT OF THE ENVIRONMENT OF IRAQ

Although there are a number of statutes, instructions, and other legal instruments that have been promulgated to regulate environmental issues in Iraq, there is only one main environment statute, and that is the Law on the Protection and Improvement of the Environment (Law 27 of 2009). As its name suggests, this is a comprehensive law aimed at "protecting and improving" the environment in a number of areas, including: water; air and noise; land use; biodiversity; hazardous waste; and petrol and natural gas extraction. To that end, it perpetuates and broadens the responsibilities of EPIC and local Provincial Environmental Protection and Improvement Committees; outlines principles for major areas of protection; establishes an environmental fund to finance cleanup; creates an enforcement team for oversight, and grants authority to the Minister of Health & Environment to force clean up of polluted sites and exact damages upon the non-compliant.

Article 18 of this Law regulates the protection of biodiversity. This section prohibits: damaging biota and their habitats; fishing, hunting, and trading of birds, wild animals, endangered and semi-endangered species; hunting, killing, possessing, or transferring of birds, wildlife and aquatic animals identified by the line Ministry; killing or harming rare, medicinal, or aromatic wild plants or their seeds; cutting certain trees in public areas within the city except by permission (trees 30 years old or older); logging without a license; introducing alien plants or animals without permission, or conducting research on genetic engineering to the detriment of the environment and biota.

IMPLEMENTATION OF ENVIRONMENTAL LAWS

Chapter 5 of Law 27 of 2009 creates an environmental enforcement team and the Minister must nominate the head of this team from the Ministry's staff. The head of the enforcement team is empowered with "state juridical power"; must be assisted by the environmental police, and has the authority to access businesses and other establishments.

Also some actions are being taken to better protect important habitats. For example, the authority over Natural Protected Areas, which were previously the responsibility of the Ministry of Agriculture were reassigned to the Iraq Ministry of Health & Environment (IMoHE) and Protected Areas regulations were developed and approved in 2014.

PARTICIPATION IN INTERNATIONAL ORGANIZATIONS/AGREEMENTS

Iraq is a signatory nation to a number of environmental conventions, which are listed in Table 4.

Table 4: International Environmental Conventions Iraq has joined

#	Convention/Agreement	Year	Responsible Ministry
1	Stockholm Convention on Persistent Organic Pollutants	2016	Ministry of Environment
2	The Cartegena Protocol on Biosafety to the Convention on Biological Diversity	Mar-2014	Ministry of Environment
3	Convention on International Trade in Endangered Species of Wild Fauna and Flora (CITES)	Mar-2014	Ministry of Environment
4	Convention of Migratory Species (CMS)	2014	Ministry of Environment
5	Convention of Conservation of African-Eurasion Waterbird Agreement (AEWA)	2014	Ministry of Environment
6	The International Treaty on Plant Genetic Resources for Food and Agriculture	2014	Ministry of Agriculture
7	The Basel Convention on the Control of Transboundary Movements of Hazardous Wastes and Their Disposal	May-2011	Ministry of Environment
8	United Nations Convention to Combat Desertification	May-2010	Ministry of Environment

#	Convention/Agreement	Year	Responsible Ministry
9	The Convention on Biological Diversity	Oct-2009	Ministry of Environment
10	United Nations Framework Convention on Climate Change	Jul-2009	Ministry of Environment
11	Kyoto Protocol - Framework Convention on Climate Change	Jul-2009	Ministry of Environment
12	Ramsar Convention on Wetlands	Feb-2008	Ministry of Water Resources
13	Convention on the Prohibition of the Use, Stockpiling, Production and Transfer of Anti-Personnel Mines	Aug-2007	Ministry of Environment
14	The Montreal Protocol and the Vienna Convention for the Protection of the Ozone Layer	Jun-2006	Ministry of Environment
15	Convention Concerning the Protection of the World Cultural and Natural Heritage (the World Heritage Convention)	1977	Ministry of Culture and Antiquities
16	International Plant Protection Convention	1954	Ministry of Agriculture

In addition, the Rotterdam Convention on the Prior Informed Consent Procedure for Certain Hazardous Chemicals and Pesticides in International Trade is under consideration but has not yet been ratified.

3.B CONSERVATION INITIATIVES IN IRAQ

KEY BIODIVERSITY AREAS PROGRAM

Previously, the only effort to identify important sites for conservation in Iraq was the work done by Evans (1994) to identify Important Bird Areas and Scott (1995) to identify key wetlands within the country, but almost no active field research had taken place since the 1970s. Little was known about the status of many sites or species during the intervening years, despite the fact that major ecosystem changes had taken place (e.g. drainage of the lower Mesopotamian Marshlands). The KBA Program was initiated in 2004 by the Iraq Ministry of Health and Environment and Nature Iraq with support from the Canada-Iraq Marshlands Initiative (CIMI) and BirdLife International. CIMI and BirdLife supported much of the initial training of the program teams and the first exploratory bird surveys in southern Iraq to test field methods. Subsequently, in 2006 major funding came from the Italian Ministry of Environment, Land and Sea. As a whole, the KBA Program represents the largest and most comprehensive survey and data collection effort focused on Iraq's biodiversity. The data set was also augmented by surveys conducted by Nature Iraq and sponsored by a number of international funding agencies that work in southern Iraq and the Kurdistan Region.

Attempting such a program in a post-conflict country that had not seen significant field research in over 20 years was a daunting task and required active management, oversight and on-going training. The program, particularly the ornithological work, benefitted greatly by having the on-going expertise of Richard Porter (BirdLife International) who advised on the methodology and reviewed and vetted all data collected since the inception of the program. As a result, the ornithological data have been quality controlled to the best existing abilities and represents some of the most in-depth and comprehensive surveys in the region. The other data generated in the KBA process did not go through as rigorous a review but staff involved in the data collection undertook annual field training and independent experts were consulted to help verify observations.

In the first two years of the program, staff had collected data during field visits but did not conduct analysis or prepare any reports on findings and it was not until 2007 that all field staff began to produce reports on the findings of the field work. The annual reports from 2008 to 2010 (now superseded by the current inventory) documented the progress of the KBA Program and represent an important capacity-building component of the process. They led to the evolution of the staff's capabilities and to the publication of this inventory – the product of more than seven years of surveys and data analysis against strict KBA criteria.

ADDITIONAL GOVERNMENT CONSERVATION ACTIONS

In addition to being the main partner in the Key Biodiversity Areas Program, the Iraqi Ministry of Health & Environment took a series of actions to address biodiversity conservation. Iraq joined the Convention on Biological Diversity in 2009 and submitted its first National Report to the CBD Secretariat on the status of the country's biodiversity in 2010[4] prior to attending the CBD Conference of Parties (COP10) in Nagoya, Japan (IMoHE, 2010). This report evaluated where Iraq stood in achieving the 2010 CBD goals and targets and found that while progress was being made there were significant information gaps and on-the-ground conservation actions to protect biodiversity were not yet in place.

Funded by the Italian Ministry of Environment, Land & Sea (IMELS) starting in 2005, the **New Eden Group** was made up of Italian and Iraqi environmental experts, with Nature Iraq as an implementing partner. It provided on-going support

4 Though this is Iraq's first report to the CBD it is entitled "The Fourth National Report on Biodiversity in Iraq." (IMoHE, 2010)

to the Iraqi Ministry of Health & Environment as well as the Iraqi Ministry of Water Resources (specifically to the Center for the Restoration of Iraqi Marshes and Wetlands (CRIM-W) and assisted the Iraqi government in developing a number of feasibility studies, management plans and pilot projects including work for the Central Marshes National Park and the Hawizeh Marsh Ramsar Site (both discussed below). It also provided extensive training and capacity building programs to a number of Iraqi ministries on environmental topics.

In 2012, the IMoHE released the **National Environmental Strategy and Action Plan (NESAP)** for Iraq. This outlines ten strategic objectives for 2013 to 2017 to protect air and water quality, the marine and coastal environment, biological diversity; handling and managing hazardous wastes; reduction of pollution and radioactive contamination, and developing an institutional and legal framework for the environmental sector (see Table 5). While all these have implications for Iraq's biological diversity, Strategic Objective V has six components specific to biodiversity protection and use.

Table 5: NESAP Strategic Objective V: Protection and sustainable use of biodiversity

Convention/Agreement	Year
Component I: Local species 5.1 Biodiversity protection	5.1.1 Establishment of the national network of natural reserves 5.1.2 Biodiversity protection in the Marshlands 5.1.3 Maintaining areas of natural heritage 5.1.4 The national strategy for biodiversity 5.1.5 Mapping of the important areas of biodiversity and birds 5.1.6 Inventory of biodiversity and lists of endemic and endangered species 5.1.7 Strategy of invasive species control 5.1.8 Natural pest control
Component II: Keeping samples of Iraqi organisms 5.2 Saving genetic samples and germplasm	5.2.1 Saving genetic germplasm of agriculture (palm groves) 5.2.2 Establishment of stations for breeding of endangered Iraqi deer 5.2.3 Establishment of nature reserves 5.2.4 Animal genetic resources 5.2.5 DNA Project 5.2.6 Museum of Natural History
Component III: Bio safety and security 5.3 Bio-safety and Bio-security in Iraq	5.3.1 The national framework for biological safety in accordance with Cartagena Protocol 5.3.2 Implementation of the National Program for Biological safety in accordance with Cartagena Protocol
Component IV: Sustainability of ecosystems 5.4 Biological diversity in curricula of research, education and training	5.4.1 Integration of biodiversity in the curricula of research
5.5 Ecotourism	5.5.1 Activation of eco-tourism
Component V: Institutional and legal frameworks 5.6 Development and enforcement of biodiversity protection legislations	5.6.1 Developing the institutional framework for biodiversity management 5.6.2 Monitoring and activation of compliance with laws 5.6.3 Biodiversity staff capacity-building
Component VI: Environmental awareness and public participation 5.7 Outreach and inventory of cultural environmental heritage	5.7.1 Outreach and inventory of cultural environmental heritage 5.7.2 Biodiversity actors award

Several of the projects and activities listed under NESAP Objective V have been undertaken or are planned in Iraq but Iraq will also need to prepare and implement a specific action plan for biodiversity conservation as required under the Convention on Biological Diversity. In its national report submitted to the CBD Secretariat in 2010, Iraq provided a roadmap forward to the development of the **National Biodiversity Strategy & Action Plan (NBSAP)**. In addition, an action plan for the implementation of a **Programme of Work for Protected Areas (POWPA)** in Iraq was later submitted by the IMoHE to the CBD Secretariat in May 2014 (this is discussed in more detail in the next section).

Also in 2012, UNEP-ROWA (United Nations Environmental Programme — Regional Office for West Asia) and the Iraq Ministry of Health & Environment (IMoHE) officially launched the Global Environment Facility (GEF)-funded project *"First NBSAP for Iraq and Development of Fifth National Report to the CBD"*. This support enabled the government of Iraq to pursue further consultation with key Iraqi stakeholders and release its Fifth National Report, submitted to the CBD Secretariat in March of 2014. This report outlines many of the actions taken towards implementation of the CBD in Iraq since 2010; provides in detail the progress towards achieving 2020 Aichi Biodiversity Targets for Iraq, and gives further information on the on-going efforts to develop Iraq's National Biodiversity Strategy and Action Plan (NBSAP).

Technical staff from the IMoHE along with Nature Iraq, UNEP and the IUCN has also been involved in the development and updating of Red Lists for globally threatened taxa in Iraq. Meetings organized by a UNEP-IUCN partnership focused on drafting a Regional Red List for the Arabian Peninsula, including Iraq, which help in the preparation of species action plans. A first national-level Red Listing assessment was carried out by a team of Iraqi experts under the supervision of the IUCN Regional Office in 2013 and focused primarily on species in the southern marshlands of Iraq.

Regional initiatives are also taking place that are significant to conservation in Iraq. These include the IUCN Freshwater Key Biodiversity Assessment Program completed in 2014 and smaller-scale, local initiatives such as the botanical garden project started in 2012 by the Sulaymaniyah Governorate, which has the main goals of establishing a garden to explore, explain, and conserve the botanical wealth of the Kurdistan region of Iraq through scientific research. The Botanical Garden project has not been activated for lack of funding, however, more recently the Kurdistan Botanical Foundation (KBF), a sister organization to Nature Iraq has undertaken surveys in the field actively collecting samples of flora. They continue the work of classifying the samples and preserving them for future studies and have publishes several publications documenting new species that have been found and classified. Much more work remains to be done and it is taking place as funding permits.

DESIGNATION OF PROTECTED AREAS IN IRAQ

The Convention on Biological Diversity defines a protected area as '*A geographically defined area, which is designated or regulated and managed to achieve specific conservation objectives.*' IUCN has developed different categories of protected areas (shown in Table 6). In Iraq,

Table 6: The categories of protected areas as defined by IUCN (2014b)

No.	Name	Description
Ia	Strict nature reserve	Strictly protected for biodiversity and also possibly geological/ geomorphological features, where visitation, use and impacts are controlled
Ib	Wilderness area	Usually large unmodified or slightly modified areas, without permanent or significant human habitation, managed to preserve their natural condition
II	National park	Large natural or near-natural areas with characteristic species and ecosystems, which support compatible spiritual, scientific, educational, recreational and visitor opportunities
III	Natural monument or feature	Areas set aside to protect a specific natural monument, e.g. a landform, sea-mount, marine cavern, cave, or a living feature such as an ancient grove.
IV	Habitat/species management area	Areas to protect particular species or habitats, where management reflects this priority. Many will need regular, active interventions to meet the needs of particular species or habitats, but this is not a requirement of the category
V	Protected landscape or seascape	Where interaction of people and nature over time has produced a distinct character with ecological, biological, cultural and scenic value: where keeping this interaction is needed to safeguard nature conservation and other values.
VI	Protected areas with sustainable use of natural resources	Areas that conserve ecosystems, together with associated cultural values and traditional natural resource management systems. Generally large, mainly in a natural condition, with a proportion under sustainable natural resource management.

there is often a perception that a protected area is land surrounded by a fence that people are restricted from using. This type of protection is known as "fortress conservation" — "preserving" areas by forcing people off their lands and/or not allowing traditional land use in these areas. The often unintended consequences are increased poverty, embittered local stakeholders and sometimes, paradoxically, actual declines in overall biodiversity. The IUCN Protected Area categories contradict this belief by showing the wide variety that exists in such designations.

The nomenclature of protected areas considers other types of designations as well such as Man and Biosphere Reserves and World Heritage Natural Sites. The former covers internationally designated, protected areas that are meant to demonstrate a balanced relationship between people and nature (encouraging sustainable development). Iraq has no Man and Biosphere Reserves presently but has been a signatory nation since 1974 to the UNESCO World Heritage Convention (WHC). The WHC allows for the designation of Cultural[5], Natural and Mixed Sites. In 2014, Iraq submitted a proposal to list the Southern Marshlands as Iraq's first World Heritage Mixed (Natural and Cultural) site, under the following four WHC Natural Site criteria (UNESCO, 2014):

(iii) "A unique or at least exceptional testimony to a cultural tradition or to a civilization which is living or which has disappeared"

5 Iraq previously had only four Cultural Sites listed as WHS. These include Erbil Citadel, Hatra, Ashur (Qal'at Sherqat), & Samara Archeological City. The latter three are inscribed on the List of World Heritage in Danger.

(v) "An outstanding example of a traditional human settlement, land-use, or sea-use which is representative of a culture (or cultures), or human interaction with the environment especially when it has become vulnerable under the impact of irreversible change"

(ix) "An outstanding example representing significant on-going ecological and biological processes in the evolution and development of terrestrial, freshwater, coastal and marine ecosystems, and communities of plants and animals"

(x) "Contains the most important and significant natural habitats for in-situ conservation of biological diversity, including those containing threatened species of outstanding universal value from the point of view of science or conservation"

In July of 2016, UNESCO formally accepted the "The Ahwar of Southern Iraq" as a cultural and natural World Heritage Site made up of seven components: three archaeological sites and four wetland marsh areas in southern Iraq (2016). These marsh sites include the Central Marshes (IQ075), Hawizeh (IQ073), and East and West Hammar (IQ077 & IQ076). Under the WHC, Iraq needs to implement management plans to protect these sites and report regularly on the state of conservation of its World Heritage sites. Combined the four marsh areas of the Ahwar of Southern Iraq covers an area of 210,899 hectares with an additional 207,643 hectares of buffer (see Map 7).

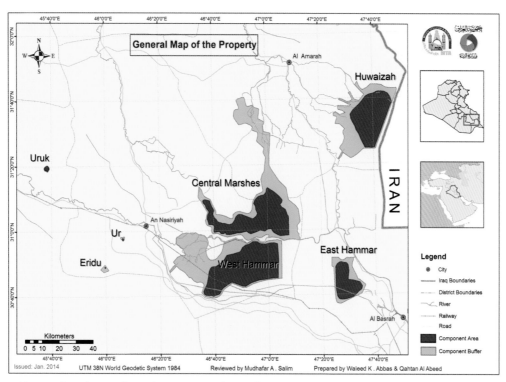

Map 7: The Ahwar of Southern Iraq: Refuge of Biodiversity and the Relict Landscape of the Mesopotamian Cities (UNESCO, 2016)

Map 8: Hawizeh Ramsar Site

But the very first official protected area designation in Iraq was assigned to the **Hawizeh Marshes Ramsar site** (part of KBA site IQ073) in 2007 when Iraq became signatory to the Ramsar International Convention on Wetlands. The Ramsar Convention requires member countries, such as Iraq, to maintain the ecological character of their "Wetlands of International Importance" and to plan for the sustainable use of all of the wetlands in their territories (Ramsar, 2014). Hawizeh Marshes are unique in Iraq as a transboundary wetland with Iran (the Iranian portion is called Hor Al-Azim) and were the only part of the Mesopotamian Marshlands that were never completely drained in the 1990s. Unfortunately, these wetlands have faced severe reductions due to drought as well as hydro/agricultural irrigation projects both inside Iraq and in upstream states. In addition, in 2011 Iran completed a 90 km embankment through the marsh along the border restricting the flow of water from Hor Al-Azim into Hawizeh. These threats led Iraq to include this Ramsar site on the

Convention's Montreux List of ecologically threatened sites in April 2010. A total of 137,700 hectares (1,377 km²) in Hawizeh is covered by its Ramsar designation as shown in Map 8, but while management plans have been developed for Hawizeh implementation of these plans has lagged behind. Additionally in 2014, 500 hectares of the Sawa Lake KBA Site (IQ071) were listed as a Ramsar Site as well as the 219,700 hectares of the Central Marshes (IQ075) and 180,000 hectares of the Hammar Marshes (including both East and West Hammar (IQ077 & IQ076)).

In 2006, a year before the designation of Hawizeh as a Ramsar site, efforts were initiated to establish Iraq's first national park in the Central Marshes. A feasibility study was developed and subsequently a management plan was developed in 2008 with an updated version completed in 2010 (New Eden Group) and submitted to the Iraq Ministry of Health & Environment (IMoHE). Several socio-economic development and educational projects in and around the park area were conducted and/or proposed throughout this period.

Planning for and establishing of a network of Protected Areas in Iraq was to continue to be of great importance to the Ministry. In 2010, a strong effort to review and elaborate on existing information collected in the extensive monitoring programs (including on KBAs) carried out since 2004 was organized in a GIS database in order to select priority sites to be considered for future protected areas status in Iraq.

In July 2013, the Iraqi Council of Ministers made an important step towards protecting part of the Mesopotamian marshlands by approving **Iraq's first National Park** in the Central Marshes (part of IQ075). See Map 9. In addition to establishing the park, as was mentioned previously, the Iraqi Council of Ministers issued an order to assign the exclusive authority for the designation of protected areas to the Ministry of Environment.

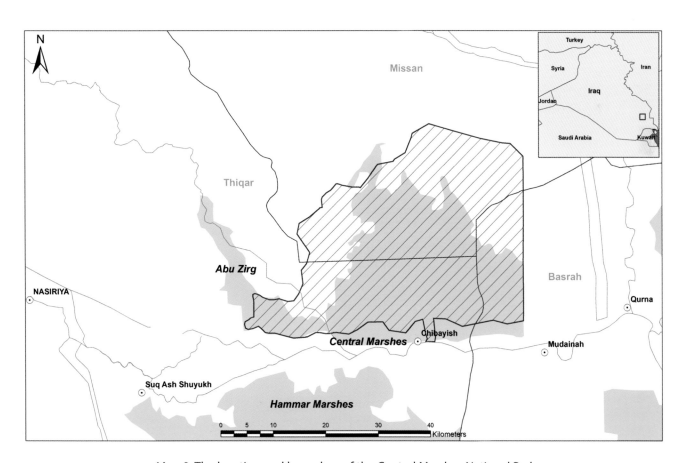

Map 9: The location and boundary of the Central Marshes National Park

In addition to these efforts, the IMoHE as part of their commitment under the CBD submitted a proposed list of protected areas (these are the first 10 sites listed in Table 7), under their "Programme of Work for Protected Areas" (PoWPA) to the CBD Secretariat in 2012.

Then in the summer of 2013, the IMoHE submitted a project proposal through UNEP-ROWA for an additional GEF-funded project entitled, "Initial steps for the *Establishment of the National Protected Areas Network*". This project had three components that include designing a Protected Areas System and institutional strengthening, implementation of the Protected Areas Network, and public awareness raising. The list of proposed protected areas developed in 2012 was then updated. Teeb (including both IQ067 & IQ068) (Site #11 in Table 7) was added to the list and later in 2014, still more sites were added to the list (Sites #12 through 19 in Table 7). In 2015, Teeb (Site #11) and Dalmaj (IQ064) (Site #1 in Table 7) were to be implemented as pilot project sites. Lack of funding has hindered the implementation of these projects.

Table 7: Original list of proposed priority sites for protected status in Iraq showing their ecoregion(s)

#	PROPOSED SITES	ECOREGION
1	Dalmaj (IQ064)	Arabian Desert and East Sahero-Arabian Xeric Shrublands (PA1303)
2	Mosul Lake (IQ009)	Middle East Steppe (PA0812)
3	Peramagroon Mt. (IQ027)	Zagros Mountains Forest Steppe (PA0446)
4	Razzaza (IQ058)	Arabian Desert and East Sahero-Arabian Xeric Shrublands (PA1303)
5	Tharthar Lake (IQ051)	Mesopotamian Shrub Desert (PA1320)
6	West Hammar (IQ076)	Tigris-Euphrates alluvial salt marsh (PA0906) (now part of the Hammar Marsh Ramsar site (2014) and the Ahwar of Southern Iraq UNESCO Site (2016))
7	Khor Az-Zubayr (IQ081)	South Iran Nubo-Sindian desert and semi-desert (PA1328)
8	Qara Dagh (IQ039)	Zagros Mountains Forest Steppe (PA0446)
9	Wadi al Weir and Sh'eeb Abu-Talha (IQ070)	Arabian Desert and East Sahero-Arabian Xeric Shrublands (PA1303)
10	Haditha Wetlands & Baghdadi (IQ050)	Mesopotamian Shrub Desert (PA1320)
11	Teeb (including Teeb Oasis & Zubaidaat (IQ067) and Teeb Seasonal Wetlands (IQ068)	South Iran Nubo-Sindian desert and semi-desert (PA1328) & Tigris-Euphrates alluvial salt marsh (PA0906)
12	Barzan Area & Gali Balnda (IQ004)	Zagros Mountains Forest Steppe (PA0446)
13	Al-Qosh (New site) – 36.75°N, 43.09°E	Zagros Mountains Forest Steppe (PA0446)
14	Bedra and Zurbatiyah (aka Jazman (Zurbatia) (IQ057)	South Iran Nubo-Sindian desert and semi-desert (PA1328), Mesopotamian Shrub Desert (PA1320), & Zagros Mountains Forest Steppe (PA0446)
15	East Hammar (IQ077)	Tigris-Euphrates alluvial salt marsh (PA0906) (now part of the Hammar Marsh Ramsar site (2014) and the Ahwar of Southern Iraq UNESCO Site (2016))
16	Marine Area (New site) – 29.75°N, 48.75°E	Arabian Gulf (90)
17	Sakran Mountain (IQ019)	Zagros Mountains Forest Steppe (PA0446)
18	Sawa Lake (IQ071)	Arabian Desert and East Sahero-Arabian Xeric Shrublands (PA1303) (now a Ramsar Site (2014))
19	Suwaibaat (or Sleibaat) (IQ074)	Arabian Desert and East Sahero-Arabian Xeric Shrublands (PA1303) & Tigris-Euphrates alluvial salt marsh (PA0906)

Table 8 provides information on the currently established and proposed protected areas as well as several other important sites in Iraq listed by type and showing their number, total area and percentage of the country.

Table 8: Current and Proposed Protected Areas in Iraq

Protected Area Type	Number	Area	% of Total Area of Iraq
National (using IUCN categories)			
II: National Park	1 (designated under category II, V & VI)	Approx. 1416 km² (141,615 ha)	0.33%
V: Protected Landscape/ Seascape	(see above)	(see above)	
VI: Managed Resource Protected Area	(see above)	(see above)	
Other			
1. National Committee of Protected Areas proposed sites (type not defined)	Proposed sites (See Table 7)	Over 13400 km² (over 1344382 ha) based on KBA boundaries	Over 3%
2. Halgurd-Sakran Park (regional proposed park)	1 proposed regional park	Approx. 322 km² (32168 ha)	0.07%
3. Barzan Area (tribal pro-tected area)	1 informal tribal protected area	Approx. 1148 km² (114822 ha)	0.26%
4. Ministry of Agriculture (MoA) Protected Areas/ Reserves	14 established	32 km²	0.01%
International Protected Area Categories			
Ramsar Sites*	4 established	Total: 5379 km²	1.24%
1. Hawizeh (2007)	Established	1377 km² (137700 ha)	
2. Sawa Lake (2014)	Established	5 km² (500 ha)	
3. Central Marshes (2014)	Established	2197 km² (219700 ha)	
4. Hammar Marshes (East & West) (2014)	Established	1800 km² (180000 ha)	
World Heritage (WH) Sites (with date of inscription)**	**5 established (Cultural Sites)**	Total: 4681.25 km²	1.08%
1. Ashur (2003)	Under criteria: (iii)(iv)	1.7 km² (70 ha with 100 ha buffer)	
2. Hatra (1985)	Under criteria: (ii)(iii)(iv)(vi)	3.2 km² (323.75 ha)	
3. Samarra Archaeological City (2007)	Under criteria: (ii)(iii)(iv)	465 km² (15058 ha with 31414 ha buffer)	
4. Erbil Citadel (2010)	Under criteria: (iv)	0.2 km² (16 ha with 268 ha buffer)	
5. The Ahwar of Southern Iraq (2016)	Under criteria: (iii)(v)(ix)(x)	4209 km² (211544 ha with 209321 ha buffer)	
WH Tentative List Sites (with data of submisson)**	10 proposed sites		
1. Ur (2000)	Cultural site criteria: (i)(iii)(iv)	Unknown	
2. Nimrud (2000)	Cultural site criteria: (i)(ii)(iii)	Unknown	
3. Ancient City of Nineveh (2000)	Cultural site criteria: (i)(ii)(iii)(iv)(v)(vi)	Unknown	
4. Fortress of Al-Ukhaidar (2000)	Cultural site criteria: (i)(ii)	Unknown	
5. Wasit (2000)	Cultural site criteria: (i)(ii)(iv)	Unknown	

Ramsar List: rsis.ramsar.org   ***World Heritage List:* whc.unesco.org/en/list/
*** *World Heritage Tentative List:* whc.unesco.org/en/tentativelists/

Table 8: Current and Proposed Protected Areas in Iraq

	Protected Area Type	Number	Area	% of Total Area of Iraq
6.	Babylon – Cultural Land-scape & Archeaological City (2003)	Cultural site criteria: (iii)(vi)	Unknown	
7.	Site of Thilkifl (2010)	Cultural site criteria: (i)(ii)(iii)(iv)(v)(vi)	Unknown	
8.	Amedy city (2011)	Mixed Cultural & Natural Site: (i)(ii)(iii) (vii)(viii)	Unknown	
9.	Historical Features of the Tigris River in Baghdad Rusafa (2014)	Cultural site criteria: (i)(ii)(iv)(vi)	Unknown	
10.	Wadi Al-Salam Cemetery in Najar (2011)	Cultural site criteria: (iii)(v)(vi)	Unknown	

Please note, official designation as a protected area as listed in Table 8 implies only a commitment to the protection of these sites and in most cases these sites have seen few if any conservation actions on the ground.

In addition to official protected sites, the Iraqi Ministry of Agriculture maintains 14 agricultural reserves that include reserves for plants, water bodies and conservation of certain wild species (see Table 9), though these areas are all relatively small in size (often less the 1 km²) and some are no longer operating.

Table 9: Ministry of Agriculture Reserves

Reserve Name	Location	Species & origins	Management	Notes
Gseibah	Baghdad	Wild animals of Iraqi origin (65 Gazelles)	State Company of forests and orchards	Operational and will receive new bird species and other animals at last report
Raodhat Al-Maha	Baghdad	-	Same	Not operational due to locals illegal seizure
Al-Nahrawan	Baghdad	-	Same	Not operational due to locals illegal seizure
Al-Dibis	Kirkuk	Iraqi plants & animals	Same	Operational at last report but current status unknown
Mandli	Diyala	Iraqi wild animals	Diala Directorate of Agriculture	Operational and will receive new species
Al-Masad	Anbar	Iraqi wild animals (>1300 Gazelles)	State Company of animal wealth	Operational at last report but current status unknown
Al-Dhabaa	Anbar	Iraqi wild animals (95 Gazelles)	State Company of animal wealth	Operational at last report but current status unknown
Sinjar	Mosul	Plants	Mosul Directorate of Agriculture	Was being rehabilitated but current status is unknown
Al-Safiyah	Basrah	Aquatic	Basrah Directorate of Agriculture	Dried out due to the Iranian embankments and is out of service now
Ali Al-Gharbi	Amara	Iraqi wild animals	Amara Directorate of Agriculture	Operational and 25 gazelles were added upon last report in 2014
Al-Samawa	Samawa	Wild animals	Samawa Directorate of Agriculture	Being rehabilitated
Karbala	Karbala	Wild animals	Karbala Directorate of Agriculture	Out of service now and proposed for investment
Najaf	Najaf	Iraqi wild animals (48 Gazelles & 21 Ostriches)	Najaf Directorate of Agriculture	Operational at last report but current status unknown
Nasiria	Thi Qar	Aquatic	Nasiria Directorate of Agriculture	Being rehabilitated

Both the Ministry of Tourism and the Tourism Board of the Kurdistan region are developing long-range plans for tourism sites throughout the country. There is interest in identifying areas to promote eco-tourism and some areas being developed for tourism are in or near KBA Sites. There are many potential synergies between tourism sites and Iraq's Protected Area Network, but many potential pitfalls as well particularly if tourism development does not ensure the protection of the species and habitats at these sites.

Additionally, it should be noted that the Kurdistan region is also interested in developing regional parks and an initiative to develop a park encompassing Halgurd and Sakran Mountains (including parts of IQ017, IQ018 & IQ019) has started with support from the Protected Areas High Committee of the Kurdistan Regional Government. A park committee was formed and an office for the park was developed in Choman, Erbil. The park has a presence on social media and the Environmental Police have attempted to curtail illegal hunting over some of the park area (Abdul-wahid H. Gwany, Personal Communication, 14 December 2013). As of January 2017, these efforts have stalled due to the government funding crisis in the Kurdistan region.

Additionally, one site in Iraqi Kurdistan, the Barzan Area (IQ004), is noteworthy for having, until recently, only informal environmental protection. Protected since the early 20th century through tribal prohibition and controls on hunting and habitat uses, this KBA site has an area of over 110,000 hectares representing approximately 4% of the Zagros Mountains Forest Steppe (PA0446) Ecoregion in Iraq. The benefit of long-term protection for this extensive area can be seen in the fact that

Map 10: Proposed Sakran-Halgurd Park shown with KBA sites (park boundary map courtesy of M. Jung, A. Oblin (AIT-Austrian Institute of Technology) & G. Loiskandl)

Plate 8: Road sign upon entering the Barzan Area (IQ004) posted by the KEPIC in Erbil ©A Bachmann/NI

this is one the few sites in Iraq where a large population of the globally Vulnerable Wild Goat *Capra aegagrus* can easily be seen. It is a noteworthy lesson that the Barzan Protected Area, which has achieved a relatively high level of protection since the beginning of the last century, was created only by local community agreement.

4.A SURVEY AREAS, SITE SELECTION AND SURVEY PERIODS

For the purpose of data collection efforts as well as the logistics and capacity reasons, Iraq was divided into three regions: The Kurdistan Region of Northern Iraq, Central Iraq and Southern Iraq.

Survey work for southern Iraq began in the winter of 2005. Surveys were then extended to the Kurdistan region of northern Iraq in the winter 2007 and in Central Iraq the surveys started in the winter of 2009.

The Important Birds Areas of Iraq chronicled in Evans (1994) and key wetland areas in Scott (1995) formed the initial list for determining survey sites. This was further developed during consultations with experts on the distribution of Iraq's national fauna and flora, and augmented by targeted field surveys of poorly known areas.

All sites were located using a *Garmin* GPS. Occasionally maps (usually at 1:100000 scale) were used to locate the area of the fieldwork. Standard site information sheets were used to capture data on fauna and flora, habitats and threats together with site location, security and logistical details, photographs taken and, if necessary, a sketch map of the site. The survey teams varied in size depending on the size of the site and the faunal groups to be studied (usually between two and five field workers).

Some sites were surveyed as a whole unit, whereas others, notably large sites, were divided into sub-sites to help manage the survey effort. These areas were surveyed generally in both summer and winter over a varying number of years between 2005 and 2010 (with additional information added from other surveys in 2011 and 2012). In some cases, sites were additionally visited in spring, notably in the Kurdistan region during the annual training program.

A total of 220 sampling sites were surveyed from 2005 – 2010 with some additional visits to some sites in 2011-2012. Some sites in southern Iraq were visited over six years with a total of twelve seasonal surveys - six summer and six winter visits, for other sites visits were less frequent.

Please see Appendix 2 for a list of the KBA sites and their survey periods.

4.B TEAM TRAINING

Workshops were held regularly throughout the program to assess the data quality, maintain a list of candidate KBAs and inform, involve and continually train the survey teams and other interested individuals. Training focused on survey methods, the use of data sheets, application of IBA, IPA, and KBA criteria, data entry onto Nature Iraq databases and reporting.

As there are few field ornithologists in Iraq, an active field-based training program was developed on bird identification and survey techniques, which led to a significant increase in the ornithological knowledge and expertise within the country. Over 40 people from the Ministry of Environment

Plate 9: Summer 2007 KBA Survey Team @NI

Plate 10: 2008 Bird Survey Training

and its district offices as well as the Kurdistan Environment Protection & Improvement Commission, universities, interested individuals, and Nature Iraq staff participated in these training courses.

4.C SURVEY METHODOLOGY

BIRDS

Bird surveys were undertaken at each site, usually in summer and winter. They were often a rapid assessment to identify presence/absence of species and count or estimate numbers. Observations were made using binoculars (usually 8x30) and telescopes (typically 30 or zoom x 65 mm) and were backed-up by photographs using a Canon 200mm and 400mm lens. An attempt was made to cover all habitats at a site as systematically as possible. In some cases small birds (passerines and non-passerines) were located by sight or voice and counted within a (very approximate) band with a width of approximately 100m; for larger birds, such as raptors, the range was much greater, up to 2-3 km.

At wetland sites, particularly areas of open water, counting all waterbirds (especially wildfowl and waders) was often possible by careful scanning from selected vantage points on the shore using binoculars and telescopes; totals were then obtained by aggregating the counts. Occasionally a motor-canoe (locally called a Shakhtoora) was used for accessing remote areas or deep water, or observations were made while wading through reed beds. Cars were used to cover the more accessible areas and often proved useful as 'hides'.

For larger birds, notably raptors, sitting and scanning from suitable vantage points for at least three hours was essential to assess numbers of potentially breeding birds present, particularly during the spring and summer when birds are displaying. The experience of the observer in understanding the habits and territorial behavior of these birds, prevents counting duplication.

During the summer surveys, strong emphasis was placed on determining the breeding status of all birds encountered. Breeding evidence was based on British Trust for Ornithology guidelines (see Appendix 3).

OTHER FAUNA SURVEYS

Efforts to conduct a more focused mammal survey was initiated in 2007, but discontinued pending more capacity building. Field teams always collected anecdotal information on mammals and other fauna including taking photographs of live animals, examination of tracks or any signs of presence in a survey site. Since many mammals are nocturnal, most species were rarely observed during the rapid assessment surveys. However, a more focused survey on mammals was undertaken in Kurdistan, northern Iraq from summer 2009 to summer 2010. This included gathering reports (often through taped interviews) of sightings from locals, and documenting the extent of hunting and trade by interviewing people at the site, at regional animal markets, and by visiting and interviewing staff at local zoos. In areas where there were minefields site access was difficult, therefore the team replied completely on taped interview with local villagers.

At animal markets taped interviews were conducted with pet shop owners and, if present, local hunters. In addition, in 2010, species and numbers were recorded in the markets, together with their capture information and the health status of species present. A photographic record was kept of these visits to markets and this proved helpful for subsequent identification of species.

During the surveys the team used a recorder to maintain a record of the interviews, (making sure to protect the anonymity of those interviewed) to allow the interviewer to collect the information freely and accurately. All data were entered into Excel datasheets, under the three categories of site visits, animal markets, and zoos (note that data from the latter two categories was not generally used in individual site assessments but provided important background information for the team).

Information collected on other fauna, notably reptiles, amphibians and large insects (the last were not included in the assessments) was also largely anecdotal and opportunistic, but where photographs were obtained later identification was often possible with the assistance of outside expertise.

Training of field staff in these, often challenging groups, and in survey methods more appropriate for these taxa is needed to enable the collection of more systematic and comprehensive fauna data in Iraq.

PLANTS & HABITATS

The botany survey was conducted at various times from late spring to summer (See Appendix 2 for the site survey periods). Initially only general observations and plant identifications were done in the field but eventually more formal waypoints were selected within the main habitat types present within a site. GPS coordinates and the elevation for each waypoint along with photographs of the waypoint were taken and a description including slope, exposure, and percentage of vegetated area was developed (the latter is described more fully below). The dominant tree, shrub, herb and grass species as well as any plants of conservation importance such as endemic species, rare plants, and plants with cultural or economic importance were noted for each waypoint. Any threats to the site were also noted. Plants were identified in the field or were collected in plastic bags and then pressed before being sent back to the office for identification.

Extensive pictures of habitat and plants were also taken during the site surveys. In 2009,

panoramic photographs of many sites were taken. Pictures of plants and habitats were taken in order to help with plant identification and to describe their status at a site. Detailed plant photos were also used to develop plant profiles (photos of plant parts to be assembled later into a complete digital profile of the plant). These aid in the identification of species and can be used to develop electronic, on-line plant guides and educational material. This method and the method of assigning herbarium numbers to individual specimens were introduced in 2009 by the CMEP/RBGE. An example of a plant profile is provided below.

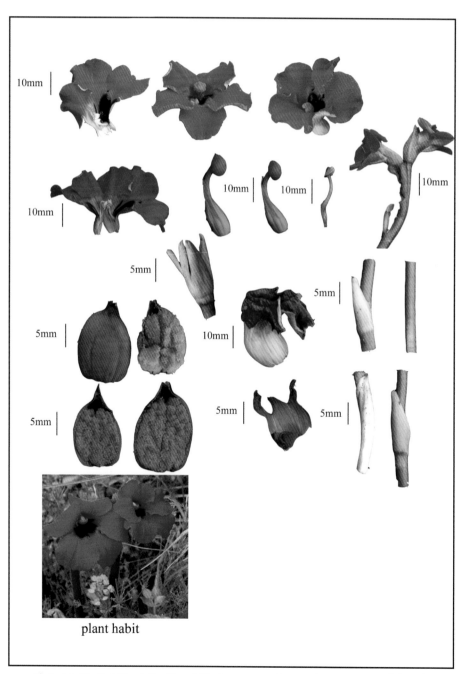

Plate 11: Partial Plant Profile of *Phelypaea coccinea* at Peramagroon (IQ027) developed with Photoshop CS3 © S Abdulrahman/NI

Vegetation cover at the site was estimated using direct observation (determining a percentage of vegetated vs. non-vegetated area). Prior to the adoption of the full threat assessment methodology the botanical team determined the ecological condition of the site based on a rating scale of 1 (least disturbed or impacted; best ecological condition and quality) to 5 (most disturbed or impacted; poorest ecological condition and quality). Though these methodologies are subjective, the goal of the survey was simply to conduct a rapid assessment of the overall plant communities as well as their habitat and the threats that they face.

REFERENCES USED

Several key references that have been used extensively to establish the framework of the program and to guide the data-gathering and analysis are listed below.

SITES

1. *Important Bird Areas in the Middle East.* (Evans, 1994). A key reference used to identify the initial sites surveyed. The Inventory of Key Biodiversity Areas attempts to update this previous publication.

2. *A directory of wetlands in the Middle East.* (Scott, 1995). An important reference on wetlands in Iraq that was also used to identify survey sites.

BIRDS

The main identification guides used in the surveys were Salim, Porter, Christensen, Schiermaker-Hanson, & Jbour (2006); Porter & Aspinal (2010), Porter, Christensen, & Hansen (1996); Mullarney, Svensson, Dan, & Grant (2001) and Allouse (1953 & 1960-62), the latter was used to review and compare the bird populations over certain areas as a whole.

In the case of birds, the BirdLife International website Data Zone (See Web Resources) have the benefits of being widely accessible and the most up-to-date documentation for most of BirdLife's science, including the Global IBA criteria (BirdLife International, 2013a). Additionally other key references included:

1. *IUCN Red List for birds* (BirdLife International, 2013b). The list of globally threatened species, updated annually by BirdLife International for IUCN; it is the source of species of global conservation concern used to select IBAs under A1 criterion.

2. *An annotated checklist of the birds of Iraq* (Salim, Al-Sheikhly, Majeed & Porter, 2012). This checklist was developed largely as a result of the KBA bird surveys.

3. *Threatened Birds of the World* (BirdLife International, 2000). This BirdLife publication identified, on behalf of IUCN, all bird species considered of global conservation concern and, together with the updated IUCN Red List, was the source of the list of bird species of global conservation concern that was used to select IBAs under the A1 criterion (see Appendix 1).

4. *Waterbird Population Estimates, 3rd and 4th Editions* (Delany and Scott, 2002 & 2006). Wetlands International has published much information on the sizes and geographical ranges of waterbird populations globally on a regional basis. They also provided access to unpublished data as well as analyses of their databases, using the 1% population thresholds for identifying IBAs under the A4i and A4iii criteria.

5. *Regional Checklist* (BirdLife International, 2010). BirdLife International provided data including species attributes, relevant IBA criteria and populations thresholds, regional and national occurrence, and taxonomy that used to aid identification of IBAs in Iraq.

Additional references used for assessments in southern Iraq included: Abed, Altaey, & Salim (2013), Kubba and Salim (2011), and Salim (2004, 2008a, 2008b, 2009a, 2009b, 2010, & 2011).

INFORMATION ON OTHER (NON-AVIAN) FAUNA

The IUCN Red List was the source of the list of mammals, reptiles and amphibians (containing species that the IUCN has evaluated) together with a draft checklist of mammals, reptiles, and amphibians of Iraq by Dr. Zuhair Amr of the Jordan University of Science & Technology as part of a review of past literature (Amr, 2009a &b). Field guides used in the survey included Murie and Elbroch (2005) and Stokes and Stokes (1986). Historical information and other sources of information on other fauna species are summarized below:

MAMMALS

1. *The Mammals of Iraq* (Hatt, 1959). This provides rich historical information on the distribution of the mammals of Iraq as well as the localities where specimens of rare species were purchased, giving a basic knowledge of choosing sites when surveying for mammals.

2. *The Mammals of Arabia* (Harrison, 1964-1972). Volume I (Insectivora, Chiroptera, Primates), II (Carnivora, Hyracoidea, Artiodactyla) & Volume III (Lagomorpha, Rodentia) - Updated reference on the mammals of this region after Hatt's book.

3. *The Mammals of Arabia, Second Edition* (Harrison and Bates, 1991). This reference is the same as Harrison's three volumes but some of the general information has been condensed and the distribution records have been updated.

REPTILES & AMPHIBIANS

1. *Handbook to Middle East Amphibians and Reptiles* (Leviton, Anderson, Adler and Minton, 1992). Published by the Society for the study of Amphibians and Reptiles

2. *The Reptile Database* (See Web Resources). Online database cataloguing all living reptile species and their classification.

3. *Gecko Fauna of the USSR and contiguous regions* (Szczerbak and Golubev, 1986). Provides a thorough description and identification keys of the Geckos.

4. *Lizards of Iran* (Anderson, 1999). Published by the Society for the Study of Amphibians and Reptiles this reference provides a thorough description and identification keys of the Iranian Lizards.

5. *Snakes of Iran* (Latifi, 1991). Published by the same society as the Lizards of Iran above this provides a comprehensive reference with descriptions and identification keys of the Iranian Snakes.

FISH

1. *Freshwater Fishes of Iraq* (Coad, 2010). Provides a good insight of the Iraqi freshwater fishes along with the conservation and economic importance of each fish species. Coad also maintains a website with updated information (See Web Resources).

2. *FishBase* (See Web Resources) and *Catalog of Fishes* (See Web Resources). Both online websites that contain information on the fresh and saltwater fishes of Iraq and provide updated scientific names on these species.

BOTANICAL INFORMATION

Botanical information on Iraq species and habitats was, during much of the survey effort, limited to the incomplete *Flora of Iraq* and *Flora Iranica*. Family limits and generic placements follow Mabberley (2008) and the *Angiosperm Phylogeny Website* (See Web Resources).

References used in identification of species include the following: Ahmad (2013a,b,c & 2014), Al-Rawi & Chakravarty (1964), Babashekh (2006), Bermani (1981), Davis (1978, 1982, & 1984), Guest (1966), Houri and Houri (2001a & b), Ghahraman (1983, 1987, 1999, 2001 and 2003), Hussain and Alwan (2008), Maahzide (2003), Mashhadani (1992), Rawi (1964), Raza and Dawd (1983), Rechinger (1964), Sardar (2003), Tohme and Tome (2002), and Townsend and Guest (1966-1985), and Ghazanfar & Edmonson (2013). General information on habitats was based on Guest (1966), Abdulhasan (2009), and Abdulhasan et al. (2009).

TAXONOMY AND NOMENCLATURE

For birds we have followed *An annotated checklist of the birds of Iraq* (Salim et al., 2012), *Birds of the Middle East* (Porter & Aspinall, 2010) and the *Ornithological Society of the Middle East Regional List* (See Web Resources), the latter giving details of taxonomic decisions. The nomenclature does not therefore always follow that adopted by BirdLife International, whose alternative names are shown in the list of bird species occurring in Iraq (Salim et al., 2012).

For mammals, we follow *Mammal Species of the World* (Wilson & Reeder, 2005) and species listings on the IUCN Red List.

For reptiles, we have followed *The Reptile Database* (See Web Resources).

For Amphibians we followed *Amphibia Web* (See Web Resources). We also utilized the species listings on the IUCN Red List (See Web Resources).

For fish, we have followed *Freshwater Fishes of Iraq* (Coad, 2010) and species listings based on the website on *Freshwater Fishes of Iraq* (See Web Resources). Marine species were checked against the website *Catalogs of Fish* (See Web Resources).

For plants and habitats, we have followed published accounts in *Flora of Iraq* (Townsend & Guest, 1966-1985 & Ghazanfar & Edmonson, 2013), and *Flora Iranica* (Rechinger, 1963-in progress) updated and added to where necessary.

Globally threatened and near threatened species were determined based on the *IUCN Red List* (See Web Resources).

4.D DATA ENTRY & VALIDATION

Upon returning from the field the staff entered and commenced the review of the data collected in the field. Initially all data was entered into Microsoft Excel spreadsheets but in 2009, individual Microsoft Access databases were developed for birds, fish and botanical information for the Kurdistan Region, Central, and Southern Iraq. In some cases, older data in Excel spreadsheets was later entered into the database (this was the case for bird data). Observations on other fauna (besides birds and fish) were originally included with bird data but later a separate Microsoft Excel datasheet was developed.

Although the survey teams were experienced in identification and survey techniques, raw data on birds was given the most thorough and regular external review by the Nature Iraq ornithological adviser Richard Porter (BirdLife International).

Other datasets from the KBA Surveys (for mammals, reptiles, plants, and fish) were reviewed by outside experts who freely provided advice and help to vet the data collected.

Plate 12: Entering field data

4.E DATA MANAGEMENT

All KBA data is maintained in the Nature Iraq Archive and Library. This includes electronic media such as spreadsheets, database files, and image files as well as hardcopies of field sheets. Copies of all program reports are with the IMoHE. For each KBA in Iraq, key data have, as far as possible, been collected on:

- location

- species

- reasons for importance

- habitats and land use

- threats

- conservation actions

In addition, BirdLife International has invested extensively in the development of information-management tools to support the activities of the BirdLife Partnership for many years. To this end all KBA bird survey data being entered on Nature Iraq's database is also provided to BirdLife International for global archiving (see *WorldBirds/Biodiversity Database* (WBDB)) and analyzed data on birds are presented on the *BirdLife website* (BirdLife Data Zone). Data is also available on *WorldBirds Middle East Birds Database*. See Web Resources for further details. Also several papers resulting from the survey work have also been published in peer-reviewed journals and these are provided in the section entitled Additional KBA Papers.

4.F SITE DELINEATION

With the exception of wetlands, which tend to be discrete, easily recognizable areas, determining boundaries of KBAs can be difficult. Therefore the sites included in this inventory have been delineated using the following criteria. A KBA site is defined, as far as possible by being:

I. Different in character and biodiversity from the surrounding area;

II. A potential (or actual) protected area, or an area that can be managed, politically and actually, for nature conservation;

III. A self-sufficient area that provides the biological requirements for the key species for which the site has been identified.

Practical considerations of how best the site may be conserved are the most important consideration. Simple, conspicuous boundaries such as roads and rivers have been used where appropriate. There is no fixed maximum or minimum size for a KBA, thus the biologically sensible has been tempered with the practical. In a few cases a pragmatic decision has been made as to whether to combine neighboring sites into one 'super KBA' or treat them separately.

All delineations in the KBA Inventory should be considered preliminary and would require further refinement based on future field studies and management decisions.

4.G THREAT ASSESSMENTS

In summer and winter 2010, the survey teams conducted a site threat assessment using the Pressure-State-Response (PSR) Model. The PSR Model relies on three major indicators:

- **Pressure** — Pressure indicators identify and track the major threats to important species and habitats. Examples include rates of agricultural expansion, over-exploitation and pollution.

- **State** — State indicators refer to the condition of the site, with respect to its important species and habitats. State indicators might be population counts of the birds or mammals that are present at the site. They might also be measures of the extent and quality of the habitat required by these species.

• **Response** — Response indicators identify and track conservation actions: for example, changes in conservation designation, implementation of conservation projects and establishment of Local Conservation Groups.

PRESSURE INDICATORS

The KBA Program always informally documented threats at survey sites, but in 2010, the teams attempted a more formal assessment, based on eight (out of 11) threat types defined by the IUCN (2014c). The threat categories and their symbol are presented in Table 10 (three threat categories could not be assessed due to lack of information):

Each threat type was evaluated based on its Timing, Scope and Severity to develop an integrated "Threat Status Score." This score allowed for the rating of the individual threat as one out of four color-coded levels (See Map 16 through Map 23 to see the color codes for threats). The full methodology was adapted from one outlined in a BirdLife International (2006) report on Monitoring Important Bird Areas.

Not all threats have equal impacts upon biodiversity. Garbage and hunting at a site might both be considered a very high threat at a site but it is the over-exploitation of species (hunting) that is likely to have a greater impact on the biodiversity of site.

Plate 13: Hunting Houbara in Western Iraq in 2010

Table 10: Threat type & their symbols used in the Inventory Assessment

Threat type	Symbol
1. **Agricultural expansion and intensification (farming and grazing regimes, aquaculture, forestry practices)**	🌿
2. **Residential and commercial development**	🏢
3. **Energy Production and mining (gravel mining, oil development, electrical towers, etc.)**	⛏
4. **Transportation & service corridors (development of roads and shipping corridors)**	🚏
5. **Over-exploitation, persecution and control (logging, hunting, over-fishing, etc.)**	🏹
6. **Human intrusions and disturbance - Effects related to non-consumption of biological resources – recreational activities, war, military exercises, work and other activities**	👥
7. **Natural systems modification (dams and changes water mgmt, filling in wetlands, drainage, dredging, canalizations**	〰
8. **Invasive or other problematic species**	Not assessed due to lack of information

Threat type	Symbol
9. **Pollution (municipal and industrial waste and garbage, noise, air, light, & thermal pollution).**	
10. **Geological events (threats from catastrophic geological events)**	Not assessed due to lack of information
11. **Climate change, severe weather, drought, floods**	Not assessed due to lack of information

It is also very important to remember that the threat assessment presented in the individual site assessments and the discussions related to these threats should only be used as an indicator for future, more in-depth assessments of the pressures on the biodiversity of a site. Most habitats in Iraq are not "natural" in the sense of being pristine: the vast majority has to a greater or lesser extent been created over thousands of years by the activities of humans. This is not necessarily a problem; sustainably used habitats such as the agro-ecosystems that have developed in Iraq are often effective in both conserving biodiversity and producing food. For this reason the state of habitat(s) needs to be taken into account when threats and management actions are discussed.

STATE INDICATORS

As stated above, understanding pressures on the biodiversity of a site are only one part of the process. The current condition or state of the site must also be assessed. Unfortunately, this could not be carried out in the KBA Program due to lack of adequate habitat and habitat/species association information.

This information, which remains a gap in the KBA Program, may include the following:

- Population sizes for one or more "trigger" species (for which there is good information) or each "trigger" species assessed individually (then applying the "weakest link" approach)

- The area and quality of the key habitats on which the "trigger" species depend, as an indirect measure, or "surrogate", for population size.

It is likely that the status of different habitats in Iraq exists along a continuum from pristine to totally destroyed. According to Tony Miller of the Center for Middle East Plants (personal communication, 3 April 2014), this continuum might be roughly divided into five types:

1. **Natural habitats** – those unaffected by the actions of man. These are rare (or may be absent) in Iraq but if they exist are probably restricted to mountaintops and cliffs, and coastal habitats).

2. **Sustainably used habitats** — these presumably covered most of Iraq in the past — examples might be coppiced oak woodlands and the reed beds of the southern marshes.

3. **Unsustainably used habitats** — these now cover large areas of Iraq where traditional management practices have broken down or been replaced by less sustainable management practices — for instance, the bulk of the oak woodlands of the Kurdistan Region.

4. **Degraded habitats** — large areas in which the former habitats have been destroyed — for instance, mountain woodland, which has been altered to steppe grassland. Some of these changes are recent and some historic. The degraded status of the habitat does not necessarily mean that it is no longer suitable for conserving biodiversity – sometimes degraded habitats develop their own unique assemblages of biota, for instance, the steppe-grasslands of Iraq.

5. **Destroyed habitats** — areas that have been destroyed and no longer of value either economically or for biodiversity.

Only the first two habitats types are at some sort of natural equilibrium. The remaining three are in an active state of (usually) decline. Agriculture, grazing, development and road construction are clearly all threats to Type 1 "Natural habitats" but may or may not be threats in the other habitats. Type 2 "Sustainably used habitats" have developed because of some sort of agricultural "interference". For instance, it might be argued (but more research is required) that in some cases coppiced oak woodlands (Type 2) have a different but equally rich or sometimes richer assemblage of biota than natural oak woodlands (Type 1). Sometimes, Type 2 habitats are reverting to Type 1 as traditional management practices change. For instance, in some areas grazing is no longer practiced because of the movement of people to cities.

It should be noted that some apparently similar habitats could have different status. For instance, a desert may be the last stage of transition from woodland to steppe-grassland to desert or it may be a natural habitat and part of an arid ecosystem.

It is clear that the situation is very complex and that the state of individual habitats needs to be understood before final management actions are recommended.

RESPONSE INDICATORS

The response indicators are described as follows:

These indicators gauge the level of response to given threats and are rated based on the level of conservation designation, management planning and conservation actions that have taken place at a given site.

Most KBA sites, except in a very few limited cases noted in the text, would score very low in terms of response to threats as there are few national, regional or local institutions, policies, or resources allocated for addressing environmental threats (pressures) in Iraq. For this reason, this part of the PSR Model assessment was not yet attempted.

4.H ANALYSIS AND IDENTIFICATION OF KBA SITES IN IRAQ

While the variety of taxa assessed required the use of different criteria systems (KBA Criteria for non-avian fauna, IBA Criteria for avian fauna, and IPA Criteria for plants and habitats), the general format for these systems is still based on the concepts of Vulnerability and Irreplaceability.

For the purposes of assessing Vulnerability, it is the presence of species that are considered threatened that we are immediately concerned with. As stated in Langhammer et al. (2007), "Vulnerability may be measured on a site basis (likelihood that the species will be locally extirpated from a site) or a species-basis (likelihood that the species will go globally extinct)."

Species that have undergone assessment under the IUCN Red Listing process are assigned to a specific category that is identified by a two-letter code (See Figure 1) based on their conservation status (IUCN, 2012).

Species threatened with extinction may be Critically Endangered (CR), Endangered (EN), or Vulnerable (VU). Lower risk categories are Near Threatened (NT) and Least Concern (LC). Species may already be Extinct (EX) or Extinct in the Wild (EW) and there may not be enough information on other species to evaluate their status and thus are considered Data Deficient (DD). Finally many species simply have not been evaluated yet (NE).

Irreplaceability is a measure of the uniqueness of a site. Here we ask the question, 'does the site contain species that occur nowhere else (or in only a few locations) or does the site attract a significant portion of a species' entire population during a part of that species' lifecycle? Such sites are irreplaceable if lost. The different criteria systems used in the KBA Program and applied to the survey data as described below are based essentially on these foundational questions.

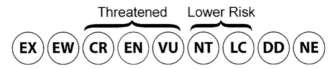
Figure 1: IUCN Categories

APPLICATION OF KEY BIODIVERSITY AREA CRITERIA FOR NON-AVIAN FAUNA

The rapid assessment survey methods used in the KBA Program did not result in data that could be used for a comprehensive assessment of criteria for non-avian fauna (see Appendix 1). However the extent to which the teams were able to establish the presence of species of conservation concern did allow two out of the six criteria and sub-criteria to be applied: Vulnerability Criterion and, to a limited extent, Irreplaceability sub-criterion Ia (restricted-range species based on global range). Species, or sub-species if the latter are IUCN-evaluated, were found to meet these criteria if observed and/or reliably reported from more than one source.

As populations of important species at a given site were impossible to determine, criteria Ib, Ic, Id, and Ie could not be assessed.

Appendix 4 provides a list of the key species whose presence did allowed for the review of KBA Criteria for Vulnerability (threatened species) and Ia (restricted-range species) and the results of the KBA assessment for these two Criteria is found in Table 16 in the next chapter. As information on conservation concern fish species became available in mid-2014, it was possible to add critically endangered and endangered fish species to the assessment under the Vulnerability criterion. A more full assessment is required as these and other fish species on the Iraq checklist may also meet other KBA criteria.

Important carnivores, even if identified as Least Concern (LC) species by the IUCN, are mentioned in the individual sites assessments when they were either observed or reported as their presence may indicate higher biological diversity at a site. Also other LC species are mentioned when they are specifically targeted at a site for persecution (for examples when hunting levels were deemed potentially unsustainable and the species could be locally eradicated).

APPLICATION OF IMPORTANT BIRD & BIODIVERSITY AREA CRITERIA

Bird data obtained at survey sites were assessed against the Important Bird & Biodiversity Area (IBA) criteria (see Appendix 1). This represents the most comprehensive and complete analysis in the KBA Inventory. This analysis relied heavily on the presence, counts and species population estimates made at each site.

CRITERIA ASSESSMENT METHODS

For a site to qualify under Criterion A1 it must regularly hold globally threatened species: the presence of a single individual of either a Critically Endangered or Endangered species or the presence of 30 individuals or 10 pairs of a Vulnerable species. For automatic qualification under Criterion A2 a site needed to hold both of the restricted-range species that occur in Iraq, namely Basra Reed Warbler *Acrocephalus griseldis* and Iraq Babbler *Turdoides altirostris,* though several sites were selected for the latter species alone in those areas of the country where only this species occurs (sites in Central Iraq and the southern edges of the Kurdistan Region). The results of the analysis for A1 and A2 Criteria can be found in Table 17 and Table 18 in the next chapter.

Criterion A3 focuses on the presence of biome-restricted species — for a site to qualify it must hold a "significant assemblage" of a group of species whose breeding distributions are largely or wholly confined to one biome. In the selection process we have attempted also to ensure a geographic spread across Iraq that support biome-restricted species. The key biomes (as defined by BirdLife International) are discussed in a previous section (see section entitled Biomes & Ecoregions of Iraq). To assist with the selection process a table was prepared listing the sites where biome-restricted species occurred (See Table 19 in the next chapter). A 30% threshold was applied to identify the sites with a "significant assemblage" that would be listed under the A3 Criterion. For example, in the Central Marshes (IQ075), nine out of 19 Sahara-Sindian biome-restricted species were found. For this biome, 30% of the 19 potential species that can be found in Iraq means that a site must have six or more Sahara-Sindian species to meet the A3 Criterion. Thus the Central Marshes is considered an A3 site, whereas Khor Al-Zubayr (IQ081) located in the same biome but with only three out of the 19 species observed did not meet the criterion.

To evaluate the field data against IBA criteria for Iraq, information on species and population thresholds was obtained from BirdLife International (2010) and Delany & Scott (2002 & 2006). Population thresholds were used specifically to determine if sites met A4i and A4ii sub-criteria for congregatory species and the values for these threshold levels are provided in Table 20 and Table 21 in the next chapter (the latter contains the results for sub-criteria A4iii as well). To compare the field data on these bird species to these threshold levels, population estimates at a given site were prepared based on the methodology listed in the next section. Based on current information, Iraq has no bottleneck sites (areas that restrict the passage of migrating flocks) thus no sites were evaluated under Criterion A4iv.

POPULATION ESTIMATES OF BREEDING BIRDS

The figures given for population estimates in the site accounts are, in most cases, a very approximate minimum number of pairs. Unless stated otherwise the numbers are based on a varying number of sample counts over very approximately defined count areas. Such counts were made as either area counts (length walked x width over which birds were counted) or spot counts (area calculated by the radius over which counts made). These sample area counts were then extrapolated to the total area of habitat in the site considered to be suitable for the species for breeding, and were then rounded downwards. This crude and simple technique was often used to give guideline figures for a rapid assessment of bird populations of a site. To use a more refined technique, such as through comprehensive transect surveys, would have been impossible given the restraints of time, number of observers and, often, the large size of the survey areas. Also, in some cases, security was a factor that had to be taken into consideration, thus preventing the surveying of the whole site.

In all cases, good knowledge of the survey area by the observers has been able to provide a degree of quality control on the estimates produced, which are presented in the site accounts as a single figure sometimes with a < notation. Nevertheless they should be regarded as very approximate estimates.

We believe that the figures produced, though approximate, are more helpful than simply describing the status of a bird as 'common' or 'rare.' Future, more detailed and comprehensive surveys are to be encouraged.

APPLICATION OF IMPORTANT PLANT AREA CRITERIA

Due to the lack of a complete flora on Iraqi plants and the limited information available in general, additional work to assess IPA Criterion A (sites with threatened species) was completed with assistance of the CMEP/RBGE under the Darwin Project for In-situ Conservation. This included the development of a list of endemics and near endemics for Iraq; geo-referencing and spatial analysis of historical data available from the Flora of Iraq and Flora Iranica, and an analysis of the relationship between this and the KBA survey data. The full list of these species developed under the Darwin Project for In-situ Conservation and also obtained during the KBA botanical surveys that was used to identify sites meeting IPA A4 Criteria is located in Appendix 7.

There is still much work to be done to identify Iraq's important plant species. For example, attempts to look at how KBA sites might overlap areas known to be important for crop wild progenitors was not possible due to a lack of information (an important area that needs more attention). Also, though some sites such as Peramagroon (IA027) and the Hawraman Area (IQ043) have full plant inventories allowing additional comparison to KBA and historical data this was not possible for the bulk of the sites for which such lists have not yet been compiled. As a result not all sub-criteria under IPA Criterion A could be assessed in the current work and the main focus was on sites that met sub-criterion A4. The sites that met IPA Criterion A4 are listed in the plant list in Appendix 7.

For IPA Criterion B (sites of botanical richness), the botanical team ranked and selected the best example of sites for each of the vegetation types. Under the KBA Program (due to insufficient information on the biogeographic zones of Iraq), sites could only be identified under Criterion B1 (the site is a particularly species-rich example of a defined habitat type) or B2b (the site is a refuge for biogeographically and bio-climatically restricted plants to 'retreat to' in the face of global climatic change). In the KBA Program, the habitat classification scheme for the Iraqi marshlands (as well as terrestrial habitats) was utilized in the evaluation of Criterion B1 and is described in details in Abdulhasan (2009) and Abdulhasan et al. (2009). An additional source for the habitat classification were the phytogeographic zones defined in Volume 1 of the Flora of Iraq (Guest, 1966), which were particularly helpful in defining habitat types in Central Iraq and the Kurdistan region. The phytogeographic zones defined in Flora included descriptions of characteristic species, altitude, and locations within Iraq. The key habitats identified at B1 sites are listed in Table 22 and the results of the B1 assessment are found in Table 23, both in the next chapter. Criterion B2b examined primarily sites that offered high elevations and varied terrain (and thus offering more potential refuge sites) or wetland sites that have historically provided more stable habitat in times of drainage or drought conditions.

For IPA Criterion C (Outstanding examples of globally or regionally threatened habitat types), the botanical team evaluated sites based on the most unique habitats within the country of global / regional significance that often are under severe threat. For example, the Mesopotamian Marshlands of southern Iraq are one of the most threatened habitats in the country because of previous drainage campaigns by the Iraqi government in the 1990s and face current issues with upstream dams and diversions, continued drainage projects and pollution. To conduct this assessment, the team evaluated the quality of the habitat(s) at a site and examined the types and severity of threats that they faced based on the Threat Assessment. The habitat classification scheme used was defined by Abdulhasan (2009), Abdulhasan et al. (2009) and the Flora of Iraq (Guest, 1966, Vol 1) as discussed above. The results of this Criterion C assessment can be found in Table 24 in the next chapter.

CHAPTER 5: **OVERVIEW OF SITES & SPECIES**

This chapter provides a comprehensive review of the findings on all sites and key species from the KBA Program. There are a total of 82 sites that meet one or more of the KBA Criteria systems that were applied in the program: KBA for non-avian fauna, IBA for avian fauna, & IPA for plants and habitats. More detailed accounts of each individual KBA site are presented in the latter half of this book (See Chapter 7).

5.A OVERVIEW OF FINDINGS

KEY BIODIVERSITY AREAS (KBAS) OF IRAQ

Of the eighty-two sites, 39 met KBA criteria for non-avian fauna, 68 met IBA criteria, and 72 met IPA criteria. The KBA sites are shown in Map 12 (a full list of these sites, their local names and their site codes can found in the Index of Sites at the end of this Report).

SURFACE AREA COVERED BY KBAS

Sites delineations provided in the Inventory as stated previously should be considered preliminary and in need of further refinement but a total of 2,838,782 hectares (28,388 km²) are included within the KBA sites in this assessment. Under the Convention of Biological Diversity (CBD), it is stated that all signatory countries should reach the goal of protecting a minimum of 17% of the major habitats within each country. The area represented by the Iraq KBAs is only 7% of the total area of the country and covers a variety of habitat types. Although the 82 sites identified here represents an important step forward in identifying priority conservation areas for Iraq, large sections of the country remain to be surveyed.

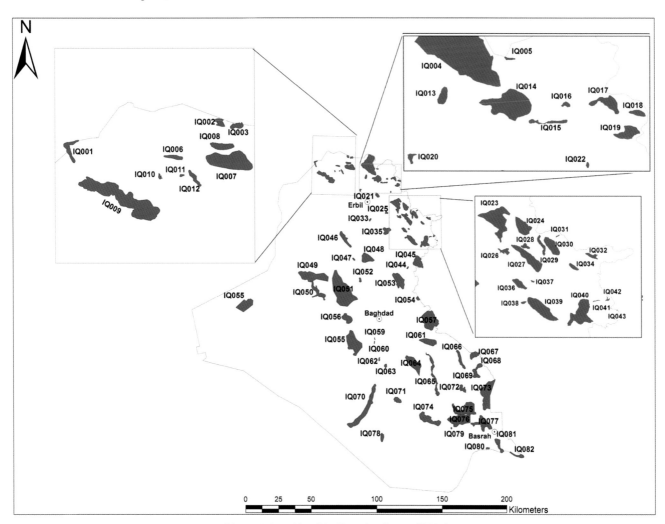

Map 12: Iraq Key Biodiversity Areas (KBAs)

KBAS BY ECOREGION

Table 11 and Figure 2 provide information for each ecoregion and the area in each ecoregion that is covered by KBAs with a comparison to currently, officially designated protected areas. Though ecoregions may have within them diverse habitats, they can serve as a rough measure of Iraq's progress in achieving the CBD goal of protecting 17% of its major habitats. Assuming that all KBAs currently identified were under some form of protection, two ecoregions (the Tigris-Euphrates alluvial salt marsh (PA0906) and the South

Iran Nubo-Sindian Desert and Semi-Desert (PA1328)) have identified KBAs that would meet this goal. As stated previously, currently only the Central Marshes (IQ075), Hawizeh Marshes (IQ072), Sawa Lake (IQ071) and Hammar Marsh (both East Hammar (IQ077) and West Hammar (IQ076)) have national protected area status. The table and figure clearly show several ecoregions (primarily desert areas of the country) where more work is needed to identify sites and some areas where significant progress has been made but additional sites should be identified (e.g. Zagros Mountains Forest Steppe (PA0446)).

Table 11: Terrestrial ecoregions with the area and percentage located within Designated Protected Areas and Iraq KBA Sites

Terrestrial Ecoregions (Ecoregion #)	Area in Iraq - km²	Designated Protected Areas - km² (%)	Area in KBA - km² (%)
Tigris-Euphrates alluvial salt marsh (PA0906)	28,795	5374 (18.7%)	6,361 (22.1%)
Arabian Desert and East Sahero-Arabian Xeric Shrublands (PA1303)	192,853	5 (0.003%)	6,424 (3.3%)
Mesopotamian Shrub Desert (PA1320)	129,995	0 (0%)	8,668 (6.7%)
Middle East Steppe (PA0812)	37,598	0 (0%)	970 (2.6%)
Eastern Mediterranean conifer-sclerophyllous-broadleaf forest (PA1207)	1,475	0 (0%)	58 (4%)
Red Sea Nubo-Sindian Tropical Desert and Semi-Desert (PA1325)	5,189	0 (0%)	0 (0%)
South Iran Nubo-Sindian Desert and Semi-Desert (PA1328)	7,993	0 (0%)	1366 (17.1%)
Gulf Desert and Semi-Desert (PA1323)	1,480	0 (0%)	30 (2.0%)
Zagros Mountains Forest Steppe (PA0446)	29,376	0 (0%)	4,468 (15.2%)
Total	434,753 km²	2,798 km² (0.64%)	28,245 km² (6.5%)*

Note that the total area differs slightly here from the total area of all KBAs because Fao (IQ082) is not completely overlapping a terrestrial ecoregion.

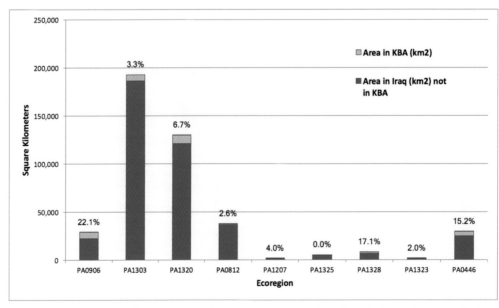

Figure 2: Area & percentage of Iraq Terrestrial Ecoregions covered by Key Biodiversity Areas (light blue)

Table 12: Freshwater & marine ecoregions with the area and percentage located
within Designated Protected Areas and Iraq KBA Sites

Ecoregions (Ecoregion #)	Area in Iraq - km²	Designated Protected Areas - km² (%)	Area in KBA - km² (%)
Freshwater Ecoregions			
Arabian Interior (440)	142,494	0 (0%)	1,978 (1.4%)
Lower Tigris and Euphrates River Basin (441)	227,497	5,379 (2.4%)	19,880 (8.7%)
Upper Tigris and Euphrates River Basin (442)	64,745	0 (0%)	5,739 (8.9%)
Orumiyeh (445)	17	0 (0%)	0 (0%)
Marine Ecoregion			
The Arabian Gulf (90)	103,716 (including 58 km of coastline and a territorial limit of 22 km)	0 (0%)	8,793 (8.5%)

Table 12 and Figure 3 provide information for the four freshwater ecoregions and one marine ecoregion that cover Iraq. The amount and percentage of area of each ecoregion that is covered by KBAs is listed with a comparison to what are officially designated protected areas currently.

With only a limited amount of territorial sea and much of the marine environment of the country heavily impacted by high nutrient loads and oil pollution, Iraq was thought not to have any coral reef habitats, which are highly sensitive and critically threatened not only in the Arabian Gulf but throughout the world as well. Then in 2012 and 2013 a team from the Basrah Marine Science Center and Scientific Diving Center at the Technical University of Freiburg in Germany discovered a living coral reef system in the turbid waters off the mouth of the Shatt Al-Arab. This reef is estimated to cover an area of approximately 28 km², which represents 0.01% of Iraq's portion of the Arabian Gulf Ecoregion (90). Though not officially part of the KBA Program, this habitat would appear to be very unique in the region and with additional survey work to fully characterize this site it may constitute an important Marine Key Biodiversity Area for Iraq.

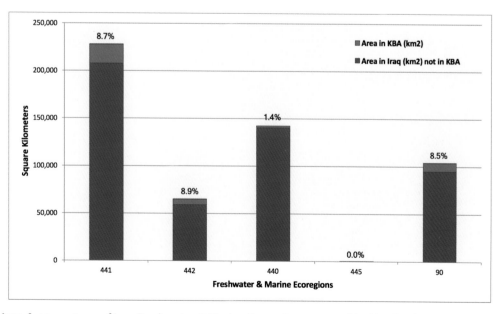

Figure 3: Area & percentage of Iraq Freshwater & Marine Ecoregions covered by Key Biodiversity Areas (light blue)

5.B CRITERIA ASSESSMENT FINDINGS

The rigorous application of quantitative KBA criteria assures that the network of these 82 identified sites is of global importance. While some sites may have met only one criterion within one system (KBA for non-avian fauna, IBA, or IPA), the majority met several criteria often within one or more of the systems. Figure 4 provides an overview of the number and grouping of sites within the different regions of country (Kurdistan (northern Iraq), Central, & Southern Iraq) as well as countrywide, which meet these criteria systems.

The importance of individual sites, as indicated by the type(s) of criteria that each fulfills, is shown in the following tables and maps.

IMPORTANT AREAS FOR OTHER (NON-AVIAN) FAUNA

Sites meeting KBA for the presence of conservation concern mammals, as well as some fish, reptiles, and amphibians make up 39 (48%) of the total number of KBA Sites. They cover over 18,660 km^2 or 4.3% of the land area of Iraq as a whole. Table 13 shows the list of these sites, what specific criterion they meet and their area. Much work remains to be done to fully characterize these and other sites throughout the country for such species as well as other taxa that were not covered in the program. Many sites could simply not be surveyed adequately and it is likely that many of the KBA sites hold important populations of conservation concern species.

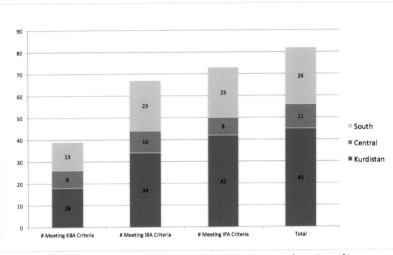

Figure 4: Number of sites meeting global criteria in each region of Iraq

Table 13: Summary of Key Biodiversity Areas identified based on specific criteria for other (non-avian) fauna

Site Code	Site names	Vulnerability	Ia. Irreplacbility Restricted Range	Area (ha)
		Criteria		Area (ha)
IQ003	Dure	✓	-	2,310
IQ004	Barzan Area & Gali Balnda	✓		114,822
IQ007	Gara Mountain and Garagu	✓	-	10,447
IQ014	Bakhma & Bradost Mountain	✓	-	21,858
IQ017	Halgurd Mountain	✓	-	6,393
IQ019	Sakran Mountain	✓	-	6,001
IQ023	Dukan Lake	✓	✓	47,253
IQ024	Assos Mountain	✓	-	20,017
IQ027	Peramagroon Mountain	✓	-	16,738
IQ029	Mawat Area	✓	-	2,699
IQ036	De Lezha	✓	-	8,110
IQ038	Sangaw Area	✓	-	1,873
IQ039	Qara Dagh	✓	-	25,345
IQ040	Darbandikhan Lake	✓	✓	36,842
IQ042	Ahmed Awa	✓	-	887

Site Code	Site names	Vulnerability	Ia. Irreplacbility Restricted Range	Area (ha)
		Criteria		Area (ha)
IQ043	Hawraman Area	✓	-	4,463
IQ044	Kalar Area	✓	-	1,130
IQ045	Maidan Area	✓	-	53,785
IQ046	Jabal Makhool	✓	✓	35,257
IQ047	Mahzam and Al-Alam	✓	✓	2,145
IQ049	Qadissiya Lake	✓	✓	145,230
IQ050	Haditha Wetlands & Baghdadi	✓	✓	48,274
IQ051	Tharthar Lake and Al-Dhebaeji Fields	✓	✓	340,573
IQ052	Samarra Wetlands	✓	✓	4,470
IQ054	Mandli	✓	-	5,386
IQ056	Habbaniya Lake	✓	✓	45,390
IQ057	Jazman (Jurbatia)	✓	-	155,095
IQ059	Musayab	✓	✓	162
IQ060	Hindiya Barrage	✓	✓	278
IQ062	Ibn Najm, North	✓	✓	1,789
IQ063	Ibn Najm	✓	✓	4,000
IQ064	Dalmaj	✓	✓	92,076
IQ065	Gharraf River	✓	✓	50,461
IQ071	Sawa Lake and Area	✓	✓	20,058
IQ072	Auda Marsh	✓	✓	19,241
IQ073	Hawizeh Marshes	✓	✓	164,023
IQ075	Central Marshes	✓	✓	131,780
IQ076	Hammar, West	✓	✓	136,326
IQ077	Hammar, East	✓	✓	82,968
Total		39	21	1,865,955

Map 13 shows the locations of KBA for non-avian fauna in relation to sites that do not meet these criteria (meeting either IBA or IPA criteria instead).

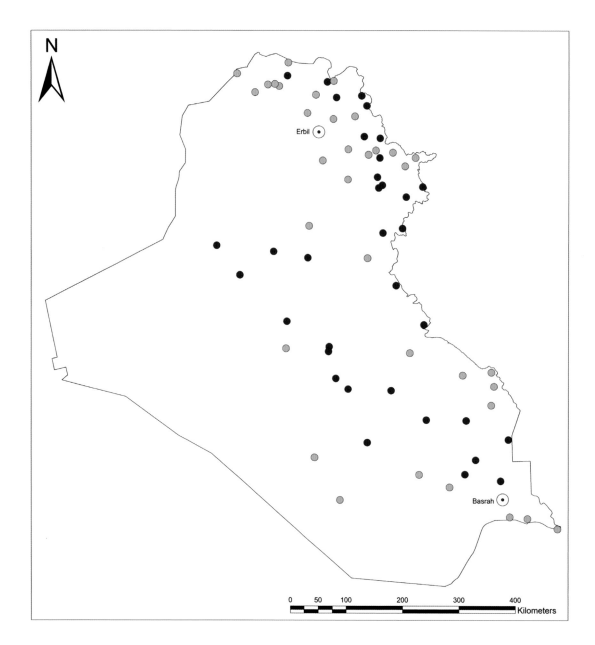

Map 13: Location of KBAs identified for non-avian fauna (dark blue)

IMPORTANT BIRD & BIODIVERSITY AREAS

Sites meeting Important Bird & Biodiversity Area Criteria (based on data obtained on birds at each site) make up 67 (82%) of the total number of KBA sites. They cover over 27,552 km² or 6.3% of the land area of Iraq as a whole. Nineteen of these sites fall within the Mesopotamian Marshlands Endemic Bird Area that encompass much of the central to southern Tigris-Euphrates River Basins within Iraq. Table 14 and the related figure provide an overview of all IBA sites and the criteria that they meet.

Table 14: Summary of Important Bird & Biodiversity Areas in Iraq
(67 sites covering c.27,660 km^2)

Site Code	Site Name	Species of global conservation concern — A1	Restricted-range — A2	Biome-restricted — A3 12	05	06	04	13	Congregations of waterbirds — A4I	Congregations of seabirds/landbirds — A4II	More than 20,000 waterbirds — A4III	Bottleneck sites — A4VI	Area (ha)
IQ001	Fishkhaboor									✓			3,315
IQ002	Benavi & Sararu			✓									1,809
IQ004	Barzan Area & Gali Balnda	✓		✓		✓							114,822
IQ006	Mangesh	✓		✓									2,287
IQ007	Gara Mountain & Garagu	✓		✓									10,447
IQ008	Amedi & Sulav	✓		✓									6,037
IQ009	Mosul Lake	✓							✓				48,128
IQ010	Dohuk Lake	✓											386
IQ011	Zawita	✓											288
IQ012	Atrush & Bania Area	✓											2,411
IQ013	Gali Zanta & Garbeesh	✓											3,540
IQ014	Bakhma & Bradost	✓		✓									21,858
IQ016	Sari Hassan Bag					✓							923
IQ017	Halgurd Mountain					✓							6,393
IQ021	Doli Smaquli & Ashab	✓		✓		✓							7,027
IQ023	Dukan Lake	✓				✓			✓				47,253
IQ024	Assos Mountain					✓							20,017
IQ025	Taq Taq	✓	✓										1,845
IQ026	Chami Razan Area	✓		✓		✓							4,530
IQ027	Peramagroon Mountain	✓		✓	✓	✓							16,738
IQ028	Sargalu	✓											5,235
IQ030	Kuradawe & Waraz	✓											7,272
IQ031	Sharbazher Area					✓							2,322
IQ032	Penjween	✓											4,035
IQ033	Altun Kopri Marsh		✓								✓		1,575
IQ034	Parazan	✓											3,617
IQ035	Chamchamal Area		✓										24,745
IQ036	De Lezha	✓				✓							8,110
IQ037	Hazarmerd	✓											607
IQ039	Qara Dagh	✓		✓		✓							25,345
IQ040	Darbandikhan Lake	✓							✓				36,842
IQ042	Ahmed Awa	✓		✓		✓							887
IQ043	Hawraman Area			✓									4,463
IQ044	Kalar Area	✓											1,130
IQ046	Jabal Makhool	✓											35,257
IQ047	Mahzam and Al-Alam		✓										2,145

Site Code	Site Name	Species of global conservation concern	Restricted-range	Biome-restricted					Congregations of waterbirds	Congregations of seabirds/landbirds	More than 20,000 waterbirds	Bottleneck sites	Area (ha)
		A1	A2	A3					A4I	A4II	A4III	A4VI	
				12	05	06	04	13					
IQ048	Albu Ajeel - Himreen Hills	✓											54,261
IQ049	Qadissiya Lake	✓											145,230
IQ050	Haditha Wetlands & Baghdadi	✓	✓					✓					48,274
IQ051	Tharthaar Lake & Dhebaeji Field	✓						✓					340,573
IQ052	Samarra Wetlands		✓										4,470
IQ053	Himreen Lake	✓											80,275
IQ055	Ga'ara	✓											89,558
IQ056	Habbaniya Lake	✓											45,390
IQ057	Jazman (Zurbatia)	✓						✓					155,095
IQ058	Razzaza Lake	✓	✓					✓	✓		✓		156,234
IQ060	Hindiya Barrage	✓	✓										278
IQ061	Shuweicha Marsh	✓											54,938
IQ062	Ibn Najm, North	✓	✓						✓	✓			1,789
IQ063	Ibn Najm	✓	✓										4,000
IQ064	Dalmaj Marsh	✓	✓					✓	✓	✓	✓		92,076
IQ065	Gharraf River	✓	✓										50,461
IQ066	Hoshiya and Saaroot	✓									✓		33,560
IQ067	Teeb Oasis & Zubaidaat	✓						✓					28,578
IQ068	Teeb Seasonal Wetlands	✓									✓		14,827
IQ069	Sinnaaf Seasonal Wetlands	✓							✓		✓		26,049
IQ070	Wadi Al-W'eir & Sh'eeb Abu-Talha	✓						✓					142,755
IQ071	Sawa Lake and Area	✓											20,058
IQ072	Auda Marsh	✓	✓										19,241
IQ073	Hawizeh Marshes	✓	✓						✓		✓		164,023
IQ074	Suwaibaat (or Sleibaat)	✓									✓		84,753
IQ075	Central Marshes	✓	✓						✓	✓	✓		131,780
IQ076	Hammar, West	✓	✓					✓	✓	✓	✓		136,326
IQ077	Hammar, East	✓	✓					✓	✓		✓		82,968
IQ078	Salman	✓											14,959
IQ081	Khor Az-Zubayr Marshes		✓										31,854
IQ082	Fao	✓	✓										16,909
Total		55	19	12	1	12	0	9	11	5	11		2,755,183

Sites are often important for a number of species, each of which qualifies under the same category. For example, several waterbird species may congregate at a wetland site in a given season, each in numbers that exceed the A4i population threshold — or several globally threatened species may have key populations at a single site, thus repeatedly meeting the A1 criterion.

In addition, a site may be important for birds that meet different criteria and sites also often qualify under more than one criterion. Figure 5 provides an overview of the number of sites meeting different IBA Criterion.

Map 14 shows the locations of IBAs in relation to sites that do not meet these criteria (meeting either KBA for non-avian fauna or IPA criteria instead).

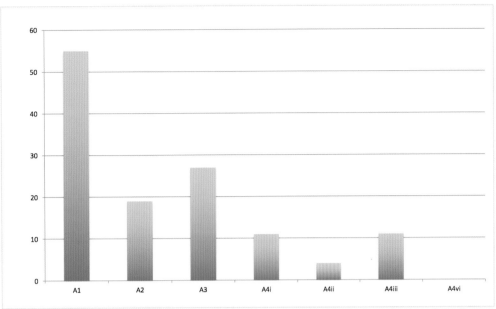

Figure 5: Number of sites meeting individual Important Bird & Biodiversity Area Criterion

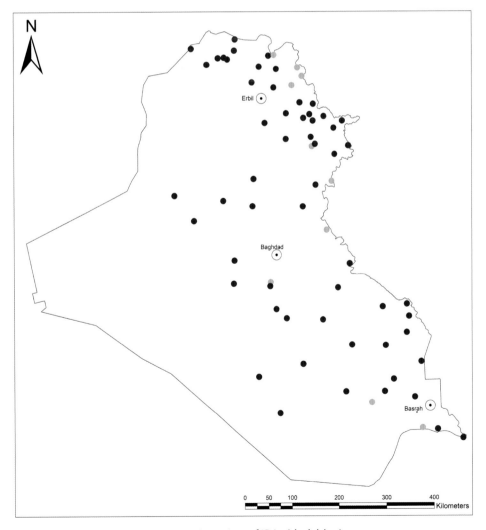

Map 14: Location of IBAs (dark blue)

Table 15: Summary of Important Plant Areas

Site Code	Site Name	Endemic & near-endemic species	Particularly species-rich example of a defined habitat	A refuge for plants to 'retreat to' in the face of climatic change.	Threatened habitat type	Area (ha)
		A4	B1	B2b	C	
IQ001	Fishkhaboor	✓	✓		✓	3,315
IQ002	Benavi & Sararu	✓	✓			1,809
IQ003	Dure	✓	✓		✓	2,310
IQ004	Barzan Area & Gali Balnda	✓	✓			114,822
IQ005	Kherazook		✓			607
IQ006	Mangesh	✓	✓			2,287
IQ007	Gara Mountain & Garagu	✓	✓	✓		10,447
IQ008	Amedi & Sulav	✓	✓			6,037
IQ010	Dohuk Lake	✓				386
IQ011	Zawita	✓	✓	✓	✓	288
IQ012	Atrush & Bania Area	✓	✓			2,411
IQ013	Gali Zanta & Garbeesh	✓	✓			3,540
IQ014	Bakhma & Bradost Mountain	✓	✓			21,858
IQ015	Rawanduz Gorge	✓	✓	✓		2,680
IQ016	Sari Hassan Bag	✓	✓			923
IQ017	Halgurd Mountain	✓	✓	✓		6,393
IQ018	Haji Omran	✓	✓			2,108
IQ019	Sakran Mountain	✓	✓	✓		6,001
IQ020	Bahrka		✓			1,306
IQ021	Doli Smaquli & Ashab	✓	✓			7,027
IQ022	Doli Plngan	✓	✓			420
IQ023	Dukan Lake	✓	✓			47,253
IQ024	Assos Mountain	✓	✓			20,017
IQ025	Taq Taq		✓			1,845
IQ026	Chami Razan Area	✓	✓			4,530
IQ027	Peramagroon Mountain	✓	✓	✓		16,738
IQ028	Sargalu	✓	✓			5,235
IQ029	Mawat Area		✓			2,699
IQ030	Kuradawe & Waraz		✓			7,272
IQ031	Sharbazher Area	✓	✓			2,322
IQ032	Penjween	✓	✓			4,035
IQ033	Altun Kopri Marsh		✓		✓	1,575
IQ034	Parazan	✓	✓			3,617
IQ035	Chamchamal Area	✓	✓			24,745
IQ036	De Lezha	✓	✓			8,110
IQ038	Sangaw Area	✓	✓			1,873
IQ039	Qara Dagh	✓	✓	✓		25,345

Site Code	Site Name	Endemic & near-endemic species A4	Particularly species-rich example of a defined habitat B1	A refuge for plants to 'retreat to' in the face of climatic change. B2b	Threatened habitat type C	Area (ha)
IQ040	Darbandikhan Lake	✓	✓			36,842
IQ041	Zalm River	✓	✓		✓	550
IQ042	Ahmed Awa	✓	✓			887
IQ043	Hawraman Area	✓	✓			4,463
IQ044	Kalar Area		✓		✓	1,130
IQ046	Jabal Makhool		✓			35,257
IQ047	Mahzam and Al-Alam Area		✓			2,145
IQ048	Ajeel Himreen Hills		✓			54,261
IQ050	Haditha Wetlands & Baghdadi	✓	✓			48,274
IQ051	Tharthar Lake & Dhebaeji Field		✓			340,573
IQ053	Himreen Lake	✓				80,275
IQ054	Mandli	✓				5,386
IQ055	Ga'ara	✓	✓			89,558
IQ057	Jazman (Zurbatia)	✓	✓			155,095
IQ058	Razzaza Lake		✓			156,234
IQ060	Hindiya Barrage		✓	✓		278
IQ061	Shuweicha Marsh				✓	54,938
IQ062	Ibn Najm, North				✓	1,789
IQ063	Ibn Najm		✓		✓	4,000
IQ064	Dalmaj Marsh		✓		✓	92,076
IQ065	Gharraf River		✓	✓	✓	50,461
IQ066	Hoshiya and Saaroot				✓	33,560
IQ067	Teeb Oasis & Zubaidaat	✓	✓			28,578
IQ068	Teeb Seasonal Wetlands		✓			14,827
IQ069	Sinnaaf Seasonal Wetlands		✓			26,049
IQ070	Wadi Al-W'eir & Sh'eeb Abu-Talha		✓	✓		142,755
IQ071	Sawa Lake and Area		✓			20,058
IQ072	Auda Marshes		✓		✓	19,241
IQ073	Hawizeh Marshes		✓	✓	✓	164,023
IQ075	Central Marshes		✓		✓	131,780
IQ076	Hammar, West		✓		✓	136,326
IQ077	Hammar, East		✓		✓	82,968
IQ079	Lehais		✓			797
IQ080	Jabal Senam		✓			2,918
IQ081	Khor Az-Zubayr		✓			31,854
IQ082	Fao		✓			16,909
		41	67	11	17	2,441,299

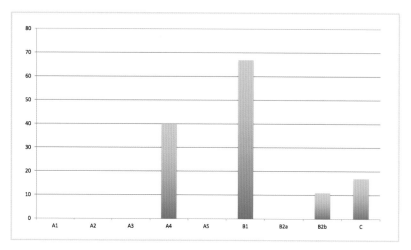

Figure 6: Number of sites meeting individual Important Plant Area Criterion

IMPORTANT PLANT AREAS

Sites meeting IPA Criteria make up 73 (89%) of the total number of KBA Sites. They cover over 24413 km^2 or 5.6% of the land area of Iraq as a whole. Table 15 (on the 2 proceeding pages) and Figure 6 provide an overview of all IPA sites and the criteria that they meet. Map 15 shows the locations of IPAs in relation to sites that do not meet these criteria (meeting either KBA for non-avian fauna or IBA criteria instead).

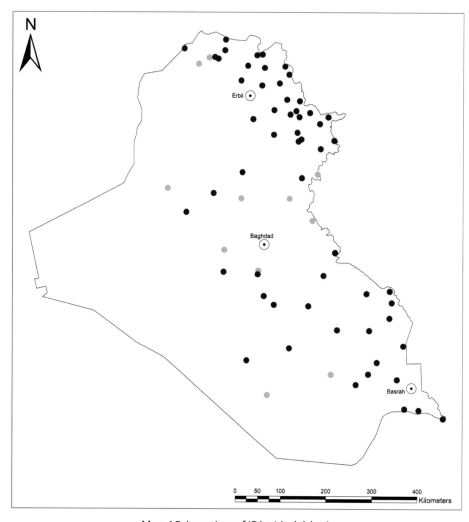

Map 15: Location of IPAs (dark blue)

Note: Sites in light blue either do not meet these criteria or were not surveyed by the botanical team

5.C SPECIES AND ISSUES IMPORTANT TO THE IDENTIFICATION OF KBA SITES IN IRAQ

NON-AVIAN FAUNA

The general KBA criteria were used to assess non-avian species, which included mammals, reptiles, amphibians and some fish species. Appendix 4 provides the checklist of all species observed and reported in the program that were used in the identification of sites based on KBA criteria for non-avian fauna.

Of the 39 sites that met KBA Criteria for non-avian fauna, 18 sites (46%) were in Kurdistan (all meeting the vulnerability criteria in addition to two sites, Dukan Lake (IQ023) and Darbandikhan Lake (IQ040), meeting the Ia criteria as well). Eight sites (21%) were in Central Iraq and 13 (33%) were in Southern Iraq (with all sites meeting the vulnerability and irreplaceability criteria, except one site, Jazman (Zurbatia) (IQ057), which only meets the vulnerability criteria). The higher number in Kurdistan may be the result of a more dedicated effort to survey sites in northern Iraq for other taxa.

As was stated in the Methodology, only Vulnerability (based on threatened species) and Irreplaceability sub-criterion Ia (Restricted-range species) could be evaluated. A site may be important for different mammal, reptile, and amphibian species yet were only qualified as KBA based on the presence of one or two species. This is mainly due to the challenges of collecting data on these taxa and this indicates a major gap in the KBA Assessments. Table 16 lists those species where some information was obtained and used to identify sites as KBAs for non-avian fauna and provides a review of the key areas where this was possible.

Table 16: Species used in KBA (non-avian) criteria assessments

Species	KBA Criteria Met	Notes & Locations
Mammals		
Smooth-coated Otter *Lutrogale perspicillata* **(Vulnerable)**	V. Vulnerability Criterion	The southern marshes lie at the center of the distribution of an isolated subspecies of smooth-coated Otter *Lutrogale perspicillata maxwelli*. Its status and distribution have been unclear due to confusion with the Eurasian Otter *Lutra lutra* (Near Threatened), which also occurs in the region. Recent surveys (Omer et al. 2012, Al-Sheikhly and Nader 2013) have confirmed the presence of smooth-coated otter in parts of the southern marshes for the first time since the 1950s-1960s. (Central Marshes, Hawizeh, Taq Taq, etc.)
Wild Goat *Capra aegagrus* **(Vulnerable)**	V. Vulnerability Criterion	Found in the mountains of Kurdistan, northern Iraq (Barzan, Peramagroon Mountain, Qara Dagh, etc.)
Persian Leopard *Panthera pardus saxicolor* **(Endangered)**	V. Vulnerability Criterion	Found in the mountains of Kurdistan, northern Iraq (Qara Dagh and reported in other sites) and one report of a hunted leopard in Mandli (Central Iraq) in 2009.
Goitered Gazelle *Gazella subgutturosa* **(Vulnerable)**	V. Vulnerability Criterion	Found or reported in Kurdistan (Sangaw and Kalar Area) and Central Iraq (Ajeel Himreen Hills, Jabal Makhool, and Mandli). Gazelles used to be found in the lower parts of Wadi Al-W'eir & Sh'eeb Abu-Talha (southern Iraq) until the last couple of decades but it is not clear whether this population still exists.
Reptiles		
Euphrates Softshell Turtle *Rafetus euphraticus* **(Endangered)**	V. Vulnerability Criterion & Irreplaceability Ia (Restricted range)	Found throughout Iraq in wetland and rivers of the Tigris-Euphrates Basin (Central Marshes, Dukan Lake, Darbandikhan Lake, etc.)
Spur-thighed Tortoise *Testudo graeca* **(Vulnerable)**	V. Vulnerability Criterion	Found in Kurdistan, northern Iraq (Qara Dagh, Peramagroon, etc.).
Amphibians		
Kurdistan Mountain Newt *Neurergus derjugini* **(Critically Endangered)**	V. Vulnerability Criterion	Found at three sites in Kurdistan, northern Iraq (Ahmad Awa, Hawraman Area, and Mawat area)
Azerbaijan Mountain Newt *Neurergus crocatus* **(Vulnerable)**	V. Vulnerability Criterion	Found at three sites in Kurdistan, northern Iraq (Gara Mountain & Garagu, Halgurd Mountain, and Sakran Mountain)

Species	KBA Criteria Met	Notes & Locations
Fish		
Leopard Barbel *Luciobarbus subquin- cunciatus* **(Critically Endangered)**	V. Vulnerability Criterion	Found at Darbandikhan and Dukan Lake
Haditha Cave Garra *Garra widdowsoni* **(Critically Endangered)**	V. Vulnerability Criterion	Found at the sinkhole at Sheikh Hadid Shrine at Haditha Wetlands and Baghdadi in the spring of 2012
Haditha Cavefish *Caecocypris basimi* **(Critically Endangered)**	V. Vulnerability Criterion	Known historically at the Sheikh Hadid Shrine sinkhole at Haditha Wetlands and Baghdadi but not observed in 2012

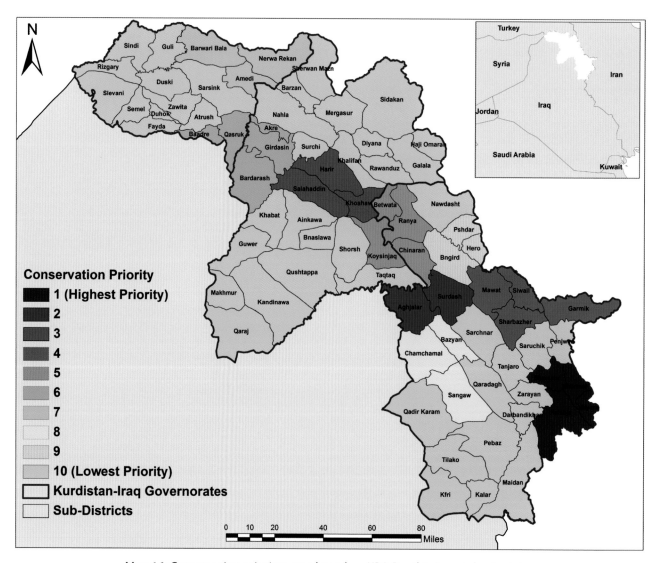

Map 16: Conservation priority areas based on KBA Benthic Invertebrate data

Though fish species were not fully evaluated (except for three critically endangered species under the Vulnerability Criterion), nevertheless fish are important economic and biodiversity resources in Iraq and additional information on important species (including several Vulnerable species) was provided at sites where fish surveys were undertaken. Importance here is broadly defined as species of economic importance for their value in fisheries as a food resource or fish that provide important ecosystem services (i.e. the mosquito fish, which is not eaten but contributes to malaria control). Problematic species are also included in the individual assessments as their presence and abundance may have an impact on the ecosystem as a whole (for example, the introduced and invasive *Tilapia zilli* may displace other species or *Silurus triostegus*, which is not fished commercially (scaleless) but is of ecological importance as a predator and its conservation status in Iraq is not well understood). Other species listed in the assessments were species that were considered rare or restricted-range species in Iraq, such as *Garra widdowsoni* (a cave-dwelling fish, assessed as Critically Endangered by IUCN) and other species for which little information are available. The checklist of these fish species is provided in Appendix 6, which provides additional details on these fish species.

For similar reasons, insects were not included in the KBA assessment though extensive collections of benthic macro-invertebrates were undertaken during the surveys from 2007 to 2010. Currently efforts are underway to describe and assign names for species found in the KBA surveys but this may take several years. Nevertheless preliminary information has indicated hotspot areas of conservation concern for aquatic insects in Kurdistan, northern Iraq based on the presence of rare and sensitive water quality bio-indicators (Mohammed Al-Saffar, Personal communication, 6 November 2013). See a preliminary map provided by M. Al-Saffar in 2013 on the proceeding page.

In addition, the previously mentioned Mediterranean-wide IUCN Freshwater Key Biodiversity Assessment Program, which was released at the end of 2014, has increased our understanding of several conservation concern freshwater species and the sites that are important for them in Iraq.

Finally, in the work on the coral reef habitat recently found in Iraq's gulf waters by scientists at the University of Freiburg and the University of Basrah's Marine Science Center a number of living stone corals and octocorals have been identified along with several ophiuroids (brittle stars). These are living in a highly turbid environment at 7-20 meters in depth and characterized by seawater temperatures ranging from between 14 and 34°C with rapidly changing salinity. According to the researchers (Pohl et al., 2014), these corals are adapted to one of the most extreme coral-bearing environments on earth and this is the first time that corals have been found in the turbid, sediment-rich coastal waters of Iraq.

BIRDS

For many key bird species, Important Bird and Biodiversity Areas (IBAs), one of the subsets of KBAs, represent the minimum network of sites needed for their conservation, at the full potential of their range, distribution and population size. This is especially true of long-distance migrants such as wildfowl and waders that breed in Western Siberia and winter in Africa or Asia. IBAs in Iraq play a vital role in providing safe feeding and roosting areas for such species and supportive networks of sites for intercontinental migrant species are a core value of IBAs.

SPECIES OF GLOBAL CONSERVATION CONCERN

A total of twelve species of global conservation concern (A1) were observed in Iraq with nine of them occurring regularly (see Table 17). Of the 67 IBAs, a total of 55 (82%) are important because they support significant numbers of one or more of these species of global conservation concern (see Appendix 5 for details on species). Nine of these species are found at IBAs in Iraq in numbers reaching or exceeding the qualifying thresholds and thus they rely on IBAs in Iraq to a greater or lesser extent for their survival. In addition, three A1 species occur at a total of seventeen IBAs either in numbers that do not reach qualifying levels or their status is currently insufficiently known to justify inclusion as qualifying species.

Table 17: Threatened species (A1) and their Sites in Iraq

Note: ✓ indicates the bird meets threshold levels at the site, + means the bird is present below the threshold levels (for status of individual species, see Appendix 5); Site codes in bold text indicate that this site qualifies under the A1 Criterion.

Site Code	Site Name	Lesser White-fronted Goose *Anser erythropus* (Vulnerable)	Red-breasted Goose *Branta ruficollis* (Endangered)	Marbled Duck *Marmaronetta angustirostris* (Vulnerable)	White-headed Duck *Oxyura leucocephala* (Endangered)	Dalmatian Pelican *Pelecanus crispus* (Vulnerable)	Egyptian Vulture *Neophron percnopterus* (Endangered)	Greater Spotted Eagle *Aquila clanga* (Vulnerable)	Eastern Imperial Eagle *Aquila heliaca* (Vulnerable)	Saker Falcon *Falco cherrug* (Endangered)	Macqueen's Bustard *Chlamydotis macqueenii.* (Vulnerable)	Sociable Lapwing *Vanellus gregarious* (Critically Endangered)	Basra Reed Warbler *Acrocephalus griseldis* (Endangered)	Total species per site
IQ003	Dure							+						1
IQ004	Barzan Area & Gali Balnda						✓							1
IQ006	Mangesh						✓							1
IQ007	Gara Mountain & Garagu						✓							1
IQ008	Amedi & Sulav						✓							1
IQ009	Mosul Lake	+					✓							2
IQ010	Dohuk Lake						✓							1
IQ011	Zawita						✓							1
IQ012	Atrush & Bania Area						✓							1
IQ013	Gali Zanta & Garbeesh						✓							1
IQ014	Bakhma & Bradost Mountain						✓							1
IQ020	Bahrka						✓							1
IQ021	Doli Smaquli & Ashab						✓							1
IQ023	Dukan Lake	✓	✓	+			✓		+					5
IQ025	Taq Taq						✓							1
IQ026	Chami Razan						✓							1
IQ027	Peramagroon Mountain						✓		+					2
IQ028	Sargalu						✓							1
IQ030	Kuradawe & Waraz						✓							1
IQ032	Penjween						✓							1
IQ034	Parazan						✓							1
IQ035	Chamchamal						✓		+					1
IQ036	De Lezha						✓							1
IQ037	Hazarmerd						✓							1
IQ039	Qara Dagh						✓							1
IQ040	Darbandikhan Lake						✓							1
IQ042	Ahmed Awa						✓							1
IQ044	Kalar Area						✓							1
IQ046	Jabal Makhool						✓							1
IQ048	Ajeel Himreen Hills						✓				+			2
IQ049	Qadissiya Lake						✓					✓		2
IQ050	Haditha Wetlands & Baghdadi			+					+			✓		3
IQ051	Tharthar Lake & Dhebaeji Field			✓										1

Site Code	Site Name	Lesser White-fronted Goose *Anser erythropus* (Vulnerable)	Red-breasted Goose *Branta ruficollis* (Endangered)	Marbled Duck *Marmaronetta angustirostris* (Vulnerable)	White-headed Duck *Oxyura leucocephala* (Endangered)	Dalmatian Pelican *Pelecanus crispus* (Vulnerable)	Egyptian Vulture *Neophron percnopterus* (Endangered)	Greater Spotted Eagle *Aquila clanga* (Vulnerable)	Eastern Imperial Eagle *Aquila heliaca* (Vulnerable)	Saker Falcon *Falco cherrug* (Endangered)	Macqueen's Bustard *Chlamydotis macqueenii.* (Vulnerable)	Sociable Lapwing *Vanellus gregarius* (Critically Endangered)	Basra Reed Warbler *Acrocephalus griseldis* (Endangered)	Total species per site
IQ052	Samarra Wetlands			+										1
IQ053	Himreen Lake	✓		+				+	+					4
IQ055	Ga'ara						✓							1
IQ056	Habbaniya Lake			✓										1
IQ057	Jazman (Zurbatia)							+	+		✓			3
IQ058	Razzaza Lake			✓				✓						2
IQ060	Hindiya Barrage			+				✓					✓	2
IQ061	Shuweicha Marsh	✓									✓			2
IQ062	Ibn Najm, North			✓									✓	1
IQ063	Ibn Najm			✓									✓	2
IQ064	Dalmaj Marsh	✓	✓	✓	✓				+		✓		✓	7
IQ065	Gharraf River			+					+				✓	3
IQ066	Hoshiya and Saaroot			+					+		✓			3
IQ067	Teeb Oasis & Zubaidaat										✓			1
IQ068	Teeb Seasonal Wetlands	✓						+	+		✓			4
IQ069	Sinnaaf Seasonal Marshes			✓					+					2
IQ070	Wadi Al-W'eir & Sh'eeb Abu-Talha						✓				✓			2
IQ071	Sawa Lake and Area			✓							✓			2
IQ072	Auda Marshes			✓				+					✓	3
IQ073	Hawizeh Marshes			✓	✓			+	+				✓	5
IQ074	Suwaibaat (or Sleibaat)			✓							✓			2
IQ075	Central Marshes			✓				+	+				✓	4
IQ076	Hammer, West			✓					+				✓	3
IQ077	Hammar, East			✓				+	+				✓	4
IQ078	Salman										✓			1
IQ080	Jabal Senam								+		+			2
IQ081	Khor Az-Zubayr Marshes							+						1
IQ082	Fao			✓		+		+					✓	4
	Total Occurrence	6	2	22	2	1	31	11	15	1	11	2	11	
	Present below A1 Threshold	1	0	7	0	1	0	10	15	1	2	0	0	
	Meets A1 Threshold	5	2	15	2	0	31	1	0	0	9	2	11	

Some of these species have limited ranges or have specific requirements that are met only by habitats found in Iraq. For example, the Mesopotamian Marshlands Endemic Bird Area of Southern Iraq represents the main, if not the only, breeding grounds for Basra Reed Warbler globally and the wetlands of southern Iraq hold the main global population of Marbled Duck. Thus disruptions to the habitats that they rely upon can have a major impact on these threatened species. Other species are under serious threat from hunting pressure such as Macqueen's Bustard, Lesser White-fronted Goose, Marbled Duck, Red-breasted Goose, and White-headed Duck. Also while most species of global conservation concern in Iraq are found at several IBAs three species were extremely rare in Iraq — Sociable Lapwing was found at only two sites:

Haditha Wetlands & Baghdadi (IQ050) and an individual was satellite tracked to an area near Qadissiya Lake (IQ049). It is listed as Critically Endangered because of a very rapid decline in its population and range. The Red-breasted Goose is an Endangered species also undergoing a rapid population decline and was recorded at only two sites in Iraq in winter: Dukan Lake (IQ023) and Dalmaj (IQ064). White-headed Duck, another Endangered species, was found only one time in Hawizeh Marhes (IQ073) in 2005 and once in Dalmaj (IQ064) in 2011. Thus conservation actions directed at these specific A1 sites is critical in protecting some of our most at risk birds though many of the A1 sites contain two or more of these species as well as species that meet other criteria.

Finally four additional conservation concern species are on the Iraq checklist but were never observed at any site during the KBA Program surveys:

1. **Northern Bald Ibis** *Geronticus eremita* CE — Formerly very rare passage migrant or vagrant in central Iraq but only recorded in the 1910s and early 1920s; in 2006 satellite signals of tagged birds from the tiny Syrian breeding colony suggest one bird may have passed close to, or through extreme western Iraq on its southerly migration. This species is the focus of a single species action plan for conservation (Armesto, Boehm, & Bowden, 2006).

2. **Great Bustard** *Otis tarda* VU — Considered a rare winter visitor to Iraq, this large bustard is facing rapid population reductions across most of its range owing to hunting and the loss of its habitat.

3. **Slender-billed Curlew** *Numenius tenuirostris* CE — Historically very rare or vagrants, last record was in 1979.

4. **Syrian Serin** *Serinus syriacus* VU — This small, long-tailed canary breeds in the mountains of Lebanon, Syria, Palestine and Jordan and appears to be declining at key sites. This species will be removed from the Iraq checklist (Porter, 2014).

RESTRICTED RANGE SPECIES

Two range-restricted species are endemic to the Mesopotamian Marshlands Endemic Bird Area located in southern Iraq. These are Iraq Babbler *Turdoides altirostris* and Basra Reed Warbler *Acrocephalus griseldis*. Iraq Babbler is found primarily in the lower Mesopotamian valley of the Tigris and Euphrates as well as in parts of southwestern Iran though during the KBA Program, it was also observed at sites along the southern edge of the Kurdistan Region including Taq Taq (IQ025), Altun Kopri (IQ033) and Chamchamal (IQ035). It is found in reedbed vegetation but it also utilizes rural habitats and irrigation canals (BirdLife, 2014). The Endangered Basra Reed Warbler visits Iraq in the summer to breed in reedbeds from Baghdad south to Basrah (in the KBA Program the northernmost sites where this species was observed was Hindiya Barrage (IQ060) and North Ibn Najm (IQ062), though here the bird does not occur in high numbers at these sites). In winter it returns to central East Africa.

Out of a total of 67 IBAs, 19 (28% of all IBA sites) have been identified on the basis that they support one or both of these restricted range species, thus meeting the A2 criterion (See Table 18). Together these 19 sites cover 13,116 km2 (46% of the total area of all KBAs in the country) with 16% occurring in the Kurdistan region, 16% in Central Iraq and the majority (68%) occurring in southern Iraq.

Table 18: Restricted Range Bird Species (A2) and their Sites in Iraq

Note: ✓ indicates the bird meets threshold levels at the site, + means the bird is present but was not selected as an A2 site (for status of individual species, see Appendix 5); Site codes in bold text indicate that this site qualifies under the A2 Criterion.

Site Code	Site Name	*Turdoides altirostris*	*Acrocephalus griseldis*
IQ025	**Taq Taq**	✓	
IQ033	**Altun Kopri Marsh**	✓	
IQ035	**Chamchamal Area**	✓	
IQ047	**Mahzam and Al-Alam**	✓	
IQ050	**Haditha Wetlands & Baghdadi**	✓	
IQ051	Tharthaar Lake & Dhebaeji Field	+	
IQ052	**Samarra Wetlands**	✓	
IQ057	Jazman (Zurbatia)	+	
IQ058	**Razzaza Lake**	✓	
IQ059	Musayab	+	
IQ060	**Hindiya Barrage**	✓	
IQ062	**Ibn Najm, North**	✓	✓
IQ063	**Ibn Najm**	✓	✓
IQ064	**Dalmaj Marsh**	✓	✓
IQ065	**Gharraf River**	✓	✓
IQ066	Hoshiya and Saaroot	+	
IQ072	**Auda Marsh**	✓	✓
IQ073	**Hawizeh Marshes**	✓	✓
IQ075	**Central Marshes**	✓	✓
IQ076	**Hammar, West**	✓	✓
IQ077	**Hammar, East**	✓	✓
IQ081	**Khor Az-Zubayr Marshes**	✓	
IQ082	**Fao**	✓	✓
Total Occurrence		23	11
Present below A2 Threshold		4	0
Meets A2 Threshold		19	11

BIOME-RESTRICTED SPECIES

A total of 43 biome-restricted species occur in Iraq's five distinct biomes. Approximately thirty have been proven to occur regularly at sites in Iraq and are breeding there (birds with confirmed or probable breeding status). Out of a total of 67 IBAs, 27 (40% of the IBAs) have been identified on the basis that they support a significant assemblage of biome-restricted species, thus meeting Criterion A3. Table 19 consists of five sub-tables (one for each biome in Iraq) that provide a list of the sites where these biome-restricted species occur and identifies the sites that were determined to have "a significant assemblage" of these species. Note that these sub-tables are roughly listed in the order in which they occur within the country, from north to south. It should be noted that a number of sites have yet to be thoroughly surveyed and may, with further research be found to support biome-restricted species.

Table 19: Biome-Restricted Bird Species (A3)
and their Sites in Iraq

Note: ✓ indicates the bird meets threshold levels at the site, + means the bird is present but the site was not selected as an A3 site (for status of individual species, see Appendix 5); Site codes in bold text indicate that this site qualifies under the A3 Criterion.

Mediterranean (ME12)

Site Code	Site Name	*Lanius nubicus*	*Poecile lugubris*	*Sitta neumayer*	*Oenanthe melanoleuca*	*Emberiza melanocephala*
IQ001	Fishkhaboor					+
IQ002	**Benavi & Sararu**	✓	✓	✓	✓	✓
IQ003	Dure		+	+	+	+
IQ004	**Barzan Area & Gali Balnda**	✓	✓	✓	✓	✓
IQ005	Kherazook	+	+		+	
IQ006	**Mangesh**	✓	✓	✓	✓	✓
IQ007	**Gara Mt. and Garagu**	✓	✓	✓	✓	✓
IQ008	**Amedi & Sulav**	✓	✓	✓	✓	✓
IQ009	Mosul Lake					+
IQ011	Zawita		+		+	+
IQ012	Atrush & Bania		+		+	+
IQ013	Gali Zanta & Garbeesh	+			+	+
IQ014	**Bakhma & Bradost Mountain**	✓	✓	✓	✓	✓
IQ016	Sari Hassan Bag Mountain				+	+
IQ017	Halgurd Mountain Area	+			+	+
IQ018	Haji Omran Mountain		+	+	+	+
IQ019	Sakran Mountain	+				+
IQ020	Bahrka					+
IQ021	**Doli Smaquli & Ashab**	✓	✓	✓	✓	✓

Mediterranean (ME12)

Site Code	Site Name	*Lanius nubicus*	*Poecile lugubris*	*Sitta neumayer*	*Oenanthe melanoleuca*	*Emberiza melanocephala*
IQ022	Doli Plngan		+		+	+
IQ023	Dukan Lake	+		+	+	+
IQ024	Assos Mountain	+		+	+	+
IQ025	Taq Taq					+
IQ026	**Chami Razan Area**	✓	✓	✓	✓	✓
IQ027	**Peramagroon Mt.**	✓	✓	✓	✓	✓
IQ028	Sargalu	+	+		+	+
IQ030	Kuradawe & Waraz	+	+		+	
IQ031	Sharbazher Area	+	+		+	
IQ032	Penjween	+				+
IQ033	Altun Kopri Marsh			+		+
IQ034	Parazan	+	+		+	+
IQ035	Chamchamal Area		+			+
IQ036	De Lezha	+	+		+	+
IQ037	Hazarmerd				+	+
IQ039	**Qara Dagh Area**	✓	✓	✓	✓	✓
IQ040	Darbandikhan Lake					+
IQ041	Zalm Area					+
IQ042	**Ahmed Awa**	✓	✓	✓	✓	✓
IQ043	**Hawraman Area**	✓	✓	✓	✓	✓
IQ044	Kalar Area					+
Total Occurrence		24	24	17	29	37
Doesn't meet threshold		12	12	5	17	25
A3 sites		12	12	12	12	12

Eurasian High-Montane biome (ME05)

Site Code	Site Name	*Tetraogallus caspius*	*Pyrrhocorax graculus*	*Montifringilla nivalis*	*Prunella ocularis*	*Anthus spinoletta*	*Serinus pusillus*
IQ004	Barzan Area & Gali Balnda		+				
IQ008	Amedi & Sulav						+
IQ017	Halgurd Mountain Area			+	+		
IQ024	Assos Mountain		+				
IQ027	**Peramagroon Mt.**		✓	✓		✓	✓
Total Occurrence		0	3	2	1	1	2
Doesn't meet threshold		0	2	1	1	0	1
A3 sites		0	1	1	0	1	1

Irano-Turanian biome (ME06)

Site Code	Site Name	Ammoperdix griseogularis	Phylloscopus neglectus	Hippolais languida	Sylvia mystacea	Sitta tephronota	Irania gutturalis	Oenanthe xanthoprymna	Oenanthe finschii	Oenanthe albonigra	Emberiza buchanani	Emberiza semenowi	Carpospiza brachydactyla
IQ001	Fishkhaboor	+			+								
IQ002	Benavi_&_Sararu					+			+			+	
IQ003	Dure					+							
IQ004	**Barzan Area & Gali Balnda**	✓		✓	✓	✓	✓	✓	✓			✓	
IQ005	Kherazook								+			+	
IQ006	Mangesh				+	+	+						
IQ007	Gara Mt. and Garagu				+	+	+		+				
IQ008	Amedi & Sulav				+	+						+	
IQ009	Mosul Lake	+											
IQ011	Zawita				+	+							
IQ012	Atrush & Bania				+	+							
IQ013	Gali Zanta & Garbeesh					+			+			+	
IQ014	Bakhma & Bradost Mountain	+			+	+			+				
IQ016	**Sari Hassan Bag Mountain**					✓	✓	✓	✓		✓	✓	
IQ017	**Halgurd Mountain Area**	✓			✓	✓	✓	✓	✓			✓	
IQ018	Haji Omran Mountain					+			+			+	
IQ019	Sakran Mountain	+		+		+						+	
IQ020	Bahrka	+			+								
IQ021	**Doli Smaquli & Ashab**	✓			✓	✓	✓		✓			✓	
IQ022	Doli Plngan					+						+	
IQ023	**Dukan Lake**	✓		✓	✓	✓	✓		✓				✓
IQ024	**Assos Mountain**	✓			✓	✓	✓	✓				✓	
IQ025	Taq Taq	+			+							+	
IQ026	**Chami Razan Area**	✓			✓	✓	✓		✓			✓	
IQ027	**Peramagroon Mt.**	✓	✓	✓	✓	✓	✓	✓	✓		✓	✓	✓
IQ028	Sargalu				+	+			+			+	
IQ030	Kuradawe & Waraz	+			+	+	+		+			+	
IQ031	**Sharbazher Area**	✓			✓	✓	✓		✓			✓	
IQ032	Penjween				+				+				
IQ033	Altun Kopri Marsh	+							+				
IQ034	Parazan				+	+							
IQ035	Chamchamal Area	+			+				+				
IQ036	**De Lezha**	✓			✓	✓	✓		✓				✓
IQ037	Hazarmerd	+			+	+			+				
IQ039	**Qara Dagh Area**	✓		✓	✓	✓	✓		✓			✓	
IQ040	Darbandikhan Lake	+				+			+				
IQ041	Zalm Area				+								
IQ042	**Ahmed Awa**	✓			✓	✓	✓		✓				✓
IQ043	Hawraman Area	+			+	+			+				
IQ044	Kalar Area					+							
IQ046	Jabal Makhool	+						+	+				
IQ055	Ga'ara	+											

Irano-Turanian biome (ME06)

Site Code	Site Name	Ammoperdix griseogularis	Phylloscopus neglectus	Hippolais languida	Sylvia mystacea	Sitta tephronota	Irania gutturalis	Oenanthe xanthoprymna	Oenanthe finschii	Oenanthe albonigra	Emberiza buchanani	Emberiza semenowi	Carpospiza brachydactyla
IQ057	Jazman (Zurbatia)	+			+								
IQ058	Razzaza Lake				+								
IQ059	Musayab				+								
IQ060	Hindiya Barrage				+								
IQ063	Ibn Najm				+								
IQ064	Dalmaj Marsh				+								
IQ067	Teeb Oasis & Zubaidaat	+			+					+			
IQ073	Hawizeh Marshes				+								
IQ082	Fao				+								
Total Occurrence		27	1	5	37	31	15	6	26	1	2	19	4
Doesn't meet threshold		16	0	1	26	18	2	1	15	1	0	10	0
A3 sites		11	1	4	11	13	13	5	11	0	2	9	4

Eurasian Steppe and Desert biome (ME04)

Site Code	Site Name	Accipiter brevipes
IQ002	Benavi & Sararu	+
IQ003	Dure	+
IQ007	Gara Mt. and Garagu	+
IQ008	Ser Amadia & Sulav Resort	+
IQ013	Gali Zanta & Garbeesh	+
IQ022	Doli Plngan	+
IQ030	Kuradawe & Waraz	+
IQ034	Parazan	+
IQ039	Qara Dagh Area	+
Total Occurrence		9
Doesn't meet threshold		9

Sahara-Sindian Desert biome (ME13)

Site Code	Site Name	Chlamydotis macqueenii	Vanellus leucurus	Cursorius cursor	Pterocles senegallus	Otus brucei	Caprimulgus aegyptius	Corvus ruficollis	Hypocolius ampelinus	Alaemon alaudipes	Ammomanes cinctura	Ammomanes deserti	Eremophila bilopha	Pycnonotus leucotis	Ptyonoprogne obsoleta	Acrocephalus griseldis	Turdoides altirostris	Oenanthe lugens	Passer moabiticus	Rhodospiza obsoleta
IQ001	Fishkhaboor																		+	
IQ006	Mangesh													+						
IQ016	Sari Hassan Bag Mountain														+					
IQ020	Bahrka											+								
IQ021	Doli Smaquli & Ashab																			+
IQ022	Doli Plngan																			
IQ023	Dukan Lake											+		+						
IQ025	Taq Taq													+			+		+	

Sahara-Sindian Desert biome (ME13)

Site Code	Site Name	Chlamydotis macqueenii	Vanellus leucurus	Cursorius cursor	Pterocles senegallus	Otus brucei	Caprimulgus aegyptius	Corvus ruficollis	Hypocolius ampelinus	Alaemon alaudipes	Ammomanes cinctura	Ammomanes deserti	Eremophila bilopha	Pycnonotus leucotis	Ptyonoprogne obsoleta	Acrocephalus griseldis	Turdoides altirostris	Oenanthe lugens	Passer moabiticus	Rhodospiza obsoletus
IQ026	Chami Razan Area													+						
IQ027	Peramagroon Mt.													+				+		+
IQ028	Sargalu													+						
IQ030	Kuradawe & Waraz													+						
IQ032	Penjween																+			
IQ033	Altun Kopri Marsh											+					+		+	
IQ035	Chamchamal Area											+		+			+			
IQ036	De Lezha					+						+		+						
IQ039	Qara Dagh Area													+						
IQ040	Darbandikhan Lake													+						
IQ044	Kalar (Diyala River)													+			+		+	
IQ045	Maidan Area											+								
IQ046	Jabal Makhool		+					+		+		+		+			+	+	+	
IQ048	Ajeel Himreen Hills			+				+	+			+		+			+			
IQ049	Qadissiya Lake											+		+						
IQ050	**Haditha Wetlands & Baghdadi**		✓	✓			✓	✓	✓			✓	✓	✓			✓		✓	
IQ051	**Tharthaar Lake & Dhebaeji Field**		✓	✓			✓		✓			✓		✓			✓		✓	
IQ052	Samarra Wetlands							+						+			+		+	
IQ053	Himreen Lake																+			
IQ054	Mandli													+			+			
IQ055	Ga'ara			+			+			+	+	+	+							
IQ056	Habbaniya Lake		+							+			+	+						
IQ057	**Jazman (Zurbatia)**				✓	✓	✓		✓	✓	✓	✓	✓	✓	✓					✓
IQ058	**Razzaza Lake**	✓	✓	✓	✓		✓	✓	✓	✓	✓	✓	✓	✓			✓		✓	
IQ059	Musayab		+			+								+			+			
IQ060	Hindiya Barrage						+		+					+		+				
IQ061	Shuweicha Marsh			+			+			+										
IQ062	Ibn Najm, North		+			+	+		+					+		+			+	
IQ063	Ibn Najm		+				+		+					+		+	+		+	
IQ064	**Dalmaj Marsh**	✓	✓	✓			✓	✓	✓	✓	✓			✓		✓	✓		✓	
IQ065	Gharraf River		+			+								+			+		+	
IQ066	Hoshiya and Saaroot		+				+		+	+							+			
IQ067	**Teeb Oasis & Zubaidaat**			✓	✓				✓	✓	✓	✓		✓					✓	✓
IQ068	Teeb Seasonal Wetlands		+	+	+		+		+	+	+	+		+						
IQ069	Sinnaaf Seasonal Wetlands		+				+			+	+	+								
IQ070	**Wadi Al-W'eir & Sh'eeb Abu-Talha**	✓		✓	✓		✓	✓	✓	✓				✓						
IQ071	Sawa Lake and Area			+			+	+		+	+	+		+						

Sahara-Sindian Desert biome (ME13)

Site Code	Site Name	Chlamydotis macqueenii	Vanellus leucurus	Cursorius cursor	Pterocles senegallus	Otus brucei	Caprimulgus aegyptius	Corvus ruficollis	Hypocolius ampelinus	Alaemon alaudipes	Ammomanes cinctura	Ammomanes deserti	Eremophila bilopha	Pycnonotus leucotis	Ptyonoprogne obsoleta	Acrocephalus griseldis	Turdoides altirostris	Oenanthe lugens	Passer moabiticus	Rhodospiza obsoleta
IQ072	Auda Marsh		+						+					+			+			
IQ073	Hawizeh Marshes		+		+		+		+					+		+	+		+	
IQ074	Suwaibaat (or Sleibaat)		+		+		+	+		+				+					+	
IQ075	**Central Marshes**		✓		✓		✓		✓					✓		✓	✓		✓	
IQ076	**Hammar, West**		✓		✓	✓	✓	✓	✓					✓		✓	✓		✓	
IQ077	Hammar, East		+				+		+					+		+	+		+	
IQ078	Salman	+		+	+			+		+	+	+								
IQ080	Jabal Senam	+		+	+		+	+		+	+	+								
IQ081	Khor Az-Zubayr Marshes		+														+			
IQ082	Fao		+				+									+	+			
Total Occurrence		5	20	11	15	7	19	11	18	19	10	21	5	37	2	9	24	2	22	4
Doesn't meet threshold		2	15	5	8	4	13	7	10	11	6	15	2	29	1	6	18	2	14	2
A3 sites		3	5	6	7	3	6	4	8	8	4	6	3	8	1	3	6	0	8	2

Together the 27 sites that meet A3 Criterion cover 15330 km² (55% of the total area of IBAs in the country). All five of the biomes are found in Iraq, although the extent of Eurasian High-Montane is restricted to high elevation sites in Kurdistan and the area of the Mediterranean biome is limited to just the extreme north of the country. The number of IBAs in each of the five biomes is shown in Figure 7 and this figure also indicates in which region of the country the biome-restricted birds are concentrated. The fact that only one site was selected based on species restricted to the Eurasian High-Montane biome (ME05) is due to the limited number of high elevation sites visited in the survey. Additional survey work in these areas is likely to show that significant populations of key species are present and that additional sites will qualify under Criterion A3. Also there is only one Iraq species associated for the Eurasian Steppe and Desert biome (ME04), Levant sparrowhawk *Accipiter brevipes*, and while this bird was seen at several sites it did not make up an assemblage of species and therefore no sites were selected for this biome.

Of the 43 biome-restricted species, most occur at one or more of the IBAs meeting A3 Criteria listed in the tables above. The following species are not represented at A3 IBAs: Caspian snowcock *Tetraogallus caspius* and Radde's Accentor *Prunella ocularis* (both associated with the Eurasian High-Montane biome); Hume's wheatear *Oenanthe albonigra* (associated with the Irano-Turanian biome), *Accipiter brevipes* (associated with the Eurasian Steppe and Desert

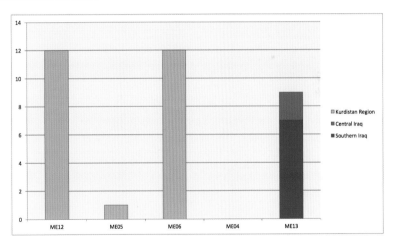

Figure 7: IBAs by Biome

biome as previously mentioned) and the Mourning wheatear *Oenanthe lugens* (associated with the Sahara-Sindian Desert biome). The Caspain snowcock was the only biome-restricted species that was never seen during any of the surveys.

While sites identified under Criterion A3 clearly form a major component (by area) of the national IBA inventory, it should be noted that each of these sites usually also meets at least one additional criterion.

CONGREGATORY SPECIES

In all, 16 sites (24% of all IBAs) qualify on the basis that they hold globally important congregations of one or more bird species, thus meeting one or more of the A4 sub-criteria and all of them fulfill at least one additional IBA criterion,

Table 20: Congregatory waterbird species (A4i) and their sites in Iraq

Species	1% Population Level*	Mosul Lake (IQ009)	Dukan Lake (IQ023)	Darbandikhan Lake (IQ040)	Razzaza Lake (IQ058)	Ibn Najm, North (IQ062)	Dalmaj Marsh (IQ064)	Sinnaaf Seasonal Wetlands (IQ069)	Hawizeh Marshes (IQ073)	Central Marshes (IQ075)	Hammar, West (IQ076)	Hammar, East (IQ077)
Greater White-fronted Goose *Anser albifrons*	150		✓									
Lesser White-fronted Goose *Anser erythropus*	110		✓									
Common Shelduck *Tadorna tadorna*	800		✓									
Ruddy Shelduck *Tadorna ferruginea*	500	✓										
Marbled Duck *Marmaronetta angustirostris*	100					✓	✓		✓	✓	✓	✓
Red-crested Pochard *Netta rufina*	2,500				✓							
Greater Flamingo *Phoenicopterus roseus*	2,400				✓							
Great Cormorant *Phalacrocorax carbo*	1000		✓									
Eurasian Coot *Fulica atra*	20,000						✓					
White-tailed Lapwing *Vanellus leucurus*	250				✓							
Kentish Plover *Charadrius alexandrinus*	1,000				✓						✓	
Black-tailed Godwit *Limosa limosa*	1,000						✓					
Collared Pratincole *Glareola pratincola*	1,000	✓										
Slender-billed Gull *Chroicocephalus genei*	1,500		✓	✓	✓			✓			✓	✓
Whiskered Tern *Chlidonias hybrida*	1,000				✓						✓	
Total species A4ii species at each site		2	5	1	6	1	3	1	1	1	4	2

Note: 1% Population threshold values obtained from Delany & Scott (2002 & 2006)

Table 21: Species and Sites meeting Criteria A4ii & A4iii in Iraq

Criteria	Species	Fishkhaboor (IQ001)	Altun Kopri Marsh (IQ033)	Ibn Najm, North (IQ062)	Dalmaj Marsh (IQ064)	Hoshiya & Saaroot (IQ066)	Teeb Seasonal Wetlands (IQ068)	Sinnaaf Seasonal Wetlands (IQ069)	Hawizeh Marshes (IQ073)	Suwaibaat (or Sleibaat) (IQ074)	Central Marshes (IQ075)	Hammar, West (IQ076)	Hammar, East (IQ077)	Total sites
A4ii	**Dead Sea Sparrow** *Passer moabiticus**	✓			✓						✓	✓		4
A4iii	**Congregatory Waterbirds (waders and waterfowl)**		✓	✓	✓	✓	✓	✓	✓	✓	✓	✓	✓	11

*1% Population Threshold value for this species is 960 (individuals)

mostly A1. Of the four A4 sub-criteria, eleven sites meet the A4i Criterion, supporting 1% or more of the biogeographical population of one or more congregatory waterbird species; four sites meet the A4ii Criterion, selected for congregatory landbirds, 11 sites meet the A4iii Criterion, holding 20,000 or more waterbirds but no site meets the A4iv Criterion, as Iraq does not have any identified, migration bottleneck sites. Table 20 and Table 21 provide an overview of these sites and the species evaluated under A4 sub-criteria.

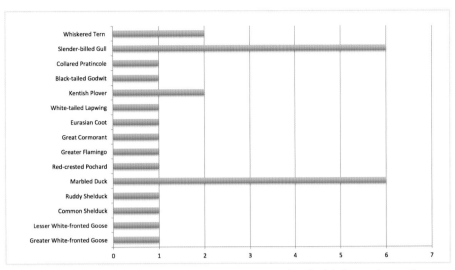

Figure 8: Number of sites meeting A4i Criterion thresholds for each species

Criterion A4i and A4iii sites are designated for waterbird populations and while five sites meeting some of the A4 sub-criteria are in the Kurdistan Region, the large complex of wetlands that make up the Mesopotamian Marshlands of southern Iraq is, by far, the most important area of the country for large congregatory species. Razzaza Lake (IQ058), Dalmaj Marsh (IQ0640) and West Hammar (IQ076) are important sites that support more than 1% of the biogeographical population of at least six, four, & five species respectively. Of the approximate 50 species of congregatory waterbirds that, theoretically, occur in Iraq, 15 species occur at IBAs in Iraq in qualifying numbers under criterion A4i and of these, 11 species occurring in these numbers were reliant on a single site. See Figure 8.

All of these congregatory waterbirds are migratory or nomadic and the individual areas within this network of sites identified are important at different stages of their annual lifecycles, especially for breeding or as migratory stopover sites. As the majority of these species are long-distance migrants it is important that there is cooperation and coordination in conservation activities between the countries through which they pass. This is particularly important for those congregatory species that are also of global conservation concern — six species in the case of Iraq: Lesser White-fronted Goose *Anser erythropus*, Marbled Duck *Marmaronetta angustirostris*, Red-breasted Goose *Anser ruficollis*, the rare migrant Sociable Lapwing *Vanellus gregarius*, and the White-headed Duck *Oxyura leucocephalus*.

Four sites — Fishkhaboor (IQ001), Dalmaj Marsh (IQ064), the Central Marshes (IQ075), and West Hammar (IQ076) have been identified under the A4ii criterion as they support 1% or more of the global population of a non-waterbird, the Dead Sea Sparrow *Passer moabiticus*.

A total of eleven sites regularly support more than 20,000 waterbirds, predominantly on migration. At the majority of sites, the number of birds is well in excess of the qualifying threshold, with sites such as the Central Marshes estimated to be used by over 70,000 birds annually. Some common waterbird species that congregate in large numbers in Iraq include Eurasian Coot *Fulica atra*, Black-tailed Godwit *Limosa limosa*, Dalmatian Pelican *Pelecanus crispus*, Greater Flamingo *Phoenicopterus roseus*, Great Cormorant *Phalacrocorax carbo*, as well as a variety of ducks such as Marbled Duck *Marmaronetta angustirostris*, herons, and various waders.

PLANTS & HABITATS

Here we will discuss another subset of KBAs, Important Plant Areas (IPAs). As part of the KBA data-collection process, information was collected on plants and habitats present at each site using several references (i.e. Guest, 1960, Abdulhasan, 2009 and Abdulhasan et al. 2009) to classify the habitats — see the "Methodology" chapter for details — and where possible the area of each was assessed. This information was then applied to the IPA criteria adapted for the Arabian Peninsula (See Appendix 1 for the full IPA criteria list). Of the 73 IPAs identified in Iraq, 42 (58% of all IPAs) were located in Kurdistan, northern Iraq, a further 8 (11%) & 23 (32%) were located in Central and Southern Iraq respectively.

According to Guest (1960, p60 & p64), Iraq is composed of five phytogeographical groups: Mediterranean; Irano-Turanian; Saharo-Sindian; Eurosiberian-Boreoamerican; and bi-and pluri-regional. Its also shows that two phytogeographical regions overlap the country: Irano-Turanian (represented by two sub-regions, the Irano-Anatolian sub-region that includes 132 endemic plants and the Mesopotamian sub-region that includes 49 endemic plants); and Saharo-Sindian (represented by the Middle Saharo-Sindian sub-region that includes 9 endemic plants). Northern Iraq, with its variety of habitats, elevations and climatic conditions is part of the Irano-Turanian phytogeographical region and is by far the most botanically rich area of the country.

As with IBA and general KBA criteria, the IPA criteria system has a specific criteria (A1-A4) that apply to the presence at a site of species that are of conservation important. In the Iraq Program, only IPA A4 criterion (sites containing national endemic, near endemic, regional endemic and/or regional range-restricted species or infraspecific taxa) could be assessed. Forty sites (55% of all IPA sites) were identified based on Criterion A4, which accounts for 7911 km² of land (1.8% of the entire country). The assessment of these forty sites was based on a list of species developed from the joint work of the KBA botanical surveys and the Centre for Middle Eastern Plants (CMEP) under the Darwin Project for In-Situ Conservation (mentioned previously in Chapter 3, section C). Appendix 7 provides a list of the plant species reported in the IPA assessments. This list is considered a partial list of Iraq's endemic and near endemic plants and contains 58 endemics and 66 near endemics but the majority (52 and 55 plants respectively) are based on historically recorded plants for the sites. Under the KBA Project, sites were also identified based on 56 plants that were considered nationally rare by the KBA botanical team.

In addition to conservation concern species, all the IPA sites were assessed based on species richness at a site (Criterion B). Two specific sub-criteria were used: B1 (a particularly species-rich example of a defined habitat type) and B2b (The site is refuge for biogeographically and bioclimatically restricted plants to 'retreat to' in the face of global climatic change).

Table 22: B1 Habitats Types found at IPA Sites

Code	Community	Habitat type
A	Alpine Zone	Alpine Zone Vegetation habitat type
B	Desert	Shrubland
C	Flooded Communities	Periodically or occasionally flooded land habitat type
D1	Herbaceous veg.	Sparsely vegetated land
D2		Steppe
E1	Inland Standing Water	Aquatic communities- Free Floating/Rooted Floating Vegetation habitat type
E2		Aquatic communities - Rooted Submerged Vegetation habitat type
E3		Ponds or lakes- Unvegetated standing water habitat type
F1	Inland running water, rivers, or canals	Riparian vegetation
F2		Submerged river or canal vegetation habitat type
G1	Marsh Vegetation	Helophytic Vegetation-Reed and Reedmace Bed habitat type
G2		Pioneer communities growing on salt or brackish habitat type
H	Moist Steppe Zone	Steppe habitat type
I1	Mountain Forest	Mountain Riverine Forest habitat type
I2		Oak Forest- Lowest, Medium, and Highest Zones habitat type
I3		Pine Forest habitat type
I4		Thorn-Cushion Vegetation habitat type
J	Riverine Forest	Riverine Forest of the Plains habitat type
K	Salt Pioneer Swards	Salt pioneer swards vegetation habitat type
L	Woodland	Shrubs

Table 23: IPA sites meeting Criterion B1 based on different habitat types.

Site Code	A	B	C	D1	D2	E1	E2	E3	F1	F2	G1	G2	H	I1	I2	I3	I4	J	K	L
IQ001																		✓		
IQ002														✓	✓					
IQ003														✓	✓					
IQ004														✓	✓					
IQ005														✓						
IQ006															✓					

Site Code	A	B	C	D1	D2	E1	E2	E3	F1	F2	G1	G2	H	I1	I2	I3	I4	J	K	L
IQ007														✓	✓					
IQ008														✓	✓					
IQ011															✓	✓				
IQ012														✓	✓					
IQ013														✓	✓					
IQ014														✓	✓					
IQ015														✓	✓					
IQ016																	✓			
IQ017	✓																✓			
IQ018																	✓			
IQ019														✓	✓		✓			
IQ020																		✓		
IQ021															✓					
IQ022														✓						
IQ023														✓						
IQ024														✓	✓					
IQ025														✓						
IQ026														✓	✓					
IQ027															✓		✓			
IQ028														✓	✓		✓			
IQ029														✓						
IQ030														✓	✓					
IQ031														✓	✓		✓			
IQ032														✓						
IQ033							✓				✓									
IQ034															✓					
IQ035																			✓	
IQ036														✓	✓					
IQ038													✓							
IQ039														✓	✓					
IQ040														✓	✓					
IQ041										✓								✓		
IQ042														✓	✓					
IQ043														✓	✓					
IQ044																		✓		
IQ046		✓			✓															
IQ047		✓							✓											
IQ048		✓			✓															
IQ050											✓									
IQ051			✓				✓				✓									
IQ055		✓			✓															
IQ057				✓																
IQ058		✓																		
IQ060									✓											✓

Site Code	A	B	C	D1	D2	E1	E2	E3	F1	F2	G1	G2	H	I1	I2	I3	I4	J	K	L
IQ063		✓									✓									
IQ064											✓									
IQ065							✓		✓											
IQ067				✓																
IQ068		✓																		
IQ069																			✓	
IQ070		✓		✓																
IQ071		✓																		
IQ072							✓				✓									✓
IQ073						✓	✓		✓		✓									✓
IQ075						✓	✓				✓									
IQ076			✓			✓	✓				✓									
IQ077						✓	✓	✓			✓	✓								
IQ079				✓																
IQ080				✓																
IQ081		✓																	✓	
IQ082							✓												✓	
Total	1	10	2	5	3	4	8	1	4	2	10	1	1	26	25	1	7	5	3	3

Sixty-seven (92% of all IPAs) were identified based on IPA Criterion B1 and these sites were identified uniformly throughout the country. The habitat types found at these sites are listed in Table 22 and the sites that were identified based on these habitats are provided in Table 23 (both listed in the proceed pages). The most commonly identified habitat types at these sites (found at 10 or more sites) were Mountain Riverine Forest, Oak Forest (Lowest, Medium, and Highest Zones), Desert Shrublands, and Marsh vegetation (Helophytic Vegetation-Reed and Reedmace Beds).

Eleven sites (15% of all IPAs) were identified based on IPA Criterion B2b (refugia sites). Though only seven sites were identified in the Kurdistan region there are likely more that may be well suited to acting as refugia because landscapes in this region offer protected slopes, canyons and valleys, higher forest cover and other features that can protect plants and their associated species assemblages. Four refugia sites were identified in the south of Iraq, several of which are wetland areas that never completely dried in the mid-1990s. These were Hindiya Barrage (IQ060), Gharraf River (IQ065), and Hawizeh Marsh (IQ073). These acted as refuge site for aquatic plants in southern Iraq. The first two sites are on important river bodies that provide source water to the Mesopotamian Marshlands, while Hawizeh is the only major wetland in southern Iraq that remained partially intact despite the intentional drainage campaign of the Mesopotamian Marshlands by the Iraqi government in the 1990s. The last refuge site identified in southern Iraq is Wadi Al-W'eir & Sh'eeb Abu-Talha (IQ070), which has protected an isolated relict of tropical vegetation that refers to a previous climate optimum.

Table 24: Summary of IPA Criterion C Sites with threatened habitats

Site Code	Site Name	Pine Forest habitat	Riverine Forest of the Plains (Al-Ahrash) habitat	Mesopotamian Marshlands	Area (ha)
IQ001	Fishkhaboor		✓		4,179
IQ003	Dure	✓			2,310
IQ011	Zawita	✓			288
IQ033	Altun Kopri Marsh			✓	1,575
IQ041	Zalm River		✓		550
IQ044	Kalar Area		✓		1,130
IQ061	Shuweicha Marsh			✓	54,938
IQ062	Ibn Najm, North			✓	1,789
IQ063	Ibn Najm			✓	4,000
IQ064	Dalmaj Marsh			✓	92,076
IQ065	Gharraf River			✓	50,461
IQ066	Hoshiya and Saaroot			✓	33,560
IQ072	Auda Marshes			✓	19,241
IQ073	Hawizeh Marshes			✓	164,023
IQ075	Central Marshes			✓	131,780
IQ076	Hammar, West			✓	136,326
IQ077	Hammar, East			✓	82,968
	17	2	3	12	781,194

Seventeen IPAs (23% of the total) were identified based in IPA Criterion C (Outstanding examples of globally or regionally threatened habitat types). Table 24 provides an overview of the habitat types that were the focus of this selection.

The majority of the sites selected under Criterion C were located in the Mesopotamian Marshlands in southern Iraq, which face decreasing water levels and drought, over-grazing, water pollution and human intrusion. They are located in the critical threatened Tigris-Euphrates alluvial salt marsh ecoregion. The Riverine Forest of the Plains (Al-Ahrash) is a key riparian habitat type in Iraq. Much of these river riparian areas are under threats from gravel mining, agricultural expansion and pollution from untreated sewage from cities and villages upstream. Only two sites, Kalar Area on the Diyala River (IQ044) and Zalm River (IQ041) were selected as examples of this habitat. Pine Forests is located in the critical threatened Zagros Mountains Forest Steppe Ecoregion of northern Iraq and outstanding examples of this habitat type were found at only two sites (Dure (IQ003) and Zawita (IQ011)). This important habitat is under threats of human intrusion and pollution caused by tourism.

5.D POTENTIAL THREATS AT KBA SITES

There are many threats that pose a real danger for species or habitats at sites in Iraq. These threats are discussed extensively in the individual site assessments. Sites in Iraq often face several threats of varying scope and severity. The following maps provide an overview of those areas of the country with the highest threat levels based on eight (out of 11) IUCN defined threat classifications that could be evaluated in the survey.

As was stated in the Methodology, these maps provide merely a rough indication of where more in-depth assessments on pressures to biodiversity and other environmental services should be concentrated. Not all high threats indicate that biodiversity is under pressure (e.g., garbage accumulation has less overall effect on biodiversity than overgrazing or hunting). Also some activities may have been listed as a low threat simply due to a lack of clear information (e.g. the team may not have been aware of upstream dams and water management policies). Certain activities are quite complex and require more training to be able to fully assess their impact on biodiversity (e.g. determining whether certain agricultural practices such as grazing or tree cutting is unsustainable or actually a traditional practice that may foster biological diversity).

The assessment of threats was primarily done in the field between 2009 and 2010 and while some newer information resulted in updates to the threat assessments for certain sites, a schedule for re-evaluating threats should be pursued at these sites. For example, the renewed conflicts in Northwestern, Western and Central Iraq that intensified in 2014 are clearly a threat to over a dozen KBA sites in these regions and the current threat assessment has not taken this into account.

Threats involving agriculture are shown in Map 17. Agriculture was mostly represented at sites by small-holder farms, livestock farming (both nomadic and small-holder grazing and animal production) and in some areas, from small-holder aquaculture. The mere presence of agricultural fields or grazing does not necessarily represent a threat and in fact, agro-ecosystems that are traditionally managed are often highly biologically diverse. The team's evaluation attempted to identify potentially unsustainable practices, such as extension of crop fields (or aquaculture farms) into stream riparian zones or other natural areas and heavy grazing pressure. But given the long history and large area devoted to crop and grazing lands within Iraq, agricultural impacts on biodiversity require much much greater study than what could be accomplished in the KBA Program.

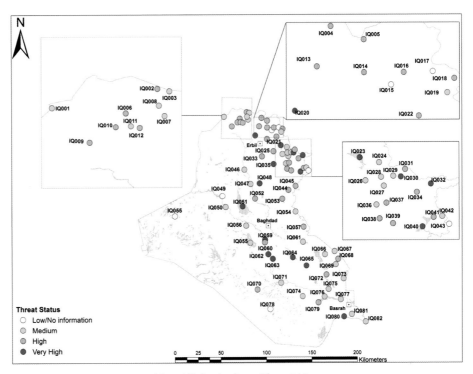

Map 17: Agriculture Threat Map

Map 18 gives a rough evaluation of where impacts from development are occurring within KBA sites. The main activities here were urban housing development, some commercial and industrial development and tourism and recreational development. The team attempted to evaluate these threats based on the degree that such developments involved the conversion of or damage to natural areas within a site.

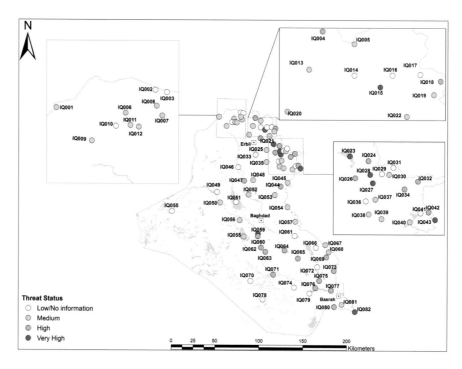

Map 18: Residential and Commercial Development Threat Map

For Iraq, energy production is largely related to oil development and production activities but mining, particularly for rock and gravel, was considered a severe threat at many sites as shown in Map 19. Oil development impacts is particularly a problem in southern Iraq but is increasing in the Kurdistan Region and gravel mining in and along rivers is often a very high threat for river riparian and in-stream habitats at many of the sites that contain these habitat types. These activities often caused additional human intrusion and pollution threats beyond simply the direct conversion and/or damage of natural habitats.

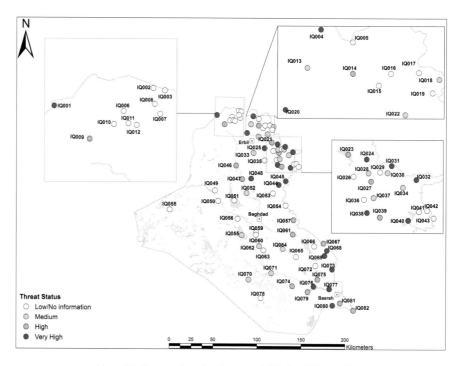

Map 19: Energy Production and Mining Threat Map

Map 20 shows assessments that were focused primarily on transportation issues such as the development of existing and new roads, which in Iraq tend to cause erosion and habitat destruction due to construction methods and the placement of these roads as well as opening up access to previously remote natural areas that lead to other types of impacts.

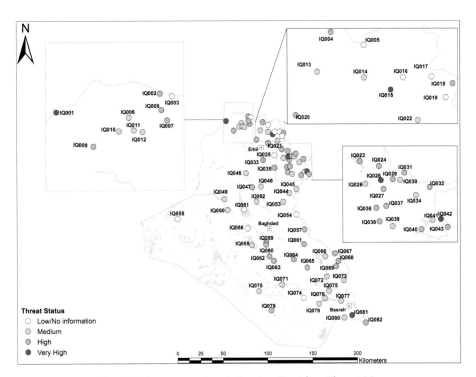

Map 20: Transportation & Service Corridors Threat Map

Hunting, over-fishing and harvesting of natural resources was sometimes difficult to adequately assess at a site due to the nature of these activities and the rapid assessment methodology. These threats, shown in Map 21, may occur at times and places where the KBA team could not be present to assess them directly, but the survey was often augmented by additional information provided by locals at or near the site. The key issues examined were uncontrolled hunting of waterfowl but also other bird, mammal and reptile species and high fishing pressure and/or the use of unsustainable fishing practices (electro-fishing and to some extent poisons and explosives). Harvesting of other natural resources such as plant resources was occasionally encountered but was largely not well assessed.

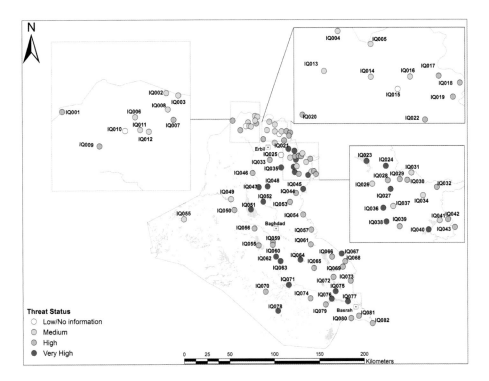

Map 21: Over-exploitation, Persecution and Control Threat Map

Map 22 focuses on threats from non-consumptive uses of biological resources including recreational activities, work,conflict and the other activities that impose human disturbance at a KBA Site. Many sites had evidence of previous wars and conflicts. For example, many sites along the Iranian border contain or are close to active mine fields. Some areas are popular places for picnicking and local tourism and are heavily used for recreation. Other areas simply have high human disturbance due to the proximity to people living and working near a site (agriculture, urban and industrial development, resource extraction activities, etc.).

In 2015, a review was done of Iraq's KBA affected by the conflict zone of the Islamic State of Iraq and Syria (ISIS). Fourteen sites in Iraq were within or near this zone including: Ajeel Himreen Hills (IQ048), Ga'ara (IQ055), Habbaniyah Lake (IQ056), Haditha Wetlands & Baghdadi (IQ050), Himreen Lake (IQ053), Hindiyah Barrage (IQ060), Jabal Makhool (IQ046), Mahzam & Al-Alam (IQ047), Mosul Lake (IQ009), Musayab (IQ059), Qadissiya Lake (IQ049), Razzaza Lake (IQ058), Samarra Wetlands (IQ052), and Tharthar Lake & Dhebaeji Field (IQ051). It should be noted that the Map 22 does not reflect the increased threat that these 14 sites face.

To the degree that the team was able to assess dams, water resource management decisions, drainage, dredging, and canalization activities that pose threats to the biodiversity of a site, these are shown in Map 23 but accurate information was often hard to obtain and very little is known about future planned activities that may impact KBA sites. In addition, this assessment is focused within the site itself and

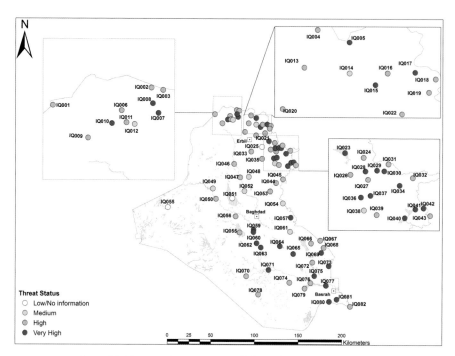

Map 22: Human intrusions and disturbance Threat Map

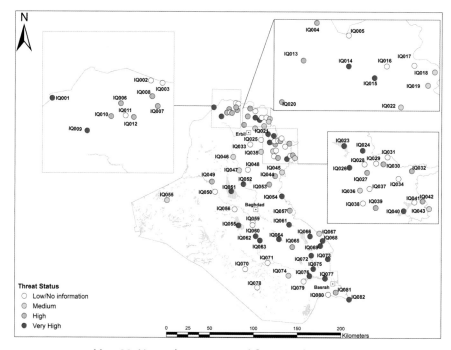

Map 23: Natural systems modification Threat Map

often such activities may be going on upstream at some distance from the area and the field teams may not have been aware or been able to properly assess the full threat to the site from these activities. As a result sites listed with a low or medium threat may actually be facing significantly higher threat from these activities.

The main pollution threats to sites that the team tried to assess, which are indicated in Map 24, were municipal and industrial waste and garbage and to some extent agricultural pollutants and noise pollution. Military waste, air, light and thermal pollution threats were difficult for the team to evaluate but are likely significant problems at many sites as well.

For a more extensive discussion of the specific pressures occurring at an individual site, see the site assessments themselves in Chapter 7 and also see the Threat Table in Appendix 8 that shows the data and calculations used to determine the threat status at each site.

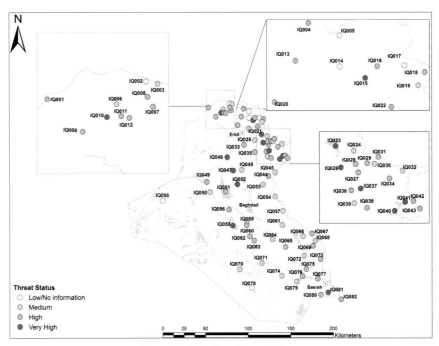

Map 24: Pollution Threat Map

5.E LIMITATIONS OF THE KBA EXERCISE IN IRAQ

Several limitations within the KBA Program have already been mentioned in proceeding chapters. The value of the KBA rapid assessment methodology is that it can provide strong indicators of which areas of the country have high biological diversity (threatened and other sensitive species, high endemism, species-richness and/or threatened habitats, etc.). The drawbacks of this approach are that only certain species and habitat types show up on the radar screen, while others that are more difficult to observe are almost entirely not represented. In addition, while the scale of the KBA Program effort spanned an impressive six years thanks to the commitment of partner organizations — financial, logistical and security constraints still limited the scope of the survey effort and left many parts of the country under-surveyed or were not visited at all. In addition, while the program had a strong capacity building component, not all necessary topics could be adequately covered within the training programs.

REMAINING GAPS

While the KBA Program has served to fill many of the gaps in our understanding that have existed since prior efforts to identify important areas in Iraq (Evans, 1994 & Scott, 1995), additional areas in central, western and southwestern Iraq remain to be surveyed and monitored. Many of these areas continue to have major problems with security and lack of access.

Even in areas that were surveyed, the effort was limited at best to two visits in a single year and while many sites were visited in all the years of the study effort, several sites were visited only once. Many of these were rejected as KBA sites, even several that were previously identified as important, because they currently did not meet any criteria but in fact, given the rapid ecological changes in Iraq, these areas may still be important sites for biodiversity and are deserving of more study and conservation efforts. These sites are mentioned in a special section at the end of the individual site assessments.

Finally, the focus of the KBA Program was on birds and to progressively lesser degrees on plants, habitats, mammals, fish, reptiles, and amphibians. Even for bird surveys, much remains to be learned about the areas that are important for key species. Continued surveys are needed including research using different survey methodologies to address information gaps for birds and other taxa. Several species and areas that require additional research and work include:

BIRD:

• More intensive surveys are needed during the migration seasons in autumn and spring to identify the best sites for migrating species.

• Identify sites with trigger species on the Iraq avifauna list, which the current list of sites doesn't adequately capture (or capture at all).

- Further surveys at sites that were borderline qualifiers but for which adequate data weren't available.
- Writing an updated Arabic field guide on the birds of Iraq.

PLANTS & HABITATS:

- The completion of the Flora of Iraq series is needed in order to have a full list and good reference on the plants in Iraq.
- There is an urgent need to evaluate the conservation status of plants in Iraq with a particular focus on the endemic species. This evaluation can follow the IUCN Red-listing model and criteria or any other global or regional criteria of plants assessment.
- Additional capacity building and training of botanists who can conduct research and develop a database on threatened plants and habitats in Iraq and how to monitor them.
- Local and national field guides as well as other publications and outreach materials are also needed as important tools to assist in plant conservation in the country.
- Biophysical studies to establish the status of the habitat(s) within the KBA sites and understand the biological diversity within the different habitat types found in Iraq (including areas that are under active human management, etc. marshlands and coppiced woodlands).
- Also develop research projects focused on threatened habitats in Iraq (e.g. addressing issues such as overgrazing, drought and declining water levels in the threatened habitats such as the marshlands in southern Iraq).
- Conduct more studies as part of the development of a database of habitats in the country with a special focus on threatened habitat types (mentioned above). This would include conducting a study to develop a full inventory of all habitats types in Iraq, their components, conservation status, and the threats to each type.

MAMMALS:

- Conduct rapid survey assessments for rodents and bats by trap and release and carnivores by camera trapping and other methods, to fill knowledge gaps on important mammal species, their area of occurrence and complete the Iraq checklists.
- Studies on population and conservation status of large mammals.

- Field research on specific flagship species (such as Persian Leopard or Syrian Brown Bear) that utilize trans-boundary territory opens up opportunities for collaborative work with researchers and conservationists in Iraq's neighboring countries. For example The Persian Wildlife Heritage Foundation, an Iranian wildlife conservation organization, has suggested collaboration on a presence/absence survey of leopard in all potential habitats, which could develop a map on the distribution of this species in Iraq and Iran and lead to participatory educational programs with locals (a preliminary step to conservation of KBAs important for these species).

ICHTHYOLOGY AND FRESHWATER BIODIVERSITY:

- Collecting and collating all distribution data on threatened freshwater biodiversity in Iraq.
- Freshwater biodiversity studies can also be accompanied by an ecosystem services study and social impact studies to understand where, for example, hydropower development will have the lowest and the highest impact and increase awareness about the geographic pattern of such impacts.
- Fisheries surveys are also a high priority to identify and provide guidance on sustainable catch limits to fisherman and regulators.

HERPTOFAUNA:

- Country-wide surveys and presence/absence studies are needed to identify high diversity areas and hot spots, locate endangered species that until now are poorly known in Iraq, identify endemic species, potentially describe new species, and complete the Iraq checklist for these taxa.

ENTOMOLOGY:

- Studies are needed to update and complete the Iraq checklist and identify conservation concern species as well as key sites important for these species.

PRESSURES-STATE-RESPONSE (PSR):

- For pressures indicators, more in-depth site-specific assessment of threats (and related capacity building for survey teams on how to conduct these assessments) with local involvement (see Stakeholders). Also coupling in-depth research with education/community engagement programs to focus on specific threats that KBAs areas suffer from, such as over-grazing, poaching or electro-fishing;

- For state indicators, socio-economic surveys should be undertaken to establish the status of individual KBA sites, examine land management (past and present) and impacts on biodiversity. In addition, more detailed research on key "trigger" species, areas of occurrence and their population within Iraq along with assessments of their habitat requirements.

- For response indicators, conservation initiatives are now beginning in Iraq and it will be important to develop standardized and consistent methods for measuring the success of these actions as a metric for progress in conserving biodiversity in Iraq

STAKEHOLDERS:

- Socio-economic surveys and outreach activities to identify and involve regional and site-specific stakeholders and identify more clearly the local actions that have an impact on biodiversity. This should occur before but as part of conservation planning and actions.

COLLABORATION WITH OTHER NATIONAL AND INTERNATIONAL INSTITUTIONS

It is clear that the protection of these KBAs and other important sites in Iraq cannot be accomplished through the work of one entity alone, even with the resources of a government ministry. Biodiversity conservation planning and management needs to be mainstreamed to other agencies and organizations throughout Iraq such as the Iraq Ministries of Water Resources, Agriculture, Higher Education, Science and Technology, Interior and others. The related local and national NGOs and local stakeholders at individual sites also need to be actively involved. International institutions are already playing a part as well and there are currently a number of on-going initiatives as was previously discussed in the section 3.C International Measures Relevant to the Conservation of Sites.

Affiliation between Iraqi institutions and international agencies and groups is also increasing within the country. For example, the Iraqi Ministry of Health and Environment is a member of the IUCN and Nature Iraq is a BirdLife International Affiliate and accredited to the United Nations Environment Programme (UNEP). These international environmental organizations offer affiliation and membership programs with partner organizations in Iraq to actively promote conservation initiatives within the country.

EXAMPLES OF COLLABORATIVE PROJECTS IN IRAQ

In a 2013 article reviewing progress on the protection of the Hawizeh Marshlands Ramsar Site, Clay Rubec, a Canadian expert who oversaw the development of the management plan for the Iraq Ramsar site in 2009, outlined some of the key first step projects needed for the implementation of the plan that require support from a variety of stakeholders. One specific initiative needed is to establish a "Hawizeh Marsh Stakeholder Advisory Committee or similar citizen policy group that represents a broad range of interests of the community" (Rubec, 2013). Through the auspices of the Convention on Wetlands Small Grant Fund administered by the Ramsar Secretariat, Nature Iraq began a small project in 2014 to implement this task involving local stakeholders from Hawizeh and members of both the National Wetlands Committee and district ministry officials. Nature Iraq conducted socio-economic surveys within the Hawizeh Marshland area; held a series of informational meetings to raise awareness locally about the Ramsar designation, and identified key, future stakeholders that will be part of this Stakeholder Advisory Committee.

Another example of collaborative projects is this book itself. The nucleus of the KBA Program started with a project sponsored by the Canadian International Development Agency (CIDA) to conduct field visits to check on previously identified IBA's and was further expanded over the years with trainings and survey trips sponsored by a number of other national government agencies and international organizations (IMELS, USAID, BirdLife, UNEP, and others). The teams included students and researchers at Iraqi universities, NGOs, international experts, and staff from the Iraqi Ministry of Environment, Kurdistan Environment Board, etc. The vetting of the data and the editing of this book included BirdLife International, CMEP/RBGE, Canadian universities, Italian academics, and so many entities that it is hard to give credit to all. The next steps to protecting biodiversity in Iraq will take the concerted and dedicated efforts of many individuals, groups and institutions both within Iraq and beyond.

The preceding chapters have outlined the nearly decade worth of work to study and catalogue the Key Biodiversity Areas in Iraq as well as provide compelling evidence of the need for site-based conservation of birds, mammals, plants, habitats and other species. This chapter identifies a set of objectives and strategic action recommendations, together with an indication of their urgency, needed to protect KBAs in Iraq and its wider biodiversity. These recommendations are organized under four topic areas as follows:

6.A RESEARCH & MONITORING

KBAs have much strength, as they are objectively defined using established, globally applicable criteria that give them weight and credibility. Regardless of their size, this means that KBAs are an effective global conservation currency, one that is becoming increasingly well recognized. Nevertheless, the 82 sites listed here are defined by today's knowledge and as yet only 6.5% of Iraq's terrestrial, inland freshwater, and coastal marine areas are included within KBA sites and no marine areas have yet been listed. The criteria used for the evaluation of these KBA's have recently been amended by the International Union for the Conservation of Nature (IUCN). Therefore the analyses presented in this book may need to be updated, however, this will be done in future publications as this is a living document and by no means represents all important sites in the entire territory of Iraq. There are many more potential sites that need to be identified by an ongoing process of field surveys. Also the KBA rapid assessment methodology, though comprehensive for bird species, leaves clear gaps for other taxa as has already been discussed in the previous chapter.

Nevertheless, this publication sets important baselines from which progress in the long-term conservation of KBAs can be monitored. It is vital that trends in the status of KBAs are identified, so as to provide early warning of change. Assessments can be made in terms of the presence and estimated population sizes of key indicator species, and in the success and failure of future management interventions undertaken. Data are required not only to review priorities, but also to understand the reasons for any change in the status of sites and the species that depend upon them. There also needs to be an increasing role for the people living close to or in KBAs in monitoring the status of these sites.

Recommended Objective A1: KBAs that are less well known are thoroughly surveyed and potential additional KBA sites are surveyed and identified; Monitoring programs established for all priority KBAs involving — whenever possible — relevant local stakeholders.

• **Strategic Action A1.1:** Financial and logistical support, as well as coordination efforts for continued field research and data sharing, should be pursued by the Ministry of Environment in a series of cooperative agreements with academic and non-governmental organizations as well as international conservation organizations.

• **Strategic Action A1.2:** The identified research and monitoring needs listed in the previous section entitled "Remaining Gaps" should be prioritized by the IMoHE in collaboration with other involved organizations and local stakeholders and an overall action agenda developed outlining the key stakeholders, areas of focus, schedule and resources needed to fill these gaps.

• **Strategic Action A1.3:** Regular monitoring becomes institutionalized as part of the annual work plan of the IMoHE and related organizations and key stakeholders.

A strong capacity-building component needs to accompany this objective to make its implementation sustainable and cost effective. The lack of qualified field researchers is a major drawback for site-based research and monitoring activities. The IMoHE, NI and their partners be they government ministries or academia, along with funding partners, have put significant effort in investing in the next generation of ornithologists, botanists, and conservationists who can take over this role but more is needed.

Plate 14: Benthic invertebrates studies in 2008 ©A Bachmann/NI

Recommended Objective A2: Capacity of conservation staff in the use of effective monitoring protocols is well established throughout the country.

- **Strategic Action A2.1:** Continue to support yearly capacity building for key field conservation personnel at the IMoHE, CRIM-W and other partner organizations with both international and local trainers. These trained personnel can in turn assist in Strategic Action A2.2 below.

- **Strategic Action A2.2:** Improve the capacity for in-country conservation training at the university and community/site level within Iraq.

- **Strategic Action A2.3:** Create departments of conservation at Iraqi universities in conjunction with international universities to systematically train future scientists in conservation methodologies and to keep track of the latest developments in the field of conservation sciences.

Plate 15: 2010 Ornithology training in Iraqi Kurdistan
©RF Porter

6.B DEVELOPMENT OF A PROTECTED AREA NETWORK

Under the Aichi Biodiversity Targets of the CBD, a goal has been set to conserve at least 17% of terrestrial and inland water, and 10% of coastal and marine areas by 2020 (CBD, 2014, Strategic Goal C, Target 11). Iraq has initiated this process with the designation of the Hawizeh Marshes, East and West Hammar Marshes, and Sawa Lake as Ramsar sites; the Central Marshes as a National Park, and the recent inclusion of the Mesopotamian marshlands as a World Heritage Site. These are all important KBA sites covered in this Inventory. Map 2 in the introduction shows the overlaps between these officially designated sites and KBA Sites, demonstrating the significant gaps in the current Protected Area Network.

As was discussed in the previous chapters, several additional areas of the country are being studied as proposed protected areas. There are also regional initiatives such as the proposed Halgurd-Sakran Park that have been the subject of previous reports (AIT, 2012 & Bachmann, 2009). Some areas have been designed for tourism and recreational parks (for example, the 2012-2025 Strategic Plan of the KRG Tourism Board includes proposed tourism development in several KBA sites) but it should be noted that while there are clear connections, there is a fundamental difference between a recreational park and a protected area. If the area is developed strictly for recreation without regard to the overall health and/or the species that utilize the site, then ultimately the potential for the area to become degraded is high. Protected areas, such as national or regional parks, nature reserves or wilderness areas, often offer a range of protective restrictions to recreational and other uses of the site that ensure that it is not degraded and is maintained for future generations.

Attempts have been made to fence areas off in Iraq in order to protect them but there is ample evidence from the region and across the globe that such measures are usually not effective in the long-term. Portions of Qara Dagh (IQ039) have been fenced in the past but these fences were simply torn down by locals. Instead, with proper management planning and active stakeholder involvement more effective means for protecting sites are possible. See a discussion about KBA Site IQ004 - Barzan Area and Gali Balnda for an example of this in the chapter on Biodiversity Protection in Iraq. Also see Case Study 2 later in this chapter.

Recommended Objective B1: KBAs protected using national and international legal instruments, with the majority of the KBAs identified in the program legally protected (or partly protected) within the national protected areas (PA) system.

- **Strategic Action B1.1:** Fast-track sites recognized under Ramsar and World Heritage Conventions for integration into the PA System of Iraq.

- **Strategic Action B1.2:** Continue the development of a list of additional priority sites that meet the objectives of protecting areas based on sites with high threat levels, priority species, geographic spread, and other considerations.

Plate 16: DeLezha KBA Site ©RF Porter

- **Strategic Action B1.3:** Lobby for the inclusion of the priority KBAs in government development plans so that they will become legally protected and recognized under categories determined by Iraqi protected area legislation across all government agencies and official departments within Iraq.

Some sites may not be amenable to legal protection under a national protected area network. For these sites other conservation measures may include community conservation initiatives and/or activities conducted by private entities, such as incorporating restricted development zoning around KBAs by industries.

Recommended Objective B2: Additional KBAs, not protected under national or international legal instruments, covered under other informal conservation actions.

- **Strategic Action B2.1:** Assess and prioritize which sites require other approaches beside formal protection.

- **Strategic Action B2.2:** Include the full list of KBA sites as proposed sites for protection on the Protected Planet and Integrated Biodiversity Assessment Tool Websites to better inform all stakeholders about the presence of KBAs in areas of Iraq where development activities are planned.

Plate 17: Open air oil dumping in a local garbage dump
©A Bachmann/NI

6.C SITE MANAGEMENT AND THE PROMOTION OF CONCEPTS FOR SUSTAINABLE NATURAL RESOURCE USE

The KBA research has provided baseline information needed to initiate management and sustainable use of natural resources. This will include actions that reduce threats and result in active preparation and implementation of site management plans.

THREATS REDUCTION

A major goal of site management is the evaluation and reduction of threats to species and habitats. For example, residential and commercial development at Peramagroon Mountain (IQ027) was evaluated as a very high threat due to the opening of a road from Sulaymaniyah to the Mergapan valley. This has increased unplanned development within this area on the northeastern side of the mountain. The severity, scope and timing of threats change over time so it is important to regularly evaluate each threat type and update these findings. This allows stakeholders to determine if the overall site is improving or becoming more degraded over time and helps to focus mitigation efforts.

In the Peramagroon Site example, since 2010, the pollution and systems modification threats have been increasing as a result of the initial residential developments in the Mergapan valley. Urban planning to restrict development zones and creation of needed infrastructure (utility and waste handling services) are

CASE STUDY 1: INTERNATIONAL PROTECTION - WORLD HERITAGE CONVENTION (WHC) NOMINATION OF THE SOUTHERN MARSHLANDS OF IRAQ AS WORLD HERITAGE SITE.

Iraq, which became signatory to the WHC in 1974, has four listed sites: Hatra (listed in 1985), Ashur (or Qal'at Sherqat, listed in 2003), Samarra Archaeological City (listed in 2007), and Erbil Citadel (listed in 2010). Two, Ashur and Samarra are inscribed on the List of World Heritage in Danger. These are all listed under the Cultural criteria. The Marshlands of Mesopotamian (proposed in 2003) along with Amedy City (proposed in 2011) are the first mixed cultural and natural sites to be put on the tentative list of 11 sites currently being considered under the Convention.

In June 2009, UNEP and UNESCO organized a meeting in Amman, Jordan on a proposed joint project entitled "Natural-Cultural Management of the Iraqi Marshlands as World Heritage". This meeting pulled together a diverse array of Iraqi and international stakeholders to provide input and advice on this issue. It covered preservation and management plan development towards World Heritage inscription as well as implementation of such plans; capacity building and awareness-raising, and international cooperation.

By 2012 a national team was assisting the IMoHE in writing a portion of the UNESCO report on the marshlands and its potential designation as a UNESCO World Heritage Site. In July 2016, UNESCO voted to accept the Iraqi files with a caveat that proper water management plans as well as site management plans be developed in cooperation with international bodies to move the acceptance forward. The consultation of local stakeholders in the process is vital for the success of this nomination so that the people directly affected can become the strongest advocates for the site.

management actions that are recommended to reduce the negative impacts from these threats.

Recommended Objective C1: KBAs are adequately and effectively managed and the natural resources of KBAs are used sustainably.

- **Strategic Action C1.1:** Continue to monitor and assess threat levels at KBA sites and track priority sites over time.

- **Strategic Action C1.2:** Undertake a re-evaluation of these KBA sites based on existing and added data using updated KBA Criteria issued by IUCN in September 2016.

- **Strategic Action C1.3:** Develop and implement threat reduction strategies that are site and threat specific at the most threatened sites with local stakeholder involvement and implement and enforce existing natural resource regulations.

MANAGEMENT PLANNING & IMPLEMENTATION

Conservation planning takes place at the site, site-adjacent and national levels. One feature of the KBAs is that networks of important sites may be identified and it is clear that the successful conservation of biodiversity involves a large number of sites. Management plans for some protected areas in Iraq have already been developed for the Central Marshes (IQ075) and Hawizeh (IQ073) but more sites require management planning. In addition, the implementation of existing plans is lagging far behind, which will likely result in declining biodiversity levels at these sites over time.

Plate 18: Hawizeh Ramsar Site Stakeholder Workshops in 2014 ©NI

- **Strategic Action C1.4:** Develop and expedite the process for developing comprehensive management plans for new priority sites.

- **Strategic Action C1.5:** Allocate secure resources (financial, logistical, and personnel) to implement existing and future management plans for long-term conservation of current protected areas.

Multi-stakeholder involvement, particularly at the local level, is the key to effective management planning. Unless local peoples are actively involved in the decision-making process any plans for effective long-term management that ensures environmental sustainability is doomed to failure. The Iraqi Ministry of Health and Environment but also regional and local government staff will play a vital role in coordinating the planning process and getting all stakeholders to the table. They are also in the best position to incentivize this process and negotiate the variety of often complex, competing uses of an area.

As most KBAs are currently outside the developing Protected Area Network in Iraq and some may never be included within this network, the protection of these sites, particularly those of conservation priority and facing the highest threats, requires approaches new to Iraq. This will include allowing local land-owners and users (i.e. farmers, hunters, fishermen, NGOs and local communities) to play active roles in the planning process; focusing on strategies that allow for the sustainable use of resources, and developing incentives for the local population that ensures that they will benefit directly from the appropriate use of natural resources.

- **Strategic Action C1.6:** Develop and institutionalize as a guiding principal, formal and informal management planning processes that incentivize multi-stakeholder participation by ensuring that stakeholders benefit from sustainable natural resources use.

CASE STUDY 2. INFORMAL PROTECTED AREAS AND PRACTICES

Although Barzan Area (IQ004) has not yet been established as a Protected Area under national legislation, the area is recognized as a regional reserve that is subject largely to tribal management and control. Located near the Turkish border on the Greater Zab River in the Kurdistan Region, the area has been locally protected since the early 1900's. A number of natural resource protections were put in place that include prohibitions against cutting fruit and other types of trees, fishing with dynamite, hunting during the mating season, etc.

These local rules were strengthened between 1961-93 when people had easy access to guns and began using them for sport. Hunting pressures grew excessive until local leaders realized that action would need to be taken to preserve the environment. Today there is a relatively strong prohibition against hunting at this site and it has been suggested that the area be designated a protected area under the UNESCO Man and the Biosphere Program.

In another example from southern Iraq, the presence at Dalmaj Marsh (IQ064) of a relatively healthy population of Bunni (Mesopotamichthys sharpeyi), which is a Vulnerable fish species native to Iraq, is likely the result of the Ministry of Agriculture controlling and providing fisheries rights to local investors who have a vested interest in protecting the resource. Here there is also a prohibition against unsustainable fishing methods such as electro-fishing and poison. Also in Abu-Zirig, a portion of the Central Marshes (IQ075), local religious sanctions against unsustainable fishing practices have also had some positive impacts.

6.D ADVOCACY & EDUCATION

Site conservation and management strategies of KBAs require a wide range of focused advocacy activities with decision-makers and other stakeholders. Advocacy ranges from providing information on specific issues such as the review of national protected area legislation, to the development of departmental policies on issues like climate change. Building close working relationships is a key component, as is the sharing of data, analysis and advice.

Plate 19: Green Festival Event in the Marshlands National Park 2013 ©RF Porter

Recommended Objective D1: KBA information is readily available to all decision-makers, planners, and stakeholders.

- **Strategic Action D1.1**: Ensure information sharing on KBAs between the CRIM-W, IMoHE and district offices.

- **Strategic Action D1.2:** Ensure information sharing on KBAs between the CRIM-W, IMoHE and other stakeholder government agencies, academic institutions, non-governmental, private and international organizations.

KBAs need to be stamped on the national consciousness as an important cornerstone of the Iraq Protected Area Network. Key to this is informing a wider audience about the

value and importance of KBAs. Information about KBAs also provides useful material for educational purposes. As well as its more formal educational aspects, the KBA process depends on developing individual and institutional capacity across a wide range of subjects. These range from tour-guiding or bird identification skills at the local level to the ability to manage an information database or legal skills for taking action at a national level to challenge decisions likely to harm specific sites.

Plate 20: Training the next generation ©RF Porter

Intensive campaigns focused on charismatic endangered species or specific threatened sites can make a significant contribution to site conservation. In addition, sustained and consistent information can be spread most effectively through the development of Local Conservation Groups (LCGs) at important sites who also can play an important part in research and monitoring, protected area development and management planning as discussed above.

Recommended Objective D2: Information on KBAs is utilized in a wide range of public forms, including specific conservation campaigns for species, habitats and sites overall.

- **Strategic Action D2.1:** Integrate and utilize KBA information in promotional and educational activities that actively promote KBA sites throughout the country.

- **Strategic Action D2.2:** Conduct specific campaigns for the protection of threatened species, habitats, and overall sites that promote their conservation, with a particular focus in high priority and/or highly threatened KBA Sites.

- **Strategic Action D2.3:** Encourage the formation of Local Conservation Groups (LCGs) at high priority KBA Sites and support them in efforts that educate the local public in their areas about the conservation of species and habitats at these sites.

Plate 21: Darwin Conservation Course Field trip ©RF Porter

CASE STUDY 3. DARWIN PROJECT FOR IN-SITU CONSERVATION

Funded by the Darwin Initiative of the United Kingdom, the Darwin Project for In-situ Conservation was a 3-year program focused on Peramagroon Mountain in the mountainous area of Kurdistan, northern Iraq, which was selected for the project because it meets KBA Criteria. The project involved Nature Iraq and experts from the Centre for Middle Eastern Plants (part of the Royal Botanic Garden Edinburgh (RBGE)) and BirdLife International.

One of the components of the project was conservation training with local university students. An online course was first offered in the spring of 2013. It featured six modules on essential topics in biodiversity as well as conservation issues aiming to raise the level of knowledge that students have about international conservation organizations and conventions, strategies, protected areas work and related topics. The course was accompanied by an open online

forum where students could discuss questions about the environment of Iraq with close participation from experts at the Royal Botanic Garden Edinburgh, BirdLife International and Nature Iraq.

Twenty-four students and university staff from the University of Sulaymaniyah as well as four additional local and regional conservation organization representatives passed the first year's course and many of the students were also able to participate in a follow-up field trip with experts from the sponsoring organizations. The course materials for the online course were refined for 2014 with more local examples and case studies on the biodiversity of Iraq and two additional modules were added. In future, it will be possible to offer this or other on-line courses to students at more universities in the country.

CHAPTER 7: KBA ASSESSMENTS

7.A DATA PRESENTATION

The site accounts that form the major part of this publication have been prepared, as far as possible, in a standardized way.

SITE ACCOUNTS

Each account consists of a header with the site name and basic geographical information; a general site description; a table showing the qualifying species/habitats and their status under each criterion; a description of the important fauna, flora and habitat findings; a threat assessment graph; an overview of conservation issues related to the site, and recommendations for further actions. These sections, along with maps, photos and other elements of the assessments are explained and illustrated below.

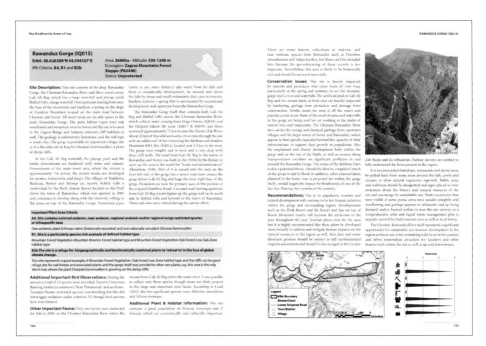

Header

Each site account has a boxed header, which gives key information about the site, including site name, site code, and coordinates. In sites with multiple sub-sites, the coordinates were for a sub-site that was usually (but not always) centrally located within the site. Also included in the header are the administrative region(s), surface area, altitudinal range, a list of the criteria by which the site qualifies as a either a IBA, a KBA for non-avian fauna and/or an IPA, ecoregion(s) and its protection status with protected area designation(s).

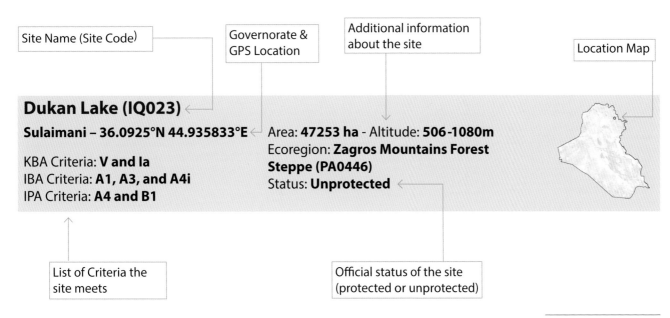

Site Name (Site Code) — Governorate & GPS Location — Additional information about the site — Location Map

Dukan Lake (IQ023)
Sulaimani – 36.0925°N 44.935833°E — Area: **47253 ha** - Altitude: **506-1080m**
Ecoregion: **Zagros Mountains Forest Steppe (PA0446)**
KBA Criteria: **V and Ia**
IBA Criteria: **A1, A3, and A4i**
IPA Criteria: **A4 and B1**
Status: **Unprotected**

List of Criteria the site meets

Official status of the site (protected or unprotected)

Site Photo(s)

A representative site photo (or photos) is provided in each assessment. Some photos are panorama images taken at the location of the site or sub-site coordinates.

Site Description

This section includes a general description of the site covering relevant information on the location, topographical and hydrological features, habitat and vegetation, proximity to other important sites and human settlements and the main land-uses, if any. In some cases, a single KBA site is quite large and multiple sub-sites were surveyed and described. Sub-sites are listed in Appendix 2.

Species/Criteria Tables

A tinted box presents data on all species known to meet global criteria at the site. The species are divided by the criteria they fulfill. Key Biodiversity Area (KBA) Criteria for non-avian fauna is listed first if such species were identified and this section provides notes on each species explaining the quality of the data used (reliable reports or direct observation) and justification for the criteria they meet.

Important Bird & Biodiversity Area (IBA) Criteria is then listed next. The nomenclature of bird species follows IUCN (2010) and Salim et al. (2012). IBA population data (counts and year) is listed under the status columns (breeding or wintering/passage). Unless stated otherwise IBA numbers are estimates, usually based on sample area counts (see the KBA Methodology).

Finally Important Plant Area (IPA) Criteria is listed at the end of the table and provides information on the individual criterion met at the site with the supporting rational.

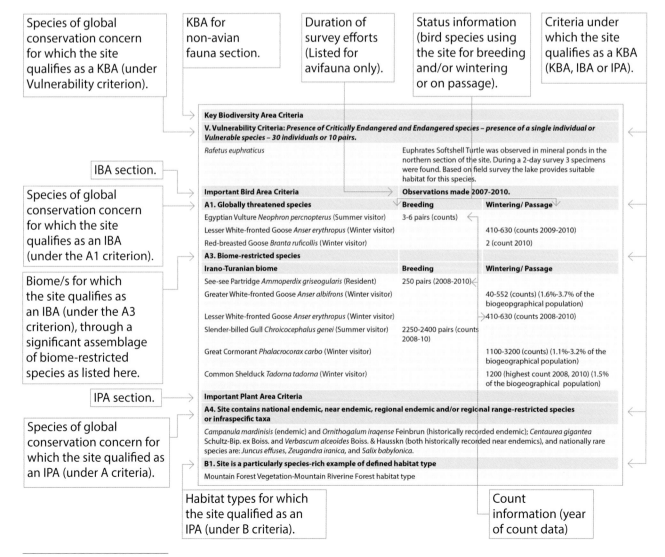

Additional Important Bird Observations

This section outlines other significant ornithological features of the site including, where known, the total number of species recorded. All species of global conservation concern that occur at the site but which do not, or are thought not to, meet IBA criteria are mentioned in the text, and reference is made to biome-restricted species that are known from the site but which do not trigger the relevant A3 criterion (Table 19 provides a complete list of the A3 species observations for each biome in Iraq). Mention is also made of congregatory species that are present in large or nationally significant numbers but which do not meet the relevant 1% thresholds. Details may also be given of the occurrence of taxa important at national or sub-regional levels. It is likely, indeed probable, that many of the less well-known sites are important for reasons additional to the justifications listed in the box above and, were further data available, some would certainly qualify by meeting additional criteria or thresholds for other species.

Other Important Fauna

This section lists any known non-avian fauna recorded from the site that are of conservation concern but do not meet non-avian KBA criteria. It also lists important carnivore species known to be present and some Least Concern species that may be facing local population declines and/or severe threat at the site. The primary discussion here covers mammals, reptiles, amphibians and sometimes fish but due to the limited nature of the surveys for non-avian fauna, this section does not fully define the biodiversity of these taxa and only provides a general indication of the relative richness found and the wider conservation importance of the site. The IUCN nomenclature has been used for both English and scientific names (IUCN Red List of Threatened Species, See Web Resources).

Additional Plant & Habitat Information

This section outlines other significant botanical about the site including species, habitats and features that are of economic, social or cultural importance at the site.

Conservation Issues & Recommendations

The Conservation Issues section includes details, where known or inferred, of the main threats (actual and potential) affecting the site, as well as further information on protection status under national law and/or international conventions, including any proposed designations. The last section of the site assessment reviews recommendations for the site. Many, if not all sites require broader environmental educational and awareness campaigns and strengthening in the application of conservation laws such as enforcement of hunting regulations but the key focus was on site-specific recommendations for monitoring activities and conservation action.

Threat Assessment Graphs

The threat assessment graph, which may appear in different locations of the site assessment as space allows, presents the threat level from 0 (low threat/No Information) to 3 (high threat) assessed for the list of eight threat types. These threat types are: Agriculture; Residential & commercial development; Energy production & mining; Transportation & service corridors; Over-exploitation, persecution & control; Human intrusions & disturbance, Natural systems mod., and Pollution. These threats were assessed as part of the rapid assessment conducted primarily during the 2010 site visits and likely do not fully represent all threats to a given site. This graph should be used to identify areas where more in-depth research on pressures to biodiversity is needed for a specific site. See the KBA Methodology Chapter for more information.

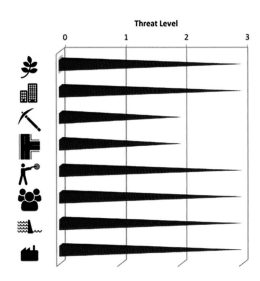

Maps

Landscapes can change enormously over time; Infrastructure such as roads and canals are built and then left to decay, wetlands and lakes were drained and agricultural or urban landscapes can expand into new areas. For Iraq information on the location of roads, wetlands and other major landscape features may be missing or inaccurate. Our maps aim to provide an overall impression of a site and its surrounding area, provide landscape condition and show the preliminary boundary of a specific site. See the Methodology section for more information on how the KBA site boundary was determined. For field trips and scientific purposes it is strongly recommended that more detailed or up-to-date maps be consulted.

The maps were produced utilizing Landsat ETM+ satellite images Band 8 (Panchromatic band at 15m spatial resolution) from 2002 (http://landsat.usgs.gov/band_designations_landsat_satellites.php) overlays a shaded relief produced from the Shuttle Radar Topography Mission SRTM (Digital Elevation Data at 90m spatial resolution) (http://srtm.usgs.gov/index.php). Additional features such as political boundaries, roads, embankments, water-bodies, wetlands, salt marshes, as well as rivers and other waterways were added to the maps where they were available and may not be consistent across the entire country. To maintain consistent map layout, each map depicts only some of the most important settlements. Where no settlement is situated near a site other features were included when possible to help orient the reader. Geographical names on the maps may not always correspond to the spelling in the text.

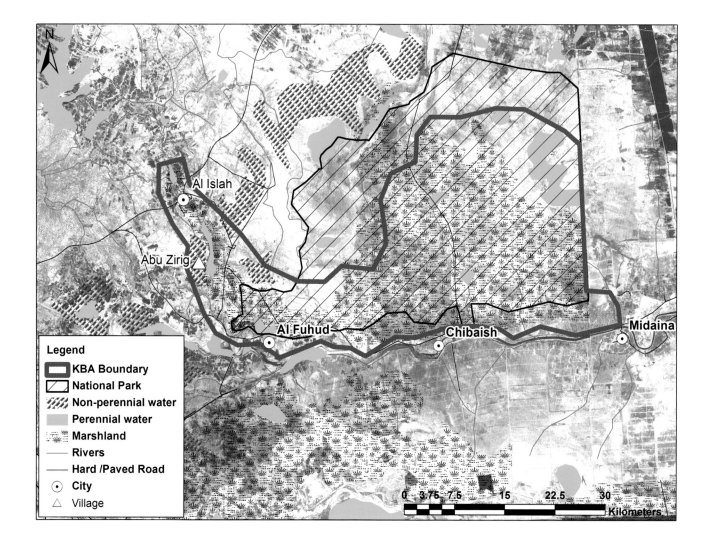

7.B INDIVIDUAL SITE ASSESSMENTS

Fishkhaboor (IQ001)

Dohuk – 37.111944°N 42.38333°E

IBA Criteria: **A4ii**

IPA Criteria: **A4, B1 and C**

Area: **4179 ha** - Altitude: **329-410 m**
Ecoregion: **Eastern Mediterranean Conifer-Sclerophyllous-Broadleaf Forest (PA1207)**
Status: **Unprotected**

Site Description: This site consists of foothills and lowlands including a narrow riparian strip along the Tigris and Fishkhaboor (Khabour) Rivers. A pontoon bridge and a bridge cross the river approximately 6.3 km downstream of the point where the Tigris River enters Iraq, which allows access to the Syrian border. This is the only KBA site that falls within the Iraqi portion of the Eastern Mediterranean Ecoregion (PA1207) as defined by the World Wildlife Fund for Nature, however Fishkhaboor does not clearly show the key features associated with this ecoregion. The main habitat here is Riverine Forest of the Plains. The site is located in the moist steppe zone, and the geology is sandstone, clay, and sandy gravels and the soil type is sandy clay. There are extensive gravel mining operations throughout the area on the Tigris and Fishkhaboor Rivers. Two sub-sites were visited in this area.

©2010 K ARARAT/NI

Important Bird Area Criteria	Observations made 2008-2010.	
A4ii. 1% or more of global population of a congregatory sea-bird or terrestrial species	**Breeding**	**Wintering/Passage**
Dead Sea Sparrow *Passer moabiticus* (Resident/summer visitor)	1400 (count 2008) (1.5% of regional population)	

Important Plant Area Criteria
A4. Site contains national endemic, near endemic, regional endemic and/or regional range-restricted species or infraspecific taxa
One nationally rare species (*Typha lugdunensis*) was found in this site.
B1. Site is a particularly species-rich example of defined habitat type
Riverine Forest of the Plains (Al-Ahrash) habitat type.
C. The site is identified as an outstanding example of a globally or regionally threatened habitat type.
This site represents a good example of the habitat type: Riverine Forest of the Plain (Al-Ahrash) habitat type, which extends throughout the country along the Tigris-Euphrates River Basins and their tributaries. Tigris-Euphrates riparian woodland is globally unique and threatened by developments and human activities along these rivers and habitats. The threats include gravel mining, cutting woodlands for different uses, and the residential and agricultural development along the rivers.

Additional Important Bird Observations: During the survey a total of 36 species were recorded. The site held breeding populations of one Mediterranean, two Irano-Turanian and one Sahara-Sindian biome-restricted species but these did not trigger inclusion under criterion A3.

Other Important Fauna: Mammal data were collected in summer 2010. According to local interviews, the significant mammals seen at the site include Grey Wolf *Canis lupus*, and Golden Jackal *Canis aureus*. Eurasian Otter *Lutra lutra* (Near Threatened) tracks were found along the river.

Fish: Surveys took place in 2007 & 2008 and the following significant species were found *Acanthobrama marmid*, *Alburnus mossulensis*, *Chondrostoma regium*, *Cyprinion kais*, *C. macrostomum*, *Garra rufa*, and *Liza abu*.

Additional Plant & Habitat Information: Two species, *Centaurea pseudosinaica* and *Brassica kaber*, were recorded in Iraq for the first time at this site.

Conservation Issues: Gravel mining along both the Fishkhaboor and Tigris Rivers has caused damage to in-stream habitats and riparian areas and transportation, especially road building in areas where most gravel mining occurs is a very high threat. This site is likely to be highly impacted by major dam construction on-going in Turkey as well. Border police reported hunting in the area and minefields are found along the Iraqi-Turkish border in areas near Tuwan such as Cheae Bekher and Ware Smaili. Sewage from the city of Zakho and other villages upstream also impact both rivers.

Recommendations: This site adjoins a Turkish IBA site (Cizre ve Silopi, TK180) and, with additional cooperation and coordination from Syria, represents a unique opportunity to develop a tri-nation "Peace Park" that could protect the biodiversity here for all three countries. A proper land use management scheme should be developed that addresses the many threats, with gravel mining and road contruction being primary issues. As this is a sensitive border area for all three countries, significant cooperation and coordination is required for any future survey and conservation work.

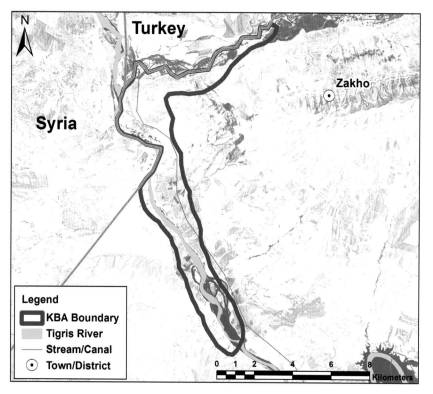

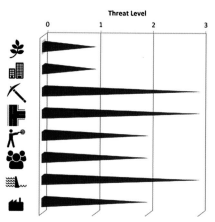

Benavi & Sararu (IQ002)

Dohuk – 37.264173°N 43.405827°E

IBA Criteria: **A3**
IPA Criteria: **A4 and B1**

Area: **1809 ha** - Altitude: **1250-1950 m**
Ecoregion: **Zagros Mountains Forest Steppe (PA0446)**
Status: **Unprotected**

Site Description: Two survey areas, Benavi and Sararu, are included in this area. Benavi includes a valley containing the village of Benavi and the adjacent mountain ridge and there is a small stream near the site. Evans (1994) originally listed Benavi as an Important Bird Area (IBA001). The main habitat types here are mountain forest vegetation (primarily oak forests) and mountain riverine forest. The slopes are rocky with interspersed grasses and rich herb flora. The site is located in the Zagros Range, where the geology is of sandstone, clay and sandy gravels, and the soil type is sandy clay. The local economy depends mainly on sheep grazing and some agriculture, which includes field crops and some walnut *Juglans regia* orchards, which are scattered around the valley.

Sararu is located in the northern part of the delineated area and has similar habitats, geology and soil types. The highest peak in the Sararu area stands about 1950 m and there is also a valley containing a large number of orchards (walnut, cherry and pomegranate) and vineyards. It is a popular area for visitors and regional tourists. The villages of Ure, Bedohe and Maye are nearby, just outside the delineation area.

Important Bird Area Criteria	Observations made 2007-2009.	
A3. Biome-restricted species		
Mediterranean biome	**Breeding**	**Wintering/ Passage**
Masked Shrike *Lanius nubicus* (Summer visitor)	10 pairs (2008-2009)	
Sombre Tit *Poecile lugubris* (Resident)	50 pairs (2008-2009)	
Western Rock Nuthatch *Sitta neumayer* (Resident)	20 pairs (2008)	
Eastern Black-eared Wheatear *Oenanthe melanoleuca* (Summer visitor)	25 pairs (2008-2009)	
Black-headed Bunting *Emberiza melanocephala* (Summer visitor)	40 pairs (2008)	
Important Plant Area Criteria		
A4. Site contains national endemic, near endemic, regional endemic and/or regional range-restricted species or infraspecific taxa		
Bellevalia kurdistanica (near endemic)		
B1. Site is a particularly species-rich example of defined habitat type		
Mountain Forest-Mountain Riverine Forest habitat type and Mountain Forest Vegetation-Oak Forest- Lowest and Medium Zones habitat type		

©2008 K ARARAT/NI

Additional Important Bird Observations: During the surveys 68 species were observed. The site also held breeding populations of three Irano-Turanian, and one Eurasian Steppe and Desert biome-restricted species.

Other Important Fauna: There were local reports of a Brown Bear *Ursus arctos* attacking a resident who lost his arm, but this could not be verified. Persian Squirrel *Sciurus anomalus* and the near endemic Zagrosian Lizard *Timon princeps kurdistanicus* were observed in this area, but brief surveys for non-avian fauna occurred only in 2008 and 2009. The area is likely rich in important species, but additional surveys are needed. A stream is present at the site but no fish observations were conducted.

Additional Plant & Habitat Information: This site contains a good population of pistachio *Pistacia eurycarp*a, which is economically and culturally important. There is also a vineyard of *Vitis vinifera*, which is important as a genetic resource and there are some grasses present that are important genetic resources, including *Aegilops columnaris*, *A. crassa*, and *Pennisetum orientale*.

Conservation Issues: There are pastures and agricultural land but the team could get little information on agricultural practices at the site, though the area is affected by overgrazing. Because this area is near the border with Turkey, it is an area of high military activity and instability. Locals indicated that there were uncleared minefields nearby. There was also road construction in the area and some hunting does occur here.

Recommendations: Mine clearance should be a priority, but ultimately the risks caused by the poor security situation along the border will require political solutions at government and regional levels. Grazing should be examined more closely. Road construction should be limited where possible.

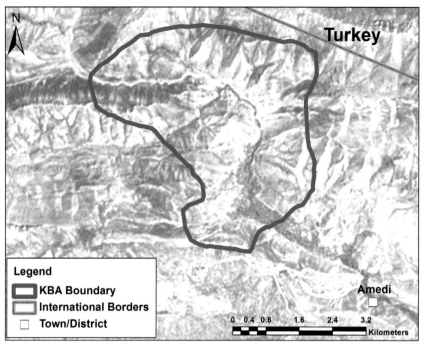

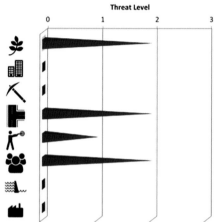

Threat Level

Dure (IQ003)

Dohuk – 37.225556°N 43.509444°E

KBA Criteria: **V**
IPA Criteria: **A4, B1, and C**

Area: **2310 ha** - Altitude: **755-1890 m**
Ecoregion: **Zagros Mountains Forest Steppe (PA0446)**
Status: **Unprotected**

©2010 K ARARAT/NI

Site Description: This site is an open area of valleys, springs, agricultural fields and woodland in the Zagros Range. The Sari Darishk, Sari Zer and Shirani Mountains form a chain close to the site. It was called Dori Serguza (IBA002) by Evans (1994), who provided incorrect coordinates for its location. The main habitat types surveyed here were mountain forest vegetation (primarily oak forests) and mountain riverine forest. *Quercus, Juniperus* and *Pinus* trees dominate the woodlands and the slopes are rocky, with grasses and isolated Quercus scrub. The geology is siltstone and limestone, and the soil type is sandy clay. The villages of Sarizer, Duskan and Barbuire and a church, which is a potentially important cultural and religious heritage site, are located near Sari Zer.

Key Biodiversity Area Criteria	Notes
V. Vulnerability Criteria: *Presence of Critically Endangered and Endangered species – presence of a single individual or Vulnerable species – 30 individuals or 10 pairs.*	
Capra aegagrus	No direct observation, reported by the border police based in the area.
Important Plant Area Criteria	
A4. Site contains national endemic, near endemic, regional endemic and/or regional range restricted species or infraspecific taxa	
Endemics are *Centaurea longipedunculata* and *Linum velutinum*; *Linum velutinum* along with three others (*Dianthus asperula, Gladiolus kotschyanus,* and *Quercus macranthera*) are nationally rare species.	
B1. Site is a particularly species-rich example of defined habitat type	
Forest Vegetation-Mountain Riparian-Forest and Mountain Forest Vegetation- Oak Forest- Medium and Highest sub-zones	
C. The site is identified as an outstanding example of a globally or regionally threatened habitat type.	
Globally important *Pinus* woodland. This represents the most easterly distribution of what was once a widespread forest.	

Additional Important Bird Observations: During the surveys 44 species were observed. Although this site does not qualify as an IBA, the following threatened species were observed: Greater Spotted Eagle *Aquila clanga* (Vulnerable) on passage and Eastern Cinereous Bunting *Emberiza semenowi* (Near Threatened) breeding. The site also held in the breeding season four Mediterranean, one Irano-Turanian and one Eurasian Steppe and Desert biome-restricted species but these did not trigger inclusion under criterion A3

Other Important Fauna: Mammal data were collected in 2010. According to border police officers, animals in the area include: Wild Goat *Capra aegagrus* (Vulnerable), Striped Hyena *Hyaena hyaena* (Near Threatened), Brown Bear *Ursus arctos*, Grey Wolf *Canis lupus*, Wild Boar *Sus scrofa*, and Persian Squirrel *Sciurus anomalus*. One near-endemic reptile

Zagrosian Lizard Timon *princeps kurdistanicus* was observed. No fish surveys were performed at this site.

Additional Plant & Habitat Information: This site contains a good population of *Pistacia eurycarpa, P. khinjuk* and *Crataegus azarolus,* which are economically and culturally important.

Conservation Issues: Human intrusion was the only issue considered a high threat at Dure. Other impacts come from farmland near the villages of Sari Zer, Duskan, and Sari Zer, hunting and pollution from picnickers.

Recommendations: The vegetation cover at Dure was very high (95%) and it is considered important for endemic and rare species of plants. It warrants some form of protected area designation and more active conservation management to reduce any threats.

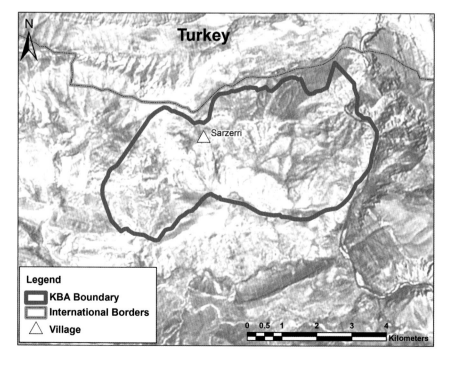

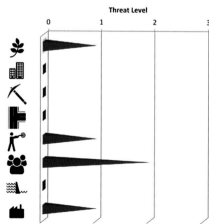

Barzan Area & Gali Balnda (IQ004)

Erbil – 36.94361°N 44.1955556°E

IBA Criteria: **V**
IPA Criteria: **A1 and A3**
IPA Criteria: **A4 and B1**

Area: **114822 ha** - Altitude: **530-2560 m**
Ecoregion: **Zagros Mountains Forest Steppe (PA0446)**
Status: **Partially protected by the Barzan Tribe**

BARZAN ©2010 S ABDULRAHMAN/NI

Site Description: This area is made up of two areas, Barzan and Gali Balnda (on Barzan's western side). Three sub-sites were visited during the survey efforts for these areas. Barzan consists of mountains and valleys covered by oak woodland that has traditionally been protected by the Barzan tribe. The most important rivers within the Barzan Area are the Rezan River and the Greater Zab River. The Rezan is fairly large (approx. 20-30 m wide) and fast flowing. It flows from Turkey through Musaka village, eventually joining the Greater Zab. The Barzan area also includes Shirin Mountain as well as other peaks, which are important for montane species. The main habitat types are mountain forest vegetation (primarily oak forests) and mountain riverine forest. The geology is sedimentary limestone and the soil type is sandy clay.

This area has often been proposed as a national park (it is one of the few places in the Kurdistan Region where it is relatively easy to see Wild Goats thanks to a long-standing hunting prohibition enforced by the Barzan Tribe) and is a popular place for picnickers. A bridge crosses the Rezan River near the access to one of the the main survey area (close to the village of Zrara) and there are remains of a much older

bridge downstream. Though the regularly visited survey site is relatively small, the overall site extends north to the Turkish border and south across the Greater Zab River. Also included within the area is the ancient Shanidar Cave, first excavated by Drs Ralph and Rose Solecki in 1971, which contained Neanderthal burial sites. Currently, the cave has been developed for tourism. It is located in a rocky gorge, with high cliffs and oak forests. A stream beneath carries seasonal winter flows to the Greater Zab River.

Gali Balnda is comprised of a valley carrying the Deraluk River, which flows from Turkey. It is surrounded by mountains and flows to the Greater Zab River. The habitat types are similar to the main Barzan area and both sites are located within the Zagros Range but the geology in Gali Balnda is sandstone, clay and gravel. There is an iron bridge (Shiladza) at the southern edge of the area that crosses the Deraluk River about 2 km from its confluence with the Greater Zab near a town of the same name. This town dumps sewage into the Greater Zab. Here the land is used for farming (mainly crops, orchards and some vineyards) but there are minefields, especially near the towns of Badan and Biav.

Key Biodiversity Area Criteria	Notes
V. Vulnerability Criteria: *Presence of Critically Endangered and Endangered species – presence of a single individual or Vulnerable species – 30 individuals or 10 pairs.*	
Capra aegagrus	Ten individuals were observed in 2007, 24 in winter 2009, 80 in summer 2010, and 12 in winter. During a training course in summer 2011, 30 individuals were counted, in addition to approximately 40 recorded by a camera trap set in the Zrara area.
Panthera pardus saxicolor	Reported reliably by locals near Rezan river.
Testudo graeca	Six individuals were seen between 2008 and 2011.

Important Bird Area Criteria	Observation made 2007-2010. Gali Balnda was visited only in summer 2009	
A1. Globally threatened species	**Breeding**	**Wintering/ Passage**
Egyptian Vulture *Neophron percnopterus* (Summer visitor)	1 pair (2009)	
A3. Biome-restricted species		
Irano-Turanian biome	**Breeding**	**Wintering/ Passage**
See-see Partridge *Ammoperdix griseogularis* (Resident)	25 pairs (2010)	
Upcher's Warbler *Hippolais languida* (Summer visitor)	1 pair (count 2011)	
Menetries's Warbler *Sylvia mystacea* (Summer visitor)	3 pairs (counts 2008 and 2010)	
Eastern Rock Nuthatch *Sitta tephronota* (Resident)	100 pairs	
Kurdistan Wheatear *Oenanthe xanthoprymna* (Summer visitor)	9 pairs (count 2011)	
White-throated Robin *Irania gutturalis* (Summer visitor)	3 pairs (count 2010)	
Finsch's Wheatear *Oenanthe finschii* (Resident)	5 pairs (2007-2008)	
Eastern Cinereous Bunting *Emberiza semenow* (Summer visitor)	11 pairs (counts 2010-2011)	
Mediterranean biome	**Breeding**	**Wintering/ Passage**
Masked Shrike *Lanius nubicus* (Summer visitor)	20 pairs (2008-2010)	
Sombre Tit *Poecile lugubris* (Resident)	20 pairs (2008-2010)	
Western Rock Nuthatch *Sitta neumayer* (Resident)	50 pairs (2011)	
Eastern Black-eared Wheatear *Oenanthe melanoleuca* (Summer visitor)	40 pairs (2007-2010)	
Black-headed Bunting *Emberiza melanocephala* (Summer visitor)	500 pairs (2008-2010)	

Important Plant Area Criteria

A4. Site contains national endemic, near endemic, regional endemic and/or regional range-restricted species or infraspecific taxa
Note: *historically recorded; **historically recorded and seen on recent surveys

Endemics at the site include:**Allium calocephalum,*Astragalus dendroproselius,*Bunium avromanum, *Cousinia acanthophysa, *C. masu-shirinensis, *Echinops nitens*, and ***Ornithogalum iraqense*; the near endemics at the site include: **Korshinskia assyriaca, *Pelargonium quercetorum Agnes,*Trigonosciadium viscidulum,*Verbascum froedinii*, and **Veronica macrostachya* var. *Schizostegia* and there are three nationally rare species: *Aristolochia paecilantha, Bromus brachystachys*, and *Michauxia tchihatchewii* (the last is a new recorded species).

B1. Site is a particularly species-rich example of defined habitat type

Mountain Forest Vegetation-Oak Forest-Lowest Zone and Medium Zone habitat and Forest Vegetation-Mountain Riverine Forest habitat

Additional Important Bird Observations: During the surveys a total of 91 species were observed (in the sub-site, Gali Balnda, a total 35 species were observed). In addition to those listed in the table above, European Roller *Coracias garrulus* (Near Threatened) was recorded breeding; the site also held a breeding population of Yellow-billed Chough *Pyrrhocorax graculus* a Eurasian High-Montane biome-restricted species. Eastern Cinereous Bunting *Emberiza semenowi* is Near Threatened.

Other Important Fauna: Data were collected from 2007 through 2010, with some additional work focused on Wild Goats (Vulnerable) in 2011. This species is thriving in Barzan, which has been a tribal protected site for many decades. The forestry police of Mergasur thought that the population of Wild Goats was around 1000, with some estimates much higher than this. In August 2010 and in 2011, an outbreak of goat plague or Peste des Petits Ruminants (PPR) led to the death of over 200 animals. In 2012, a local from Barzan succesfully photographed a Eurasian Lynx *Lynx lynx* and in the same year another local photographed a Roe Deer *Capreolus capreolus*. The globally Endangered Persian Leopard *Panthera pardus saxicolor* was also reported by locals who have seen and heard the animal in and around the Rezan River valley. Two Otters (likely Eurasian Otter *Lutra lutra*, Near Threatened) were sighted with flashlights at night in the summer of 2011. In the Zrara area, the team camera trapped Brown Bear *Ursus arctos*, Golden Jackal *Canis aureus*, Indian Creasted Porcupine *Hystrix indica* and Red Fox *Vulpes vulpes*. Zagrosian Lizard *Timon princeps kurdistanicus* was observed. Many locusts were seen near the Rezan in summer 2010, consuming large quantities of plant material.

Fish: Information was collected from 2007 through 2009 and eight species were observed. Significant species, as defined by Coad (2010) were: *Alburnoides* sp., *Alburnus mossulensis, Cyprinion macrostomum, Garra rufa, Hypophthalmichthys molitrix, Luciobarbus esocinus* (Vulnerable) and *Paracobitis* sp.

GALI BALANDA ©2010 S ABDULRAHMAN/NI

During a 2012 survey with Jörg Freyhof from the Leibniz Institute of Freshwater Ecology and Inland Fisheries, Germany *Oxynoemacheilus bergianus*, a new species for Iraq, was observed.

Additional Plant & Habitat Information: This site contains a good population of *Crataegus azarolus, Pistacia eurycarpa,* and *P. khinjuk,* which are important economically and culturally. Also there is a *Vitis vinifera* vineyard, which is important as a genetic resource.

Conservation Issues: The key environmental threat is oil development but also agriculture and grazing, urban construction, road building, tourism, landmines and pollution (both village and tourism waste) all constitute high

threats. In 2010 there was an attempt to build an earthen dam or dike at the mouth of the Rezan River where it meets the Greater Zab. According to a local, the purpose was to create a recreation area here, where a major restaurant and tea shop stop are located along the road. Winter rains washed this dam away but by the summer of 2012, a permanent structure, the Rezan River Weir Project, was under construction at the same location. There are other dam projects in the region but little was known about their extent and locations. Tourism has led to accumulation of garbage and easy access to the river has allowed car-washing to become a common practice in at least the main survey area. Though there is a hunting prohibition in the Barzan area that is largely effective in the region, poaching does still occur.

Some areas in Gali Balnda are affected by human intrusion, improper use of sewage water for irrigation, livestock grazing and agriculture in general.

Recommendations: Although this area is managed traditionally as a protected area by the local tribe there may be some pressures to remove or roll back some of the protection the site receives. Oil development must be closely monitored to avoid negative impacts to the region and dam construction should also be restricted. An official management plan should be developed and implemented and local stakeholders may help to strengthen protection of the site, particularly if oil development is occurring within the area.

Providing training and capacity building to the already-active environmental police on conservation management issues is also encouraged. The National Protected Area Committee (NPAC) identified this site in 2014 as a proposed protected area and officially recognizing the area as a national park is recommended and due to the size and richness of the area more comprehensive surveys should be conducted. For example the Gali Balnda area was only visited once during the KBA surveys and certainly requires more investigation. Minefield clearance should also be a priority for this large area.

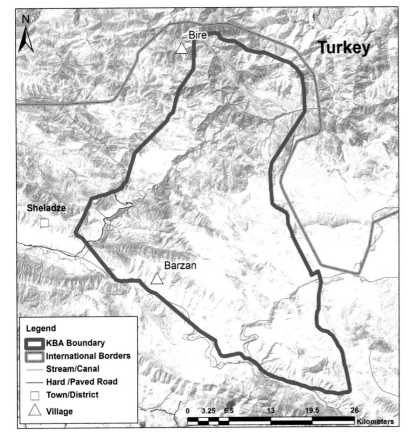

Legend
- KBA Boundary
- International Borders
- Stream/Canal
- Hard /Paved Road
- Town/District
- Village

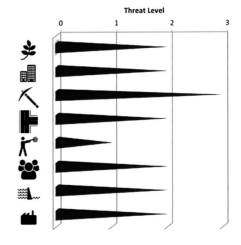

Threat Level

Kherazook (IQ005)

Erbil – 36.959167°N 44.324722°E

IPA Criteria: **B1**

Area: **607 ha** - Altitude: **576-814 m**
Ecoregion: **Zagros Mountain Steppe
Forest (PA0446)**
Status: **Unprotected**

©2007 K ARARAT/NI

Site Description: This is an open area surrounded by mountains near Kherazook Village. The key habitat examined was mountain riverine forest. There are poplar farms and riparian willow habitat on the west bank. The Kakla Stream, a tributary of the Rezan River, passes through the site.

The delineated site extends to just before the confluence with the Rezan, which flows into the Barzan Area & Gali Balnda (IQ004) about 11 km to the west. The stream in this area is about 8 m wide and flows rapidly, with a stony bottom and trees along the bank. The area of Kherazook Village is occasionally bombarded by the Turkish military.

Important Plant Area Criteria

B1. Site is a particularly species-rich example of defined habitat type

Forest Vegetation-Mountain Riverine Forest habitat type

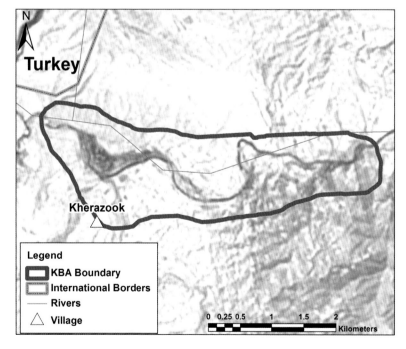

Additional Important Bird Observations: During the surveys 52 species were recorded. The site did not qualify as an IBA but held two breeding Irano-Turanian and three Mediterranean biome-restricted species though this did not trigger inclusion under A3 criteria. The Eastern Cinereous Bunting *Emberiza semenowi* (Near Threatened) was observed in the breeding season.

Other Important Fauna: Data were not collected for other fauna except fish, which were collected at this site in 2008 only, when six species were reported.

Fish: According to Coad (2010), the following were significant species: *Carassius auratus*, *Leuciscus vorax*, *Luciobarbus esocinus* (Vulnerable), and *Silurus triostegus*. The last two species found, *Barbus lacerta* and *Capoeta damascina*, have an unknown conservation status in Iraq.

Additional Plant & Habitat Information: This site contains a good population of *Aegilops crassa*, *Hordeum glaucum*, and *Hordeum bulbosum*, which are important genetic resources.

Conservation Issues: The highest threat to the site is due to its proximity to the Iraq-Turkish border and its bombardment by the Turkish military. Despite the obvious threat to humans and disturbance to wildlife, this causes more frequent forest fires in the area. The area is also impacted at a high level by agricultural activities and overgrazing, as people in the villages depend mainly on agriculture and livestock. Fishing was also seen in the Kakla stream.

Recommendations: It is important that bombardment of this area be stopped through negotiated agreement between

Turkey, the Kurdistan Regional Government (KRG), and Iraq. Additionally management of agricultural and grazing practices should be closely examined in the areas around the village to mitigate their impacts on the stream and vegetation of the area.

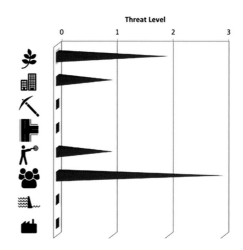

Mangesh (IQ006)

Dohuk – 37.03°N 43.071389°E

IBA Criteria: A1 and A3
IPA Criteria: A4 and B1

Area: **2287 ha** - Altitude: **850-1400 m**
Ecoregion: **Zagros Mountains Forest Steppe (PA0446)**
Status: **Unprotected**

Site Description: This site contains a mountain ridge running east to west, surrounding foothills and the valley below. The main habitat here is mountain forest vegetation with oak woodlands and also some cultivated lands lie nearby. The site is located in the Zagros mountain range, where the geology is siltstone and sandstone, and the soil type is sandy clay. It was previously surveyed under the KBA project as two sub-sites. There is a Christian archeological site located on the north side of the mountain.

Important Bird Area Criteria	Observations made 2007-2010.	
A1. Globally threatened species	**Breeding**	**Wintering/ Passage**
Egyptian Vulture *Neophron percnopterus* (Summer visitor)	1 pair (count 2007 & 2009)	
A3. Biome-restricted species		
Mediterranean biome	**Breeding**	**Wintering/ Passage**
Masked Shrike *Lanius nubicus* (Summer visitor)	15 pairs (2008)	
Sombre Tit *Poecile lugubris* (Resident)	1-2 pairs (2007)	
Western Rock Nuthatch *Sitta neumayer* (Resident)	5 pairs (2008)	
Eastern Black-eared Wheatear *Oenanth melanoleuca* (Summer visitor)	15 pairs (2008)	
Black-headed Bunting *Emberiza melanocephala* (Summer visitor)	130 pairs (2008-2009)	
Important Plant Area Criteria		
A4. Site contains national endemic, near endemic, regional endemic and/or regional range-restricted species or infraspecific taxa		
One endemic species, *Onosma albo-roseum* var. *macrocalycinum*, and two nationally rare species, *Delphinium kurdicum* and *Quercus macranthera*, were found.		

B1. Site is a particularly species-rich example of defined habitat type
Mountain Forest Vegetation-Oak Forest-Lowest & Medium Zones habitat type.

Additional Important Bird Observations: During the survey period, 56 species were recorded. In addition, the site held breeding populations of three Irano-Turanian and one Sahara-Sindian Desert biome-restricted species but these did not trigger inclusion under criterion A3.

Other Important Fauna: No additional information was collected on terrestrial fauna or fish (there were no major streams in the site).

Additional Plant & Habitat Information: This site contain a good population of *Crataegus azarolus*, which is commercially important, and *Hordeum bulbosum*, which is important as a genetic resource.

Conservation Issues: Agriculture and overgrazing (on the north side of the site) and human intrusion (from picnickers on both the north and south sides) are the highest threats at Mangesh. Building of houses and road construction as well as hunting and pollution from picnickers also have a moderate impact on the site. Streams are also being channelized here for irrigation purposes.

Recommendations: Agricultural practices and grazing land management needs more attention and better management practices. Environmental mitigation are needed in the planning and construction of roads and other development projects, including use of environmental impact assessments, zoning and introduction of construction standards to mitigate erosion and other environmental damage.

©2009 S ABDULRAHMAN/NI

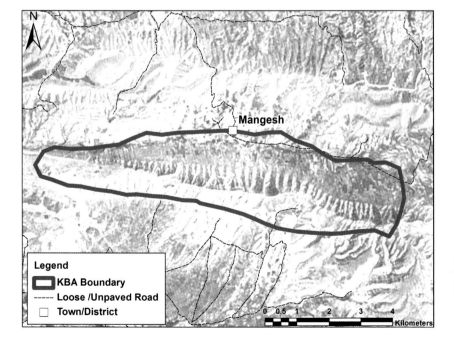

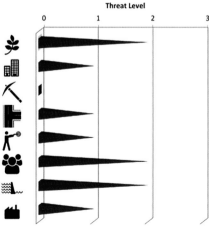

Gara Mountain & Garagu (IQ007)

Dohuk–37.010833°N 43.365°E

KBA Criteria: **V**
IBA Criteria: **A1 and A3**
IPA Criteria: **A4, B1, and B2b**

Area: **10447ha** - Altitude: **716-2053 m**
Ecoregion: **Zagros Mountains Forest Steppe (PA0446)**
Status: **Unprotected**

Site Description: Two sub-sites are contained in this area Gara Mountain and Garagu. The overall main habitat types found were mountain forest vegetation (primarily oak forests) and mountain riverine forest. The site is located in the Zagros mountain range, where the geology is siltstone and sandstone, and the soil type is sandy clay. About 60% of the area near the streams was unvegetated.

The town of Chamanke, located in an area of valleys and farmlands, is within the western edge of the delineated site. The Mani Maze Stream flowing near Chamanke is used as a source of water by the villages of Meze, Baramuinke, Kharinke and Bebad Islam and for the Shirin Bottled Water factory in Chamanke. The site is a popular tourist destination for residents of the nearby city of Dohuk. The towns of Tajilka and Ashawa are located in the northwest, where there are two streams (both former water quality sampling sites). Ashawa is a public resort. One of Saddam Hussein's former palaces is located inside the western edge of the delineation, and according to media reports in 2012 there is interest in developing the palace as a tourist site.

Garagu consists of a gorge, with a stream flowing through it, in a rocky part area of the Gara Range. The villages of Garagu, Babire, KaniBalar, Kala Shikho and Warmel are nearby. It is a popular picnicking site.

©2009 S ABDULRAHMAN/NI

Key Biodiversity Area Criteria	Notes	
V. Vulnerability Criteria: *Presence of Critically Endangered and Endangered species – presence of a single individual or Vulnerable species – 30 individuals or 10 pairs.*		
Neurergus crocatus	Observed and photographed by team at Garagu.	
Important Bird Area Criteria	**Observations made 2007-2010.**	
A1. Globally threatened species	**Breeding**	**Wintering/ Passage**
Egyptian Vulture *Neophron percnopterus* (Summer visitor)	1 pair	
A3. Biome-restricted species		
Mediterranean biome	**Breeding**	**Wintering/ Passage**
Masked Shrike *Lanius nubicus* (Summer visitor)	25 pairs (2009-2010)	
Sombre Tit *Poecile lugubris* (Resident)	60 pairs (2007, 2009 and 2010)	
Western Rock Nuthatch *Sitta neumayer* (Resident)	170 pairs (2008-2010)	
Eastern Black-eared Wheatear *Oenanthe melanoleuca* (Summer visitor)	150 pairs (2007 and 2010)	
Black-headed Bunting *Emberiza melanocephala* (Summer visitor)	350 pairs (2007, 2009 and 2010)	

Important Plant Area Criteria

A4. Site contains national endemic, near endemic, regional endemic and/or regionalrange-restricted species or infraspecific taxa
Note: *historically recorded; **historically recorded and seen on recent surveys

Endemics recorded in this area include: **Centaurea longipedunculata, *Delphinium micranthum, **Linum velutinum, Onosma albo-roseum* var. *macrocalycinum*, and **Satureja metastashiantha*

Near endemics recorded in this area include: **Astracantha crenophila, *Campanula radula, *Centaurea gigantea, *Globularia sintenisii, *Picris strigosa* subsp. *kurdica, *Stachys kotschyii.,*Viola pachyrrhiza, *Ziziphora clinopodioides* subsp. *kurdica*

Five nationally rare species include: *Asyneuma amplexicaule, Briza minor, Cosinia cymbolepis, Linum velutinum* (also endemic), *Quercus macranthera* and *Rubus caesius.*

B1. The site is a particularly species-rich example of a defined habitat type

Mountain Forest Vegetation-Oak Forest-Lowest & Medium Sub-zones habitat type and Mountain Forest Vegetation-Mountain Riverine Forest habitat type

B2b.The site is a refuge for: biogeographically and bioclimatically restricted plants to 'retreat to' in the face of global climate change.

This site represents a good example of Mountain Forest Vegetation- Oak Forest-Low, Medium, and Highest Zone habitat types. The top of mountain and some gorges/cliffs in the mountain can act as a refuge site for Oak forests and associated plants in the case of significant climate change.

Additional Important Bird Observations: A total of 60 species was observed during the surveys. Also four Irano-Turanian and one Eurasian Steppe and Desert biom.

Other Important Fauna: Data were collected only in 2010. Significant mammals reported by locals at the site were Syrian Brown Bear *Ursus arctos syriacus*. Significant reptile found at the site was Anatolian Lizard *Apathya cappadocica urmiana*. A significant amphibian found at the site was the globally Vulnerable Azerbaijan Newt *Neurergus crocatus*. No fish samples were collected.

Additional Plant & Habitat Information: This site contains a good population of *Crataegus azarolus* and *Pistacia eurycarpa*, which are economically and culturally important.

Conservation Issues: Tourism has a very high impact and related pollution was high. Agriculture is not extensive and mainly concentrated near Ashawa and Tajilka town and the valleys near the streams. There are small dams on both these streams, forming small lakes that have impacted the ecological conditions of the area. At Ashawa sub-site, where water comes from a spring on Gara Mountain, a village above is dumping sewage to the waterway. There are also landmines in the Chamanke area. Over the last three years road construction has occurred over a large part of the Amedi valley (IQ008) to the northeast and transportation corridors leading to and within the site were a high threat.

Recommendations: Many endemic and rare plant species are present and give this site special priority as a potential protected area. Trash and recycling services should be extended to this site, particularly to popular picnic areas and awareness-raising activities can be focused in these places. Minefield removal remains a priority for the area as well as better transportation planning and construction methods.

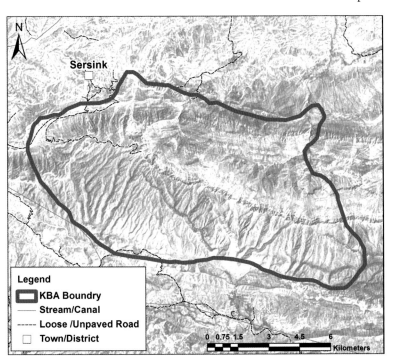

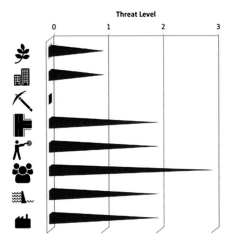

Amedi & Sulav (IQ008)

Dohuk – 37.108056°N 43.480833°E

IBA Criteria: **A1 and A3**
IPA Criteria: **A4 and B1**

Area: **6037 ha** - Altitude: **873-1563 m**
Ecoregion: **Zagros Mountains Forest Steppe (PA0446)**
Status: **Unprotected**

Site Description: Evans (1994) listed the area as an Important Bird Area (IBA003). The KBA surveys looked at two main areas here around the town of Amedi and the Sulav Resort. The main habitat types were oak woodlands and mountain riverine forest. The site is located in the Zagros Range and the geology is siltstone and sandstone. The soil type is sandy clay.

Amedi is an ancient town and historical site with about 34 archaeological features, some of which go back to the time of the Assyrians, Medeans, and different Kurdish periods. The town sits atop a rocky promontory approximately 1 km x 500 m in size and the surrounding areas features cliffs and gorges, including a ridge running east to west to the north of the town, a valley to the south and another east-west ridge at the southern extent of the site.

The Sulav area is situated on the south-facing slope of the ridge to the north of Amedi, which descends from rocky cliffs with thin scrubs to oak and juniper shrub forest that has a rocky and grassy floor. Numberous springs and cascading streams flow from here into the valley to the south, many lined by mature walnut trees, and the resort area is located just northwest of Amedi. There are many vineyards and orchards throughout the area. One of Saddam's former palaces is located inside the western edge of the delineation as well.

©2011 S ABDULRAHMAN/NI

Important Bird Area Criteria	Observations made 2007-2010.	
A1. Globally threatened species	**Breeding**	**Wintering/ Passage**
Egyptian Vulture *Neophron percnopterus* (Summer visitor)	1 pair (2009-2010)	
A3. Biome-restricted species		
Mediterranean biome	**Breeding**	**Wintering/ Passage**
Masked Shrike *Lanius nubicus* (Summer visitor)	30 pairs (2008-2010)	
Sombre Tit *Poecile lugubris* (Resident)	100 pairs	
Western Rock Nuthatch *Sitta neumayer* (Resident)	50 pairs (2008-2010)	
Eastern Black-eared Wheatear *Oenanthe melanoleuca* (Summer visitor)	20 pairs (2008, 2010)	
Black-headed Bunting *Emberiza melanocephala* (Summer visitor)	90 pairs (2008)	
Important Plant Area Criteria		
A4. Site contains national endemic, near endemic, regional endemic and/or regional range-restricted species or infraspecific taxa *Note: *historically recorded*		
Endemics at this site include: *Campanula mardinisis*, **Centaurea foreolata*, *Cousinia leatherdalei*, *Eryngium hainesii*, *Linum velutinum*, and *Onosma albo-roseum* var. *macrocalycinum*; Near endemics at the site were: **Grammosciadium cornutum*, **Salvia kurdica*, and **Stachys kotschyii*		
B1. Site is a particularly species-rich example of defined habitat type		
Mountain Forest- Mountain Riverine Forest habitat type and Mountain Forest Vegetation- Oak Forest- Lowest and Medium Zones habitat type.		

Additional Important Bird Observations: A total of 74 species were recorded. The site also held three Irano-Turanian, one Eurasian Steppe and Desert and one Eurasian High-Montane biome-restricted species. Eastern Cinereous Bunting *Emberiza semenowi* is Near Threatened.

Other Important Fauna: Mammal data were collected in 2010 only. One local reported a sighting of the globally endangered Persian Fallow Deer *Dama mesopotamica* in 2006, but this remains unverified. Persian Squirrel *Sciurus anomalus* was observed, which though a least concern species has a declining population trend and is heavily persecuted for the pet trade in Iraq. There are streams and rivers in the area but no fish survey was conducted.

Additional Plant & Habitat Information: This site contains a good population of pistachios *Pistacia eurycarpa*, which are economically and culturally important. It also held species important as genetic resources species such as *Avena fatua* and *Poa bulbosa*.

Conservation Issues: Tourism intensification was considered a very high threat to the area. Sulav is a major resort and the entire area is relatively close to Dohuk, so many people come for picnicking, resulting in the accumulation of garbage and potential sewage problems. Alot of the streams in the resort and picnic areas have also been hardened. Road construction has continued over a large area of the valley during recent year but most other development is focused in the towns and resort areas.

Recommendations: Several endemic and near endemic plant species as well as important bird species occur here and the claim that Persian Fallow Deer might be found in this area warrants more detailed surveys. Given the popularity of this area, its cultural importance and its proximity to the town of Dohuk, tourism and recreational use of this area will be one of the most important issues to address. Tourism plans are in place for the area but it is recommended to incorporate more protection for natural resources. The Gara Mountain & Garagu (IQ007) KBA site is located just to the southwest of this area and thus some integrated management planning for these two sites would be advisable.

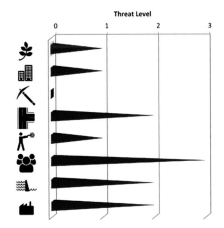

Mosul Lake (IQ009)

Dohuk & Nineveh – 36.741111°N 42.786111°E

IBA Criteria: A1 and A4i

Area: **48128 ha** - Altitude: **287-380 m**
Ecoregion: **Middle East Steppe (PA0812)**
Status: **Unprotected**

Site Description: This site consists of a large freshwater lake reservoir created by the Mosul Dam southwest of Dohuk city. This earthen embankment dam was completed in 1984 but has been under constant repair as it is built upon gypsum, a material that dissolves in water. A US Army Corps of Engineers report (Paley, 2007) stated that due to the possibility of the foundation eroding, "Mosul Dam is the most dangerous dam in the world."

Due to security concerns, the survey focused on the eastern (Dohuk Governorate) side of the lake in a limited area. The area around the lake here consists of moist steppe habitat with approximately 50% non-vegetated (similar habitat is likely on

©2010 K ARARAT/NI

the western side of the lake). The features of area were a gentle slope and sandy clay soils and the geology is a mix of siltstone and sandstone.

There is little human activity other than gravel mining, wheat and barley farming, and some net fishing on the lake. Access to the survey area was from a water station that serves Dohuk City.

Important Bird Area Criteria	Observations made 2007-2010.	
A1. Globally threatened species	**Breeding**	**Wintering/ Passage**
Egyptian Vulture *Neophron percnopterus* (Summer visitor)	7-15 pairs (counts 2009-2010)	
A4i. 1% or more of biogeographical population of a congregatory waterbird species		
	Breeding	**Wintering/ Passage**
Ruddy Shelduck *Tadorna ferruginea* (Winter visitor)		1200-9000 (counts); (2.4%-18% of regional population)
Collared Pratincole *Glareola pratincola* (Summer visitor)	500-1200 (counts 2008 and 2010); (0.5%-1.2% of regional population	

Additional Important Bird Observations: During the 2007-2010 surveys, 87 species were observed. European Roller *Coracias garrulus* (Near Threatened) was breeding and Lesser White-fronted Goose *Anser erythropus* (Vulnerable) occurred in winter, but at levels that did not meet IBA criteria.

In winter the site held up to 2500 of the armenicus race of Yellow-legged Gull *Larus michahellis*.

Other Important Fauna: Data were only collected for fish in 2007 and 2008, when 14 species were reported. Significant

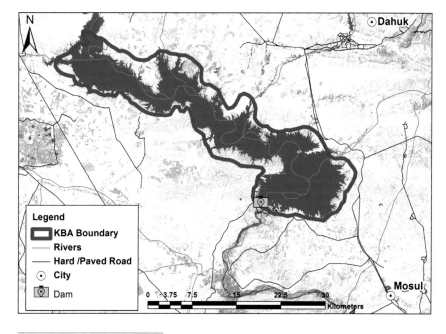

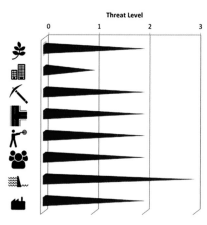

species, according to Coad (2010), were: *Acanthobrama marmid*, *Arabibarbus grypus* (Vulnerable), *Carassius auratus*, *Cyprinion kais*, *Cyprinion macrostomum*, *Cyprinus carpio* (Vulnerable), *Heteropneustes fossilis*, *Liza abu*, *Luciobarbus esocinus* (Vulnerable), *L. xanthopterus* (Vulnerable), and *Silurus triostegus*.

Conservation Issues: The Mosul Dam is under constant maintenance. The proposed solution for the dam is both expensive and untried. Agriculture and livestock production/grazing, gravel mining, and fishing are have significant impacts on Mosul Lake. In addition, road development and human disturbance due to boat, auto and tractor movement in the area were also considered high. Runoff from the surrounding lands also means that pollution due to the accumulation of trash and other pollutants is a problem as well.

Recommendations: It is imperative that a long-term, viable solution is found to address the stability of Mosul Dam. All potential solutions, including dam removal, should be considered to determine what is most cost effective. Regardless of the final actions, biodiversity protection should be taken into account as the site currently supports several important species. Regulations and controls regarding agriculture and grazing, fishing and boating (and other human disturbance) in and around the lake need to be implemented, but some of this will largely be dependent on improved security.

Dohuk Lake (IQ010)

Dohuk–36.885833°N 43.0075°E

IBA Criteria: **A1**
IPA Criteria: **A4**

Area: **386ha** - Altitude: **575-639 m**
Ecoregion: **Zagros Mountains Forest Steppe (PA0446)**
Status: **Unprotected**

Site Description: This is a small lake fed by springs from the surrounding rocky hills and mountains. The lake is actually a reservoir created by an earth-fill embankment dam on the Dohuk River just north of the city of Dohuk. The dam, which is 50 m high, holds 52 million m3 of water, and has a maximum discharge of 81 m3, was completed in 1988 with the primary purpose of providing water for irrigation and for the city. At normal operation level the reservoir is about 4 km long and 1.7 km wide. According to an Iraq and U.S. Army Corps of Engineers report (2003) there were no actions required for the dam except some short-term measures such as collection of data for evaluating trends and making some repairs. Human activity was low and fishing was not noted, but below the dam (outside of the delineated KBA site), the Dohuk River is an important center for recreational activities.

©2008 NI

Additional Important Bird Observations:
During the surveys, a total of 17 species were recorded. The European Roller *Coracias garrulus* (Near Threatened) was observed in the breeding season.

Important Bird Area Criteria	Observations made 2007-2010.	
A1. Globally threatened species	**Breeding**	**Wintering/ Passage**
Egyptian Vulture *Neophron percnopterus* (Summer visitor)	1 pair	
Important Plant Area Criteria		
A4. Site contains national endemic, near endemic, regional endemic and/or regional range-restricted species		
A historically recorded endemic for this site is *Asperula asterocephala*.		

Other Important Fauna: No survey was conducted for mammals or reptiles but fish data were collected in 2007 and 2008, during which eight species were reported. According to Coad (2010), significant species found were: *Alburnus mossulensis, Arabibarbus grypus* (Vulnerable), *Carasobarbus luteus, Carassius auratus, Cyprinion macrostomum,* and *Garra rufa.*

Conservation Issues: The 2003 Iraq and US Army Corps of Engineer report noted some maintenance issues with the dam and recommended that it be monitored for "signs of dam movement." Agriculture and grazing are a high threat to the area around the lake. The area is heavily impacted by human activities and garbage due to its close proximity to the city of Dohuk and the attraction of the area for picnicking.

Recommendations: The dam should be regularly monitored for safety and water management and releases from the dam should take into consideration the health of natural resources at the site. Agriculture and grazing should also be managed to protect native habitats and vegetation. The site will likely always be popular for recreation but adequate waste management facilities need to be provided and awareness-raising initiatives targetted in this area would likely reach many people and benefit other areas as well.

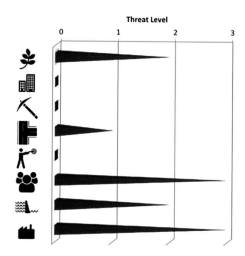

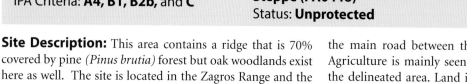

Zawita (IQ011)

Dohuk – 36.896667°N 43.146667°E

IBA Criteria: **A1**
IPA Criteria: **A4, B1, B2b,** and **C**

Area: **288 ha** - Altitude: **870-1260 m**
Ecoregion: **Zagros Mountains Forest Steppe (PA0446)**
Status: **Unprotected**

Site Description: This area contains a ridge that is 70% covered by pine *(Pinus brutia)* forest but oak woodlands exist here as well. The site is located in the Zagros Range and the geology is siltstone and sandstone and the soil type is sandy clay. A seasonal stream passes through the valley beside the main road between the towns of Zawita and Sarsang. Agriculture is mainly seen near the town of Zawita outside the delineated area. Land is used for field crops (wheat and barley) and vineyards and orchards cover the foothills. An area inside the delineation is used as a resort.

©2008 S ABDULRAHMAN/NI

Important Bird Area Criteria	Observations made 2007 and 2008.	
A1. Globally threatened species	**Breeding**	**Wintering/ Passage**
Egyptian Vulture *Neophron percnopterus* (Summer visitor)	3 pairs (count 2008)	
Important Plant Area Criteria		
A4. Site contains national endemic, near endemic, regional endemic and/or regional range-restricted species or infraspecific taxa		
One endemic was found during the survey effort here: *Onosma albo-roseum* var. *macrocalycinum* but historically the following endemics were recorded: *Asperula asterocephala*, *Erysimum boissieri*, and *Linum mucronatum* subsp. *pubifolium*		
B1. Site is a particularly species-rich example of defined habitat type		
Mountain Forest Vegetation -Pine Forest habitat type and Mountain Forest Vegetation- Oak Forest- Lowest and Medium Zones habitat type		
B2b.The site is a refuge for: biogeographically and bioclimatically restricted plants to 'retreat to' in the face of global climate change.		
This site represents a good example of Mountain Forest Vegetation -Pine Forest habitat type. The top of the mountain can work as a refuge site for these pine forests and their associated plants if there are severe climate changes.		
C. The site is identified as an outstanding example of a globally or regionally threatened habitat type.		
This site represents a good example of Mountain Forest Vegetation- Pine Forest, which is the most easterly of what was once a wide-spread forest. Also this important habitat is under threats of human intrusion and pollution caused by tourism.		

Additional Important Bird Observations: During the surveys 45 species were recorded. The site also held two Irano-Turanian and four Mediterranean biome-restricted species that were breeding but these did not trigger inclusion under criterion A3. No additional significant observations were made for non-avian fauna.

Additional Plant & Habitat Information: This site contains a good population of *Crataegus azarolus*, *Pistacia*

eurycarpa, and *P. khinjuk*, which are economically and culturally important plant species.

Conservation Issues: The site's proximity to Dohuk promotes human usage for picnicking at Zawita, which results in trash accumulation as well as potential sewage problems due to a lack of waste management facilities.

Recommendations: Due to the unique pine forest at this site and its already heavy use and importance for tourism and recreation, the area should receive more focused efforts for biodiversity conservation. Designating and providing facilities for picnic areas, waste management services and awareness programs such as interpretive trails and other educational initiatives will help to protect the flora and fauna here.

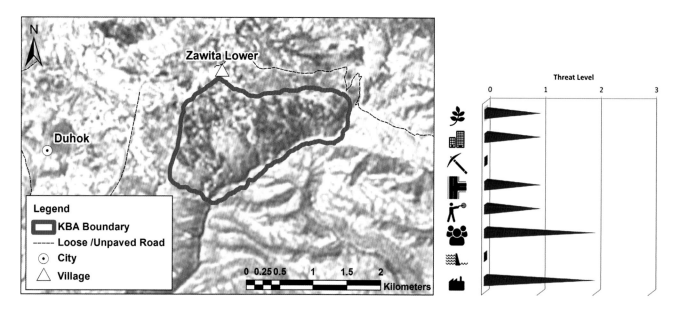

Atrush & Bania Area (IQ012)

Dohuk – 36.867222°N 43.232778°E

IBA Criteria: **A1**
IPA Criteria: **A4 and B1**

Area: **2411 ha** - Altitude: **560-917m**
Ecoregion: **Zagros Mountains Forest Steppe (PA0446)**
Status: **Unprotected**

Site Description: The Atrush Area is the central and southern part of the delineation and includes diverse habitats, including a gorge, open land covered by poplar trees, and surrounding foothill covered in woodlands. Bania (Sub-site D3C) is a mountainous area that includes the wooded and rocky valley to the north of the Atrush area. Deze and Balete villages lie near the Bedol stream that flows roughly north to south through the site. Near the center of the area (Atrush), where a smaller stream enters from the west, there are some archeological remains in a cliff overlooking the stream.

Overall the key habitats here are forest vegetation with oak forest and riverine woodland. The site is located in the Zagros range, where the geology is sandstone, clay, and sandy gravel and the soil type is sandy clay. There was road construction near the site during the time of the survey. Agriculture includes vineyards, apple orchards and wheat fields. The Bania area also has some aquaculture (carp farm). There is considerable local tourism and picnicking in the Atrush Area. Three sub-sites were surveyed here.

Important Bird Area Criteria	Observations made in 2007-2008.	
A1. Globally threatened species	**Breeding**	**Wintering/ Passage**
Egyptian Vulture *Neophron percnopterus* (Summer visitor)	2 pairs	
Important Plant Area Criteria		
A4. Site contains national endemic, near endemic, regional endemic and/or regional range-restricted species or infraspecific taxa		
One endemic species (*Pimpinella kurdica*) and two locally rare plants (*Sideritis libanotica* and *Trachomitum venetum* subsp. *sarmatiense)* were found at the site.		
B1. Site is a particularly species-rich example of defined habitat type		
Mountain Forest-Mountain Riverine Forest and Mountain Forest Vegetation-Oak Forest-Lowest & Medium Zone		

Additional Important Bird Observations: During the survey period a total of 53 species was seen. European Roller *Coracias garrulus* (Near Threatened) was breeding though at a level that did not meet IBA thresholds. In addition the site had breeding populations of three Mediterranean and two Irano-Turanian biome-restricted species but these did not trigger inclusion under criterion A3.

Other Important Fauna: No mammal or reptile surveys were conducted but five important fish species were observed during the one fish survey that took place in 2008: *Alburnus mossulensis*, *Leuciscus vorax*, *Luciobarbus xanthopterus* (Vulnerable), *Carassius auratus*, and *Garra rufa*.

Additional Plant & Habitat Information: *Scleria ciliata* and *Stachus kermanshahensis* were recorded for the first time in Iraq at Atrush. This site contains a good population of *Pistacia eurycarpa*, an economically and culturally important plant; *Hordeum bulbosum*, which is important as a genetic resource, and *Anchusa italica*, which is important as a traditional food.

Conservation Issues: Agriculture and grazing and pollution from tourist and picnicking activities, as well as changes in water management through channelization and irrigation, adversely impact this site.

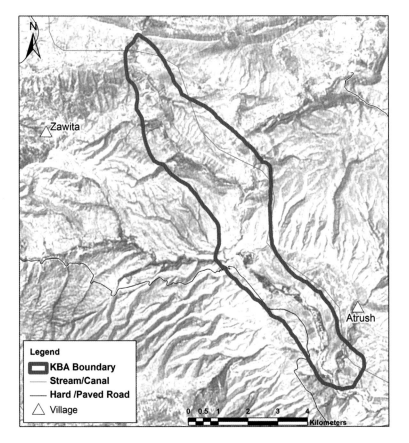

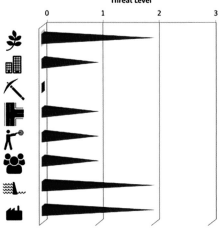

Recommendations: It is recommended that the use of land for agriculture and grazing be more tightly controlled and monitored to maintain and protect the habitats and species at the site. Areas that are popular with tourists, such as Atrush, should be the focus of educational initiatives and are also where public facilities for waste management should be provided.

Gali Zanta & Garbeesh (IQ013)

Dohuk – 36.741111°N 43.972222°E

IBA Criteria: **A1**
IPA Criteria: **A4 and B1**

Area: **3540 ha** - Altitude: **509–1175 m**
Ecoregion: **Zagros Mountains Forest Steppe (PA0446)**
Status: **Unprotected**

©2009 K ARARAT/NI

Site Description: This is a mountainous area with a narrow, winding gorge at its southern end (Gali Zanta); a broad upper valley in the center holding several villages, and above this the Garbeesh Area at the northern edge of the site. The higher ridge of Pires Mountain runs roughly northwest to southeast.

The main habitat types were mountain forest vegetation (primarily oak forests) and mountain riverine forest. The site is located in the Zagros range, where the geology is limestone and marls, and the soil type is sandy clay. Several streams drain the area towards Gali Zanta and it is covered in deciduous oak forest but about 50% of the area is non-vegetated. Picnickers frequently use the areas near the streams. There are also some orchards and vineyards. Two sub-sites were surveyed in this area.

Important Bird Area Criteria	Observations made 2007-2010.	
A1. Globally threatened species	**Breeding**	**Wintering/ Passage**
Egyptian Vulture *Neophron percnopterus* (Summer visitor)	2 pairs (counts 2008 and 2009)	
Important Plant Area Criteria		
A4. Site contains national endemic, near endemic, regional endemic and/or regional range-restricted species or infra-specific taxa		
One endemic species, *Cousinia acanthophysa* (historically recorded in this area), and two nationally rare plants, *Anogramma leptophylla* and *Bromus brachystachys*, have been found in this site.		
B1. Site is a particularly species-rich example of defined habitat type		
Mountain Forest – Mountain Riverine Forest habitat type and Mountain Forest Vegetation – Oak Forest - Lowest Zones habitat type		

Additional Important Bird Observations: During the surveys, 65 species were recorded. The site also held three Irano-Turanian, three Mediteranean and one Eurasian Steppe and Desert biome-restricted breeding species but these did not trigger inclusion under criterion A3. Eastern Cinereous Bunting *Emberiza semenowi* and European Roller *Coracias garrulus*, both Near Threatened species, were recorded during the breeding season.

Other Important Fauna: No data were collected except for fish, except in the year 2008 for the stream flowing in Gali Zanta only, when three significant species were observed: *Alburnus mossulensis*, *Carassius auratus* and *Silurus triostegus*.

Additional Plant & Habitat Information: This site contains a good population of *Crataegus azarolus*, *Pistacia*

eurycarpa, and *P. khinjuk*, which are economically and culturally important.

Conservation Issues: Agriculture and grazing affect the area, as do human intrusion (picnicking) and pollution. Picnickers leave trash and unsightly toilet facilities were located in the stream bed at Gali Zanta. Water pollution caused by the common practice of washing cars in any stream that can be accessed is also a high threat, especially at Garbeesh as was natural system modification (creating canals for irrigating farmlands). Urban and commercial development (shops for picnickers), gravel mining (in the streams in the Gali Zanta area), and hunting were all considered medium threats.

Recommendations: Agriculture (including irrigation planning) and grazing should be further examined to limit and/or mitigate any environmental problems these activities may cause. An education and enforcement program should target the problem of car washing at local streams where this is occurring. Development of picnic sites needs better planning and ongoing management particularly for waste disposal. Efforts need to be made to decrease the impacts of gravel mining and restore old mining sites.

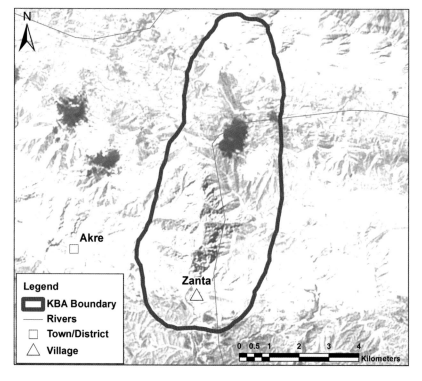

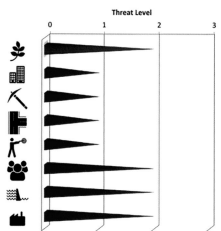

Bakhma & Bradost Mountain (IQ014)

Erbil – 36.703889°N 44.278056°E

KBA Criteria: **V**
IBA Criteria: **A1 and A3**
IPA Criteria: **A4 and B1**

Area: **21858 ha** - Altitude: **400-2010 m**
Ecoregion: **Zagros Mountains Forest Steppe (PA0446)**
Status: **Unprotected**

Site Description: This site includes Bradost Mountain and the Bakhma Area to the west. The latter, originally identified as an Important Bird Area (IBA004) by Evans (1994), is downstream of where the Greater Zab and Choman-Rawanduz Rivers meet and was the location of a dam that was

partially constructed during the Saddam Regime. The main habitat types are mountain forest vegetation (primarily oak forests) and mountain riverine forest. The site is located in the Zagros range where the geology is limestone and soil types are clay and sandy clay.

The overall site consists of deep valleys and high mountain bluffs in a region of limestone characterized by steep areas of woodland where oak trees grow in low areas and juniper trees grow on the slopes. The bulk of the delineation area contains the Bradost Mountain sub-area, a large mountain ridge that overlooks Bakhma from the east. The Lat and CorkMountains rise to the south, and to the north and northeast are a number of high hills and ridges. Vegetation is often sparse: largely oak-steppe with some occasional tamarisk along the riverbanks and farmed stands of poplar in the village.

The Bakhma dam project was started in the gap in the ridge running southeast to northwest through which the Greater Zab River flows (downstream of the confluence with the Choman-Rawanduz River). Construction was abandoned in 1991 but today the the river still flows through the diversion tunnels of the unfinished dam. There have been discussions on completing the project, which would place the first large scale barrier on what has largely remained Iraq's last wild river basin. Various other construction has been undertaken that has had a significant impact on the area adjacent to the dam site. Two sub-sites were visited during the survey effort.

©2010 K ARARAT/NI

Key Biodiversity Area Criteria	Notes	
V. Vulnerability Criteria: *Presence of Critically Endangered and Endangered species – presence of a single individual or Vulnerable species – 30 individuals or 10 pairs.*		
Capra aegagrus	30 individuals reported by the forestry police on Bradost Mountain. No direct observation has been made by the survey team.	
Important Bird Area Criteria	**Observations made 2007-2010.**	
A1. Globally threatened species	**Breeding**	**Wintering/ Passage**
Egyptian Vulture *Neophron percnopterus* (Summer visitor)	1-2 pairs (counts 2009-2010)	
A3. Biome-restricted species		
Mediterranean Biome	**Breeding**	**Wintering/ Passage**
Masked Shrike *Lanius nubicus* (Summer visitor)	2 pairs (count 2010)	
Sombre Tit *Poecile lugubris* (Resident)	5 pairs (count 2010)	
Western Rock Nuthatch *Sitta neumayer* (Resident)	8 pairs (count 2010)	
Eastern Black-eared Wheatear *Oenanthe melanoleuca* (Summer visitor)	6 pairs (count 2010)	
Black-headed Bunting *Emberiza melanocephala* (Summer visitor)	10 pairs (count 2010)	
Important Plant Area Criteria		
A4. Site contains national endemic, near endemic, regional endemic and/or regional range-restricted species or infraspecific taxa		
One nationally rare species *Plumbago europaea* occur here.		
B1. Site is a particularly species-rich example of defined habitat type		
Mountain Forest-Mountain Riverine Forest habitat type and Mountain Forest vegetation Oak Forest- medium zone habitat type		

Additional Important Bird Observations: During the surveys 75 species were observed. In addition to those listed in the table, Europen Roller *Coracias garrulous* and Eastern Cinereous Bunting *Emberiza semenowi* (both Near

Threatened) were breeding. Five Irano-Turanian and four Mediterranean biome-restricted species were also found breeding.

Other Important Fauna: Mammal data were collected in 2007 and 2010. Golden Jackal *Canis aureus* were observed. During an interview survey with the forestry police on Bradost Mountain, the following additional species were reported: Wild Goats *Capra aegagrus* (Vulnerable), sometimes seen in flocks of 30 individuals down the mountain near the main road; Grey Wolf *Canis lupus*; Persian Squirrel *Sciurus anomalus*; Indian Crested Porcupine *Hystrix indica*; Syrian Brown Bear *Ursus arctos syriacus*, seen two weeks before the interview, but only present in low numbers; Striped Hyena *Hyaena hyaena*; and Wild Cat *Felis silvestris*, which are reported in plentiful numbers, as well as some more common species. No significant reptile observation were made.

Fish: Data were collected in 2007 and 2008 from the Choman-Rawanduz River. Six significant species were found as defined by Coad (2010): *Alburnus mossulensis*, *Cyprinion kais*, *Garra rufa*, *Glyptothorax kurdistanicus*, and *Luciobarbus xanthopterus* (Vulnerable) and *Silurus triostegus*.

Additional Plant & Habitat Information: This site contains a good population of *Pistacia eurycarpa*, *P. khunjuk* and *Crataegus azarolus*, which are economically and culturally important plants. It also contains a vineyard of *Vitis vinifera*, which is important as a genetic resource.

Conservation Issues: Bakhma is relatively distant from large human populations. Although commercial development is still at a low level compared to other sites in the Kurdistan Region others threats include overgrazing and agriculture as well as gravel mining in many places in and along the Greater Zab and Choman-Rawanduz Rivers. It is likely that hunting may also be adversely affecting species and landmines are present on Bradost Mountain (and potentially other areas within the delineated site) but it was not clear how extensive these two threats are.

The greatest future threat to the site would be the completion of the Bakhma Dam. The initial construction, including roads to and from the construction area, already affects the site, but a complete dam would have far-reaching effects both upstream and downstream. These impacts include, but would not be limited to, flooding of many important cultural, recreational, agricultural and environmentally-sensitive areas upstream and eliminating the natural hydrological pulse, sediment transport and available water resources of the river downstream. The future plans for a dam are currently unclear.

Recommendations: Agriculture and livestock grazing, which is practiced throughout the area from the rivers' edge to the top of Bradost Mountain, should be examined for sustainability and potential impacts. As the Greater Zab and the Choman-Rawanduz are some of Iraq's last remaining wild rivers, they offer many development opportunities for tourism, recreation and fisheries if left undisturbed. The potential for developing the area for boating and fishing opportunities should be investigated to take advantage of these largely free-flowing rivers for sustainable ecotourism and recreation development. Gravel mining within the site and upstream should be more closely managed to mitigate and repair damage to the rivers and landmine clearance is also a priority.

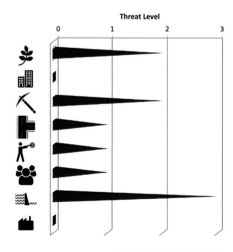

Rawanduz Gorge (IQ015)

Erbil–36.626389°N 44.594167°E

IPA Criteria: **A4, B1** and **B2b**

Area: **2680ha** - Altitude: **530-1248 m**
Ecoregion: **Zagros Mountains Forest Steppe (PA0446)**
Status: **Unprotected**

Site Description: This site consists of the deep Rawanduz Gorge, the Choman-Rawanduz River and three resort areas: Gali Ali Beg, which has a large waterfall and plunge pool; Bekhal Falls, a large waterfall 7 km upstream issuing from near the base of the mountain; and Jundyan, a spring on the slope of Handren Mountain located on the main road between Choman and Soran. All resort areas are on side-spurs to the main Rawanduz Gorge. The main habitat types were oak woodlands and mountain riverine forest and the site is located in the Zagros Range and features extensive cliff habitats as well. The geology is sedimentary limestone, and the soil type is sandy clay. The gorge is probably an important refugia site as it is the only site in Iraq for *Dionysia bornmuelleri*, a plant of damp cliffs.

At the Gali Ali Beg waterfall, the plunge pool and the banks downstream are hardened with stone and cement. Downstream of the main resort area, where the stream is approximately 7m across, the stream banks are developed for picnics, restaurants, and shops. The villages of Majidawa, Balakian, Basher and Belangr are nearby. Bekhal Falls is overlooked by the Pank Tourist Resort located on the bluff above the town of Rawanduz, which was opened in 2007 and continues to develop along with the relatively rolling or flat areas on top of the Rawanduz Gorge. Numerous pipes

(some in use, some defunct) take water from the falls and there is considerable development, in, around and above the falls by shops and small restaurants that cater to tourists. Jundyan features a spring that is surrounded by recreational development, well upstream from the Rawanduz Gorge.

The Rawanduz Gorge itself that contains both Gali Ali Beg and Bekhal Falls carries the Choman-Rawanduz River, which collects water coming from Hagi Omran (IQ018) and the Halgurd-Sakran Mt areas (IQ017 & IQ019) and flows westward approximately 72 km to join the Greater Zab River. About 20 km of this wild and scenic river runs through the site with an additional 18 km traversing the Bahkma and Bradost Mountain KBA Site (IQ014) located just 3.4 km to the west. The gorge runs roughly east to west and is very deep with sheer cliff walls. The road from Gali Ali Beg to the towns of Rawanduz and Soran was built in the 1930s by the British to open up the area to the north for "trade and administration" (Hamilton, 1930). Part of it is carved into the rock on the river-left side of the gorge but a newer road now crosses the gorge below Gali Ali Beg and hugs the river-right base of the gorge. Picnickers are now the primary uses of this portion of the original Hamilton Road. A second road starting upstream from Gali Ali Beg travels higher up the gorge wall on its south side to Bekhal Falls and beyond to the town of Rawanduz. Three sub-sites were visited during the survey effort.

Important Plant Area Criteria
A4. Site contains national endemic, near endemic, regional endemic and/or regional range restricted species or infraspecific taxa
One endemic plant *Echinops nitens* (historically recorded) and one nationally rare plant *Dionysia bornmuelleri*.
B1. Site is a particularly species-rich example of defined habitat type
Mountain Forest Vegetation-Mountain Riverine Forest habitat type and Mountain Forest Vegetation-Oak Forest-Low Sub-Zone habitat type
B2b. The site is a refuge for: biogeographically and bioclimatically restricted plants to 'retreat to' in the face of global climate change.
This site represents a good example of Mountain Forest Vegetation- Oak Forest Low Zone habitat type and the cliffs can be good refuge site for oak forests and associated plants and the gorge itself may provide for other rare plants, e.g. this area is the only site in Iraq where the plant Dionysia bornmuelleri is growing on the damp cliffs.

Additional Important Bird Observations: During the surveys a total of 13 species were recorded. Eastern Cinereous Bunting *Emberiza semenowi* (Near Threatened) and an Irano-Turanian biome-restricted species) was breeding but this did not trigger inclusion under criterion A3, though bird surveys here were limited.

Other Important Fauna: Only one survey was conducted for fish in 2008 on the Choman-Rawanduz River where the

stream from Gali Ali Beg enters the main river. It was possible to collect only three species though more are likely present in this large and important river basin. According to Coad (2010) the two significant species were *Alburnus mossulensis* and *Silurus triostegus*.

Additional Plant & Habitat Information: This site contains a good population of *Pistacia eurycarpa* and *P. khinjuk*, which are economically and culturally important.

There are many historic collections of endemic and near endemic species from Rowanduz such as *Dianthus ravandusiana* and *Tulipa kurdica*, but these can't be included here because the geo-referencing of these records is too imprecise. Nevertheless, this area is likely to be botanically rich and should be surveyed more fully.

Conservation Issues: This site is heavily impacted by tourists and picnickers who come from all over Iraq, particularly in the spring and summer, to see the dramatic gorge and is river and waterfalls. The artificial pool at Gali Ali Beg and the stream banks at both sites are heavily impacted by hardening, garbage from picnickers and damage from construction. Similar issues are seen at all the resort and popular picnic areas. Some of the small streams and waterfalls in the gorge are being used for car washing in the midst of tourist sites and restaurants. The Choman-Rawanduz River also carries the sewage and dumped garbage from upstream villages and the larger towns of Soran and Rawanduz, which appear to have greatly expanded beyond the capacity of their infrastructure to support their growth in population. Also the unplanned and chaotic development both within the gorge and on the top of the bluffs, as well as erosion along transportation corridors are significant problems in and around the Rawanduz Gorge. The status of the Bekhma Dam is also a potential threat. Should the dam be completed much of the gorge would be flood. In addition, other planned dams planned in the basin (one is proposed for within the gorge itself), would negatively impact the biodiversity of one of the last free-flowing river systems in the country.

Recommendations: Due to its popularity, tourism and related development will continue to be key human activities within the gorge and surrounding region. Developments such as the Pank Resort and Ski Resort and Spa on top of Korek Mountain nearby will increase the attraction of the area throughout the year. Tourism plans exist for the area but it is highly recommended that these plans be developed more broadly to address and mitigate human impacts on the natural resources of the region as well. Also dam and water diversion projects should be subject to full environmental impacts assessments and should be discouraged in the Greater

Zab Basin and its tributaries. Further surveys are needed to fully understand the biota present in the region.

It is recommended that shops, restaurants and picnic areas be pulled back from some areas around the falls, pools and streams to allow natural vegetation regrowth. Public areas and walkways should be designated and signs placed to raise awareness about the history and natural resources of the site and encourage its sustainable use. Waste receptacles that were visible at some picnic areas were usually unsightly and overflowing and garbage appears to ultimately end up being dumped and/or burned within or near the site anyway, so a comprehensive solid and liquid waste management plan is urgently needed for both tourism sites as well as local towns.

The Choman-Rawanduz River itself represents a significant opportunity for sustainable eco-tourism development in the region as this is one of few remaining wild rivers in the country and offers tremendous attraction for kayakers and other boaters both within the site as well as up and downstream.

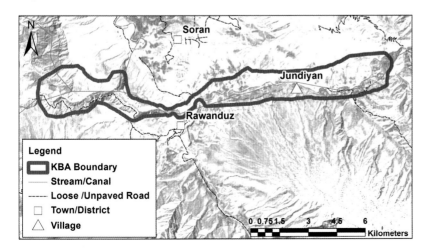

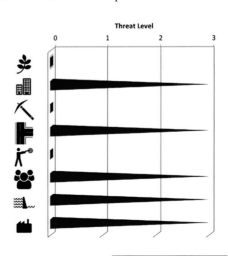

Sari Hassan Bag (IQ016)

Erbil – 36.71917°N 44.64583°E

IBA Criteria: **A3**
IPA Criteria: **A4 and B1**

Area: **923 ha** - Altitude: **1545-2495 m**
Ecoregion: **Zagros Mountains Forest Steppe (PA0446)**
Status: **Unprotected**

Site Description: This is a mountainous/subalpine area northeast of the town of Soran and was, according to the Flora of Iraq, a rich site for endemic and rare plant species in the country. The presence of minefields made botanical survey of the area difficult and thus only the thorn cushion vegetation area was examined. The site is located in Zagros Range and the geology is of basic igneous rocks, radiolarian chert, siliceous and calcareous shale, and metamorphic schist and limestone of unknown age. The soil type is sandy clay and the non-vegetated area was about 50%.

There is a small area of vineyards in the low forest mountain zone. There is picnicking in small areas near the subalpine zone and an old road leads to the top of the mountain. In addition to minefields there are traces of an abandoned airbase. The foothills suffer from overgrazing.

©2009 S ABDULRAHMAN/NI

Important Bird Area Criteria	Observations made 2009.	
A2. Biome-restricted species		
Irano-Turanian biome	**Breeding**	**Wintering/ Passage**
Eastern Rock Nuthatch *Sitta tephronota* (Resident)	2 pairs (counts 2009)	
White-throated Robin *Irania gutturalis* (Summer visitor)	1 pair (counts 2009)	
Kurdish Wheatear *Oenanthe xanthoprymna* (Summer visitor)	12 pairs (counts 2009)	
Finsch's Wheatear *Oenanthe finschii* (Resident)	2 pairs (counts 2009)	
Grey-necked Bunting *Emberiza buchanani* (Summer visitor)	1 pair	
Eastern Cinereous Bunting *Emberiza semenowi* (Summer visitor)	2 pairs (counts 2009)	
Important Plant Area Criteria		
A4. Site contains national endemic, near endemic, regional endemic and/or regional range-restricted species or infraspecific taxa		
Endemics found during recent surveys at this site were: *Astragalus helgurdensis* and *Tulipa kurdica*; Historically recorded endemics for this site were: *Cousinia mazu-shirinensis,* and a historically recorded near endemic was *Nepeta wettsteinii*		
B1. Site is particularly Species-rich example of defined habitat type		
Mountain Forest Vegetation- Thorn-Cushion Vegetation habitat type		

Additional Important Bird Observations: During the survey 24 species were recorded. The site also held two breeding Mediterranan and one Sahara-Sindian Desert biome-restricted species. The European Roller *Coracias garrulus* and Eastern Cinereous Bunting *Emberiza semenowi* were observed in the breeding season and both are Near Threatened. No other observations were obtained on non-avian fuana and there were no major streams or rivers to conduct a fish survey.

Conservation Issues: Agricultural practices, over-grazing and human activities specifically picnickers and the garbage they leave behind have a high impact on the site. There is a serious threat from mines, which cover the area, which seriously hinder any conservation activities or ecological studies. Hunting was considered a moderate threat at the site.

Recommendations: This site is located between the Halgurd Mountain (IQ017) and Rawanduz Gorge (IQ015) KBA sites and conservation actions needed here are similar and could perhaps be integrated into regional planning.

Agriculture in and around Sari Hassan Bag and grazing require further examination and stronger management. Environmental awareness should focus on areas that are popular with picnickers but involve locals as well. Mine clearance is clearly also a priority to allow for further use of the site (for research, recreational and other sustainable uses).

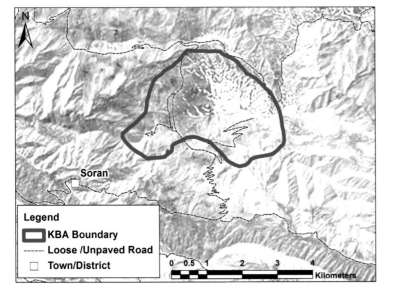

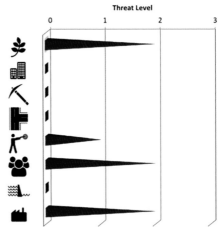

Halgurd Mountain (IQ017)

Erbil – 36.729722°N 44.884444°E

KBA Criteria: **V**
IBA Criteria: **A3**
IPA Criteria: **A4, B1, and B2b**

Area: **6393 ha** - Altitude: **1272-3613 m**
Ecoregion: **Zagros Mountains Forest Steppe (PA0446)**
Status: **Unprotected**

©2009 S ABDULRAHMAN/NI

Site Description: Halgurd Mountain is considered by many to be Iraq's highest peak at approximately 3613 m though a nearby peak Cheekha Dar or Black Tent (36.775278°N 44.918611°E) may also be a contender (Wikipedia, 2012; CIA 2012). Halgurd is part of the Hasarost (or Hasār-i Rōst) Mountain Range (itself part of the Zagros Range) near the Iranian border. The mountain retains some snow throughout the summer. Despite the fact that many places on the mountain are heavily mined, the slopes are used for summer grazing. A number of villages at lower elevations grow vegetables and raise livestock.

Two key habitats surveyed here are mountain forest vegetation — thorn-cushion vegetation and alpine zone vegetation. The geology of the mountain consists of basic igneous rock, radiolarian chert, siliceous and calcareous shale, and metamorphic schist and limestone of unknown age, and the soil types are serpentine, sandy clay, and clay. The non-vegetated area was about 50% of the site.

Key Biodiversity Area Criteria	Notes	
V. Vulnerability Criteria: *Presence of Critically Endangered and Endangered species – presence of a single individual or Vulnerable species– 30 individuals or 10 pairs.*		
Testudo graeca	One pair (also seen previously in 2008 & 2009) and there is suitable habitat for this species.	
Neurergus crocatus	Observed in 2009	
Important Bird Area Criteria	**Observations were made in Summer 2009 only**	
A3. Biome-restricted species		
Irano-Turanian biome	**Breeding**	**Wintering/ Passage**
See-see Partridge *Ammoperdix griseogularis* (Resident)	3 pairs (count 2009)	
Menetries's Warbler *Sylvia mystacea* (Summer visitor)	2 pairs (count 2009)	
Eastern Rock Nuthatch *Sitta tephronota* (Resident)	5 pairs (count 2009)	
White-throated Robin *Irania gutturalis* (Resident)	8 pairs (count 2009)	
Kurdistan Wheatear *(Oenanthe xanthoprymna)*	4 pairs (count 2009)	
Finsch's Wheatear *Oenanthe finschii* (Resident)	2 pairs (count 2009)	
Eastern Cinereous Bunting *Emberiza semenowi* (Summer visitor)	6 pairs (count 2009)	
Important Plant Area Criteria		
A4. Site contains national endemic, near endemic, regional endemic and/or regional range-restricted species or infraspecific taxa *Note: *historically recorded; **historically recorded and seen on recent surveys*		
Endemics recorded for this site include: **Alchemilla kurdica, *Allium. calocephalum, *Alyssum penjwinense, **Astragalus helgurdensis, *A. porphyrodon, *Cousinia carduchorum, *Delphinium micranthum, *Leutea rechingeri, *Onosma cornuta, *Rhynochorys elephas* subsp. *carduchorum, *Satureja metastashiantha, *Scrophularia atroglandulosa, *S. gracilis, **Tulipa kurdica,* and **Vitis hissarica* subsp. *rechingeri.* Near endemics found at the site include: **Allium arlgirdense, *Bunium cornigerum, *Carex iraqensis, *Centaurea urvillei* subsp. *deinacantha, *Colpodium gillettii, *Cousinia algurdina, *C. leptolepis, *Nepeta elymaitica, *N. wettsteinii, *Rosularia rechingeri, *Tragopogon bornmuelleri,* and **Veronica davisii.* Nationally rare species found at this site are: *Primula auriculata, Ranunculus bulbilliferus, Scilla siberica, Silene pungens, S. rhynchocarpa, S. commelinifolia, Tulipa buhseana,* and *Veronica beccabunga*		
B1. Site is a particularly species-rich example of defined habitat type		
Mountain Forest Vegetation—Thorn-Cushion Vegetation habitat type and Alpine Zone Vegetation habitat type		
B2b. The site is a refuge for: biogeographically and bioclimatically restricted plants to 'retreat to' in the face of global climate change.		
This site is represents a good example of the Thorn-Cushion Vegetation habitat type. The top of the mountain can provide refuge for Thorn Cushion plant species. Also some gorges/cliffs on the mountain can provide refuge for Oak forests and associated plants in the case of climate change.		

Important Bird Observations: During the survey 33 species were observed. The site held breeding populations of three Mediterranean and one Eurasian High-Montane biome-restricted species but did these did not trigger inclusion under criterion A3. No additional non-avian fauna observations were made and while there are important alpine and mountain streams no fish surveys were conducted.

Additional Plant & Habitat Information: This site contains a good population of *Allium akaka*, which is important as a traditional food as well as a good population of *Rheum ribes*, which is economically important.

Conservation Issues: Although threats to the area were generally assessed as low, the extensive presence of land mines is considered a high threat. There is also a high level of hunting. There are vegetable farms in the villages surrounding Halgurd Mountain operating on a small scale that could prove a threat if expanded. Garbage and trash dumps have been reported and may represent a higher threat than was observed in the KBA Surveys.

Recommendations: Halgurd Mountain is a very important for plants because it includes several regional endemic, near endemic and locally rare species. Their presence gives this

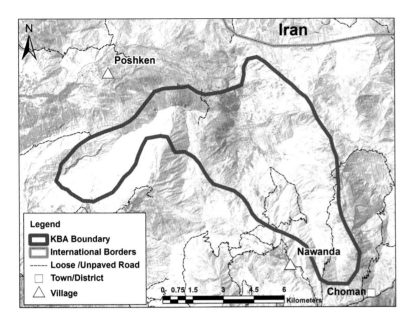

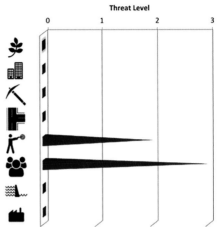

site special priority. Much of the area is part of the propose Halgurd-Sakran Park and park staff as well as other local stakeholders should be involved in management planning for this site.

Haji Omran Mountain (IQ018)

Erbil–36.666944°N 45.05°E

IPA Criteria: A4 and B1

Area: **2108 ha** - Altitude: **1455-2259 m**
Ecoregion: **Zagros Mountains Forest Steppe (PA0446)**
Status: **Unprotected**

Site Description: The site is located in the Zagros range, where the geology is of basic igneous rocks, radiolarian chert, siliceous and calcareous shale, and metamorphic schist and limestone of unknown age, and the soil type is serpentine. Several mountains including Mt. Sakran to the south and Mt. Halgurd to the west surround it. One key habitat type was examined, mountain forest with thorn-cushion vegetation and overall there is high level of plant diversity. The villages of Shiwa Rash, Mawatan, Zinwe and Megula are located within the delineated site. The town of Haji Omran, one of the main border crossings between Iraq and Iran, is also nearby. In addition there are vegetable farms (tomato, eggplant, sweet pepper, and okra) in the area.

Important Plant Area Criteria
A4. Site contains national endemic, near endemic, regional endemic and/or regional range-restricted species or infraspecific taxa *Note: *historically recorded; **historically recorded and seen on recent surveys*
Endemics at this site include: **Cousinia carduchorum, C. odontolepis*, *Echinops rectangularis*, *Erysimum boissieri*, *Ornithogalum iraqense*, *Scrophularia atroglandulosa*, and *Tulipa kurdica*. Near endemics at this site include: *Allium iranicum*, *Astracantha crenophila*, *Bellevalia kurdistanica*, *Scorzoner kurdistanica*, and *Tragopogon rechingeri*, and *Iris barnumae* is a nationally rare species
B1. Site is a particularly species-rich example of defined habitat type
Mountain Forest Vegetation- Thorn-Cushion Vegetation habitat type

Important Bird Observations: During the surveys 60 species were recorded. Three Irano-Turanian and four Mediterranean biome-restricted species were seen in the breeding season, including Eastern Cinereous Bunting *Emberiza semenowi* (Near Threatened) but these did not trigger inclusion under criterion A3.

Other Important Fauna: Mammal data were collected in 2010. Locals reported the presence of Brown Bear *Ursus arctos* and Grey Wolf *Canis lupus*. They also claimed that Eurasian otter *Lutra lutra* is present and that they had been hunted. A significant reptile found at the site was Diadem Snake *Spalerosophis diadema* and a significant amphibian

©2010 K ARARAT/NI

photographed was the Lemon-yellow Tree Frog *Hyla savignyi*. This site was not visited by the fisheries expert but the Choman-Rawanduz River as well as mountain streams flows through the site are likely very important for fish.

Conservation Issues: Agriculture intensification and grazing have a high impact on this area. New settlements and road construction may have an impact on wildlife by interfering with the passage routes of some large animals and may even lead to the death of animals by vehicles. There is extensive traffic through the site due to the border crossing

and oil spills have occurred into the river in the past due to tanker truck accidents. Hunting threatens some important species such as the Brown Bear and Grey Wolf. There are also several uncleared minefields near the Iran-Iraq border.

Recommendations: Mine clearance is a clear priority for the safety of wildlife and villagers. Efforts to develop the Halgurd-Sakran Park incorporates part of this region and Park staff are a major stakeholder in the region who can work with other local stakeholders in the development of comprehensive land management planning, educational initiatives and controls on hunting and transportation and general development.

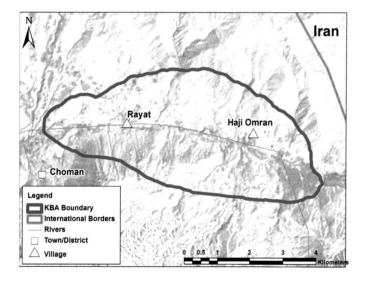

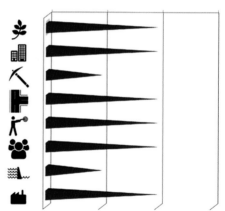

Sakran Mountain (IQ019)

Erbil–36.573889°N 44.986111°E

KBA Criteria: **V**

IPA Criteria: **A4, B1 and B2b**

Area: **6001ha** - Altitude: **1560- 3587 m**
Ecoregion: **Zagros Mountains Forest Steppe (PA0446)**
Status: **Unprotected**

Site Description: This mountainous area is part of the Qandil Range (a part of the Zagros Range) and is located in Choman District. The highest peak, Haji Ibrahim, reaches

3587 m and is surrounded by several other mountains, including Mt. Sakran to the south, Mt. Halgurd to the west and Mt. Gardamn to the north. The mountain peaks are covered

©2009 K ARARAT/NI

in snow year-round. There are several streams and small rivers throughout the area. The mountain contains oak woodlands, mountain riverine forest and thorn-cushion vegetation (sub alpine area) habitats. The geology is sedimentary limestone, and the soil type is sandy clay.

Villages include Weza, Sakran, Basan, Rezi and Ene. Its location close to the Iranian border means visits must be coordinated with the local government. It still has many minefields. This area has been locally proposed for consideration as a National Park.

Key Biodiversity Area	Notes
V. Vulnerability Criteria: Presence of Critically Endangered and Endangered species – presence of a single individual or Vulnerable species– 30 individuals or 10 pairs.	
Capra aegagrus	Reported by locals. No direct observation were made by the team.
Neurergus crocatus	Recorded in 2009
Important Plant Area Criteria	
A4. Site contains national endemic, near endemic, regional endemic and/or regional range restricted species or infraspecific taxa *Note: *historically recorded; **historically recorded and seen on recent surveys*	
Endemics noted for this site include: **Alyssum singarense*, *Astragalus helgurdensis*, **Centaurea elegantissima*, ***Cosinia carduchorum*, ***C. kurdica*, **Delphinium micranthum*, ***Leutea rechingeri*, *Onosma albo-roseum* var. *macrocalycinum*, **Scilla kurdistanica*, and *Tulipa kurdica* Near endemics for this site are: **Acantholimon blackelockii*, **Bunium cornigerum*, **Centaurea gigantea*, **Cousinia leptolepis*, **Carex iraqensis*, **Rosularia rechingeri*, **Tragopogon bornmuelleri*, **Trigonosciadium viscidulum*, **Veronica davisii*, and **Ziziphora clinopodioides* subsp. *kurdica* Nationally rare species are: *Aristolochia paecilantha* and *Fibigia suffroticosa*	
B1. Site is a particularly species-rich example of defined habitat type	
Mountain Forest Vegetation- Oak Forest- Medium Zone & Highest Zone habitat type; Mountain Forest Vegetation- Thorn-Cushion Vegetation habitat type and Mountain Forest Vegetation- Mountain Riverine Forest habitat type	
B2b.The site is a refuge for: biogeographically and bioclimatically restricted plants to 'retreat to' in the face of global climate change.	
This is a good example of the Thorn-Cushion Vegetation habitat type. The top of mountain can work as refugia for these Thorn-Cushion plants. Also some gorges/cliffs in the mountain can provide refuge for the Oak forests and associated plants in case of climate change.	

Additional Important Bird Observations: During the surveys a total of 39 species were recorded. European Roller *Coracias garrulus* (Near Threatened) was breeding but in sub-IBA-threshold numbers; in addition the site held breeding populations of four Irano-Turanian and two Mediterranean biome-restricted species, but this did not trigger inclusion under criterion A3. Eastern Cinereous Bunting *Emberiza semenowi* (Near Threatened) was also found breeding.

Other Important Fauna: Based on statements by locals, important mammal species can be found in good

numbers at Sakran Mountain, including Wild Goat *Capra aegagrus* (Vulnerable), Striped Hyena *Hyaena hyaena* (Near Threatened), and Syrian Brown Bear *Ursus arctos syriacus*. Based on interviews, hunting is practiced intensively here. It was claimed that 22 Wild Goats had been hunted in 2009. Kurdistan Viper *Montivipera raddei kurdistanica* (Near Threatened), a subspecies restricted to the mountains of Kurdistan, is also found. In Iraq, this species is known to occur only in a small region of the northeast (Alessandrini, 2010).

Additional Plant & Habitat Information: This site contains a good population of *Allium akaka, Crataegus azarolus*, and *Rumex ribes*, which are economically important plants harvested by locals to sell as food.

Conservation Issues: The most critical threats are hunting and the prevalence of minefields, which remain a danger to humans and wildlife. That said, the moderately disturbed ecological condition of this site, combined with the presence of many rare and endemic plants, make this site a strong candidate for protected status (local government efforts have already been made towards the establishment of a park that covers portions of this area).

Recommendations: It is strongly recommended that all mines be removed from the site and efforts strengthened to control and monitor hunting in the area. In the summer of 2012, a group of locals and internationals came together to form the Sakran-Halgurd National Park Committee, with the proposal to develop this area (but also parts of Halgurd Mountain (IQ017) and Hagi Omran (IQ018) areas) into a national park. For its part, in 2014 the National Protected Area Committee (NPAC) has identified Sakran Mountain as one if its proposed protected areas. Staff have already been hired for the park and efforts to protect the area continue to be pursued with the support of the Erbil Governorate. Further environmental surveys and management actions should also take place, including (but not limited to) additional botanical and fauna surveys, awareness-raising activities, socio-economic surveys, protection of cultural sites and resolution to any land ownership issues related to the park development.

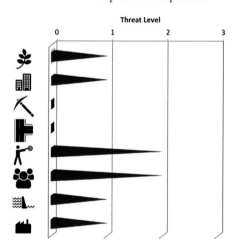

Bahrka (IQ020)

Erbil – 36.453611°N 43.810278°E

IPA Criteria: B1

Area: 1306 ha - Altitude: **280-370 m**
Ecoregion: **Middle East Steppe (PA0812)**
Status: **Unprotected**

Site Description: Bahrka means 'Small Sea' in Kurdish and Arabic. The main habitat type examined here was Riverine Forest of the Plains. The site is located in the moist steppe zone, and the geology is gravel, silt, sands, and alluvium. The area is composed of steppes and highlands, with Dara Mamz Mountain to the south. The land is mostly used for agriculture but about 40-50% is unvegetated.

The braided Greater Zab River, which comes from Turkey but also includes waters from the Rawanduz River coming from Iran and the mountains in northern Iraq, flows through the site. The river banks (riparian zone) and some uplands

are also included within the delineation. The Greater Zab continues to the southwest extending approximately 70 km to where it enters the Tigris River. The villages of Kharok, Ifraz Miran and Ifraz Kamal lie to the northeast of the survey area. There is gravel mining along the river within the site (as well as in many places up and downstream), which has disturbed parts of the riparian zone. The currently discontinued Bakhma Dam construction site, part of the Bakhma and Bradost Mountain KBA Site (IQ014), is approximately 48 km upstream.

Important Plant Area Criteria

B1. Site is a particularly species-rich example of defined habitat type

Riverine Forest of the Plains habitat type

Additional Important Bird Observations: During the surveys a total of 32 species were recorded. The site also held breeding Irano-Turanian (two), Mediterranian (one) and Sahara-Sindian Desert (one) biome-restricted species but these did not trigger inclusion under criterion A3. Note that European Roller *Coracias garrulus*, a Near Threatened species, was also observed.

Other Important Fauna: No significant observations were made for other fauna except for fish. During the 2007 & 2008 fish survey, 13 species were observed. Significant species, according to Coad (2010) were: *Acanthobrama marmid*, *Alburnus mossulensis*, *Carassius auratus*, *Chondrostoma regium*, *Cyprinion macrostomum*, *Cyprinus carpio* (Vulnerable), *Heteropneustes fossilis*, *Hypophthalmichthys molitrix*, *Luciobarbus xanthopterus* (Vulnerable) and *Silurus triostegus*.

Conservation Issues: This area is highly impacted by gravel mining, and much of the upper riparian zone has been converted to agricultural use. Also sewage from villages nearby (Ifraz Miran and Ifraz Kamal) drains into the river. There is no picnicking but there is significant human intrusion from villagers and workers involved in gravel mining and narrow unpaved roads lead to the agricultural land and through the gravel mining areas. Upstream is the unfinished Bakhma Dam and other dams are proposed at upstream sites. Should these projects proceed, they will seriously impair the ecology and function of the river and the flora and fauna at Bahrka.

Recommendations: The natural areas that remain along the river should be protected and those that have been damaged should be restored to protect river function and ecology. Gravel mining, particularly in-stream and in riparian areas, needs to be much more strongly controlled to stop, minimize and/or mitigate impacts. Any future work plans for dam construction that affect this site must consider and mitigate the full ecological impacts to the river and its biota.

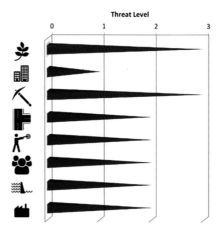

©2010 K ARARAT/NI

Doli Smaquli & Ashab (IQ021)

Erbil – 36.363611°N 44.322778°E

IBA Criteria: **A1 and A3**
IPA Criteria: **A4 and B1**

Area: **7027 ha** - Altitude: **815-1360m**
Ecoregion: **Zagros Mountains Forest Steppe (PA0446)**
Status: **Unprotected**

Site Description: This site consists of the valley and slopes between two mountain ridges that run roughly northwest to southeast; part of the Zagros Range. The ridge on the northeast side is Safin Mountain with the resort town of Shaqlawa located just beyond. Several streams flow through the area and it contains mountain forest vegetation with dense oak woodlands (also *Prunus* sp. and *Pistatia* sp. and many species of herbs) as well as agricultural fields. The geology is limestone and the soil type is sandy clay.

The villages of Tawska, Gorasheri Sarw, Gorasheri Khwarw and Qalasnji Khwarw are nearby. A new road was under construction during the 2010 survey. Agriculture is focused on grain crops (wheat and barley) and vineyards as well as some orchards especially in the valley. Housing construction was seen at the survey site. Two sub-sites were visited during the survey effort.

Important Bird Area Criteria	Observations made 2007-2010.	
A1. Globally threatened species	**Breeding**	**Wintering/ Passage**
Egyptian Vulture *Neophron percnopterus* (Summer visitor)	1-2 pairs (counts)	
A3. Biome-restricted species		
Irano-Turanian	**Breeding**	**Wintering/ Passage**
See-see Partridge *Ammoperdix griseogularis* (Resident)	40 pairs (2009-2010)	
Menetries's Warbler *Sylvia mystacea* (Resident)	90 pairs (2008-2010)	
Eastern Rock Nuthatch *Sitta tephronota* (Resident)	290 pairs	
White-throated Robin *Irania gutturalis* (Summer visitor)	25 pairs (2009-2010)	
Finsch's Wheatear *Oenanthe finschii* (Resident)	40 pairs (2009-2010)	
Eastern Cinereous Bunting *Emberiza semenowi* (Summer visitor)	10 pairs (2007)	
Mediterranean biome	**Breeding**	**Wintering/ Passage**
Masked Shrike *Lanius nubicus* (Summer visitor)	55 pairs	
Sombre Tit *Poecile lugubris* (Resident)	100 pairs	
Western Rock Nuthatch *Sitta neumayer* (Resident)	50 pairs	
Eastern Black-eared Wheatear *Oenanthe melanoleuca* (Summer visitor)	130 pairs (2007, 2009-2010)	
Black-headed Bunting *Emberiza melanocephala* (Summer visitor)	630 pairs (2008-2009)	
Important Plant Area Criteria		
A4. Site contains national endemic, near endemic, regional endemic and/or regional range-restricted species or infra-specific taxa		
Endemics here are *Bunium avromanum* and *Onosma albo-roseum* var. *macrocalycinum* and a nationally rare species was *Aristolochia paecilantha*.		
B1. Site is a particularly species-rich example of defined habitat type		
Mountain Forest Vegetation–Oak Forest–Lowest and Medium Zone habitat type		

Additional Important Bird Observations: A total of 69 species were observed. The site also held one Sahara-Sindian Desert biome-restricted species. The Eastern Cinereous Bunting *Emberiza semenowi* and European Roller *Coracias garrulus* (both Near Threatened) were obseved in the breeding season.

Other Important Fauna: Mammal data were collected at this site only in 2010. A local claimed that he had seen a brown bear *Ursus arctos* in 2009. He also claimed that the Endangered Persian Leopard *Panthera pardus saxicolor* was present before the beginning of the civil war in 1994. A key reptile observed was Urmia Rock Lizard *Apathya cappadocica urmiana*. No fish survey was made because the streams were not large enough.

Additional Plant & Habitat Information: This site contains a good population of pistachio (*Pistacia eurycarpa*, and *P. khinjuk*), a plant that is important both economically and for local heritage. Also *Gundelia tournefortii*, which is important as a traditional food and *Hordeum bulbosum* and *Vitis vinifera*, which are important genetic resources.

Conservation Issues: Although livestock production/grazing, road construction, tourism and changes to nearby land use as a result of agricultural expansion and housing construction remain high threats, the ecological condition of this area was only moderately disturbed, with high vegetation cover and several important plant species.

Recommendations: This site, along with Bahrka (IQ020), is relatively close to Erbil and also close to the town of Shaqlawa, which recieves extensive tourism. As a result, environmental education campaigns focused in this area have the potential of reaching a large audience. Agricultural expansion and grazing requires additional monitoring and control as does residential development in this area.

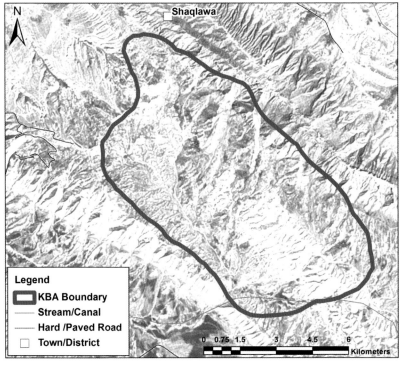

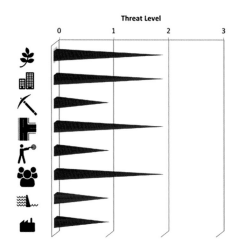

Doli Plngan (IQ022)

Sulaimani – 36.405833°N 44.751944°E

IPA Criteria: **A4 and B1**

Area: **420 ha** - Altitude: **880-1320 m**
Ecoregion: **Zagros Mountains Forest Steppe (PA0446).**
Status: **Unprotected**

Site Description: This is a valley located within a mountainous area and contains oak woodlands and mountain riverine forest. The site is located in the Zagros range, where the geology is limestone and the soil type is sandy clay. There is a small spring near Diman village where tourism and picnicking occur. Hartal and Betwata villages are nearby. There are some orchards and vineyards in the villages located within the delineated area.

Important Plant Area Criteria
A4. Site contains national endemic, near endemic, regional endemic and/or regional range-restricted species or infra-specific taxa
One nationally very rare species, *Juncus effusus*, was found at the site.
B1. Site is aparticularly species-rich example of defined habitat type
Mountain Forest – Mountain Riverine Forest habitat type

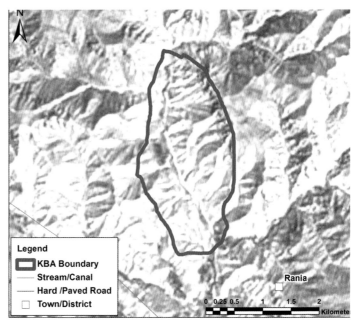

Legend
- KBA Boundary
- Stream/Canal
- Hard /Paved Road
- Town/District

Rania

0 0.25 0.5 1 1.5 2 Kilomete

Additional Important Bird Observations: During the survey a total of 31 species were recorded. The site held breeding Irano-Turanian (two), Mediterranean (three) and Eurasian Steppe and Desert (one) biome-restricted species, but this did not trigger inclusion under criterion A3. Note that Eastern Cinereous Bunting *Emberiza semenow*i and European Roller *Coracias garrulus*, both Near Threatened species, were recorded breeding. This site did not meet IBA criteria but was only visited in one survey. No additional information was collected on other (non-avian) fauna.

Conservation Issues: The high threats to this site were agriculture and livestock production/ grazing; hunting; human instrusion in the form of tourism, and pollution from the associated garbage and waste. The environment was in a poor condition in the picnicking areas. Significant amounts of trash were found,

©2009 S ABDULRAHMAN/NI

particularly close to the spring. Tourists also introduce a significant amount of noise pollution, which may be detrimental to some species.

Recommendations: Doli Plngan needs significantly more interest and environmental awareness from the local government with involvement from local stakeholders in order to be improved. Issues of human use such as agriculture, tourism and hunting require better local and regional planning and education respectively. Picnic areas should be officially designated and rubbish collection services should be made available there. Additional surveys to more fully assess this site are recommended.

Dukan Lake (IQ023)

Sulaimani – 36.0925°N 44.935833°E

KBA Criteria: **V and Ia**
IBA Criteria: **A1, A3, and A4i**
IPA Criteria: **A4 and B1**

Area: **47253 ha** - Altitude: **506-1080** m
Ecoregion: **Zagros Mountains Forest Steppe (PA0446)**
Status: **Unprotected**

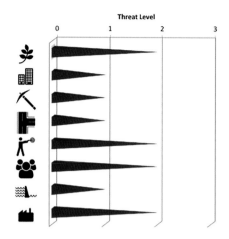

©2010 K ARARAT/NI

Site Description: This site, which was listed as an IBA by Evans (1994), is a large reservoir of about 25,000 ha that is fed by the Lesser Zab River from the northeast and the Hizop stream from the northwest. The lake is formed by the Dukan dam, built in 1959 upstream of the town of the same name. The rivers are fed by rainfall and snowmelt, leading to peak discharge in spring and low water in summer and early fall. The lake itself is divided into two parts; a larger lake to the north and a smaller lake to the south separated by a winding gorge (see map). The Rania Plain, where the lake is located, is the largest valley in the Lesser Zab Basin. Most tributaries join the Lesser Zab upstream of Dukan, the largest being the Baneh River and Qala Chwalan. A number of smaller streams join the Lesser Zab in the Rania Plain, which is now partly inundated by the lake. More than 65 villages and six towns are located around the lake, including Rania, Chwar Qurna, and Qala Dza. Much of the land around the lake is used for agriculture and grazing is practiced extensively. There are some small mineral ponds between Rania and the lake on the north side, close to Qurago and Bemushen villages.

The main habitat surveyed was mountain riverine forest and, in some areas, cliff habitat around the reservoirs edge but about 60% of the area was non-vegetated. The site is located in the foothills of Zagros Range, where the geology is sedimentary limestone, and the soil type is sandy clay. Sara and Qarasird Mountains sit to the southeast of the lake, Assos Mountain (IQ024) is situated to the northeast, Kosrat Mountain is to the southwest, and Barda Rash Mountain is to the northwest. There are also rolling hills and lowland areas characterized by oak forests and steppe ecosystems.

Key Biodiversity Area Criteria		
V. Vulnerability Criteria: *Presence of Critically Endangered and Endangered species – presence of a single individual or Vulnerable species – 30 individuals or 10 pairs.*		
Rafetus euphraticus	Euphrates Softshell Turtle was observed in mineral ponds in the northern section of the site. During a 2-day survey 3 specimens were found. Based on field survey the lake provides suitable habitat for this species.	
Luciobarbus subquincunciatus	Critically Endangered fish species	
Ia. Irreplaceability Sub-criterion: Restricted-range species based on global range		
Rafetus euphraticus	See above	
Important Bird Area Criteria	**Observations made 2007-2010.**	
A1. Globally threatened species	**Breeding**	**Wintering/ Passage**
Egyptian Vulture *Neophron percnopterus* (Summer visitor)	3-6 pairs (counts)	
Lesser White-fronted Goose *Anser erythropus* (Winter visitor)		410-630 (counts 2009-2010)
Red-breasted Goose *Branta ruficollis* (Winter visitor)		2 (count 2010)
A3. Biome-restricted species		
Irano-Turanian biome	**Breeding**	**Wintering/ Passage**
See-see Partridge *Ammoperdix griseogularis* (Resident)	250 pairs (2008-2010)	
Upcher's Warbler *Hippolais languida* (Summer visitor)	6 pairs (2009)	
Menetries's Warbler *Sylvia mystacea* (Summer visitor)	10 pairs (2010)	
Eastern Rock Nuthatch *Sitta tephronota* (Resident)	250 pairs (2008-2010)	
White-throated Robin *Irania gutturalis* (Summer visitor)	15 pairs (2010)	
Finsch's Wheatear *Oenanthe finschii* (Resident)	30 pairs (2008-2009)	
Pale Rockfinch *Carpospiza brachydactyla* (Summer visitor)	20 pairs (2008-2011)	
A4i. 1% or more of biogeographical population of a congregatory waterbird species		
	Breeding	**Wintering/ Passage**
Greater White-fronted Goose *Anser albifrons* (Winter visitor)		40-552 (counts) (1.6%-3.7% of the biogeopgraphical population)
Lesser White-fronted Goose *Anser erythropus* (Winter visitor)		410-630 (counts 2008-2010)
Slender-billed Gull *Chroicocephalus genei* (Summer visitor)	2250-2400 pairs (counts 2008-10)	
Great Cormorant *Phalacrocorax carbo* (Winter visitor)		1100-3200 (counts) (1.1%-3.2% of the biogeographical population)
Common Shelduck *Tadorna tadorna* (Winter visitor)		1200 (highest count 2008, 2010) (1.5% of the biogeographical population)
Important Plant Area Criteria		
A4. Site contains national endemic, near endemic, regional endemic and/or regional range-restricted species or infraspecific taxa		
Campanula mardinisis (endemic) and *Ornithogalum iraqense* Feinbrun (historically recorded endemic); *Centaurea gigantea* Schultz-Bip. ex Boiss. and *Verbascum alceoides* Boiss. & Hausskn (both historically recorded near endemics), and nationally rare species are: *Juncus effuses, Zeugandra iranica,* and *Salix babylonica*.		
B1. Site is a particularly species-rich example of defined habitat type		
Mountain Forest Vegetation-Mountain Riverine Forest habitat type		

Additional Important Bird Observations: During the surveys 181 species were recorded. The following were observed at levels that did not meet the IBA criteria: in winter Marbled Duck *Marmaronetta angustirostris* (Vulnerable), Ferruginous Duck *Aythya nyroca* (Near Threatened), Eastern Imperial Eagle *Aquila heliaca* (Vulnerable), Pallid Harrier *Circus macrourus* (Near Threatened) and Little Bustard *Tetrax tetrax* (Near Threatened); breeding European Roller *Coracias garrulus* (Near Threatened). The site also held breeding populations of four Mediterranean and one Sahara-Sindian

Desert biome-restricted species. In winter the site also held up to 1500 (over 2%) of the armenicus race of Yellow-legged Gull *Larus michahellis.*

Other Important Fauna: Data on mammals are only for 2008-2010, with more detailed information collected in 2010. Local fishermen reported observations of two globally Near Threatened species: Eurasian Otter *Lutra lutra* and Striped Hyaena *Hyaena hyaena.* One local reported the hunting of an Otter in the year 1996. There also have been reports of wolf attacks on local animal herds.

Fish: Data were collected for the years 2007 and 2008 only, during which 21 species were reported. According to Coad (2010), additional significant species observed were: *Acanthobrama marmid, Alburnus caeruleus, Alburnus mossulensis, Arabibarbus grypus* (Vulnerable), *Carassius auraatus, Cyprinion macrostomum, Cyprinus carpio* (Vulnerable), *Hypophthalmichthys molitrix, Heteropneustes fossilis, Gambusia holbrooki, Luciobarbus esocinus* (Vulnerable), *L. xanthopterus* (Vulnerable), *Squalius cephalus, Silurus triostegus.* To protect fish spawning in the area the Kurdistan Ministry of Agriculture prohibits fishing from mid-May to the end of July. Fishing, when allowed, is usually done with nets whose mesh size ranges from 30 mm to 80 mm. Individual anglers are also frequently observed.

Additional Plant & Habitat Information: This site contain a good population of *Anchusa italica* and *Gundelia tournefortii,* which are important as a traditional food; as well as a good population of *Linum usitatissimum,* which is important as a genetic resource.

Conservation Issues: Hunting and fishing constitute a very significant threat to the area. Hunters were observed during winter in Dukan, especially in the northern part of

the lake near the place where Lesser White-Fronted Goose was observed. In Rania district, several pools have been constructed with bird hides to attract water birds. Dukan is one of the most popular picnic sites for people coming from Sulaimani and the surrounding area, particularly during spring and summer and parts of this site are heavily impacted by picnickers and tourist activities and the trash they generate.

The road to the lake from the town of Dukan was widened in 2010/2011, which caused a high level of erosion into the lake itself. The dam continues to raise environmental concerns downstream due to fluctuating water levels, the entrapment of sediments and the low sediment transport downstream, which increases the erosion potential and decreases biodiversity below the dam.

Pollution from local industries, towns and villages and from agriculture is another very significant threat. Farmers in the northeastern part of the site near Rania were observed using poison to kill mice in their fields. In 2009, just in the Rania district alone (on the north side of the lake and containing an area important for birds, especially for raptors) farmers were provided with 120 kg of bromadioline a rodenticide, and 997 L of malathion, another pesticide (both are slightly to moderately toxic to many birds).

There are many gravel mines along the Lesser Zab River above and one below the lake within the delineated area impacting the instream and riparian habitats. Overgrazing and urban development also affect areas around the lake and along the Lesser Zab River.

Recommendations: In order to protect the lake fishery, the timing of fishing prohibitions may require further study to improve spawning success and the maturation of fish eggs (Ararat et al., 2008). Local fish hatcheries are releasing carp and potentially other non-native species to the lake and this process needs more research to understand its utility and affect on the biota of the lake and river. It is recommended that educational campaigns and outreach programs are focused in areas with high public use (recreational areas along the Lesser Zab River in Dukan, at tourism areas on the lower lake tourism areas and the upper lake near Rania).

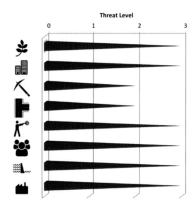

Legend
KBA Boundary
Stream/Canal
Hard /Paved Road
Town/District

Assos Mountain (IQ024)

Sulaimani – 35.988911°N 45.213889°E

KBA Criteria: **V**
IBA Criteria: **A3**
IPA Criteria: **A4 and B1**

Area: **20017 ha** - Altitude: **548-2400 m**
Ecoregion: **Zagros Mountains Forest Steppe (PA0446)**
Status: **Unprotected**

Site Description: This is a mountainous area and its habitat is mountain forest vegetation with oak forest and mountain riverine forest. The latter is found along the streams formed by snowmelt from the top of the mountain as well as along the Lesser Zab River. Assos Mountain, which is part of the Zagros Range, is south of the Lesser Zab, which flows northwest through the site towards Dukan Lake (IQ023). Here the river is fast flowing, moving through a dramatic, narrow gorge as the river leaves the Iranian border. The geology of the area consists of basic igneous rocks, radiolarian chert, siliceous and calcareous shale, and metamorphic schist and limestone of unknown age, and the soil type is sandy clay.

©2010 K ARARAT/NI

There are several villages (Isawe, Ashkana, Barozh, Priska and Gwaran to the north of the mountain and Kurkur, Kani Tu, Awazhe, Loter, Sofian and Bardashan to the south). Livestock grazing and small-scale agriculture are the main land management practices.

Key Biodiversity Area Criteria	Notes
V. Vulnerability Criteria: *Presence of Critically Endangered and Endangered species – presence of a single individual or Vulnerable species– 30 individuals or 10 pairs*	
Capra aegagrus	Wild goat was not observed but was reliably reported by locals.
Panthera pardus saxicolor	One Persian leopard was killed on Assos Mountain in January 2011, recorded on video, after reportedly attacking livestock. Also one was observed by locals previously in 2009.

Important Bird Area Criteria	Observations were made in 2010.	
A2. Restricted-range species	**Breeding**	**Wintering/ Passage**
See-see Partridge *Ammoperdix griseogularis*	80 pairs	
Eastern Rock Nuthatch *Sitta tephronota* (Resident)	140 pairs	
Menetries's Warbler *Sylvia mystacea* (Summer visitor)	30 pairs	
White -throated Robin *Irania gutturalis* (Summer visitor)	30 pairs	
Kurdistan Wheatear *Oenanthe xanthoprymna* (Summer visitor)	50 pairs	
Eastern Cinereous Bunting *Emberiza semenowi* (Summer visitor)	30 pairs	
Important Plant Area Criteria		
A4. Site contains national endemic, near endemic, regional endemic and/or regional range-restricted species or infra-specific taxa		

Hesperis kurdica var unguiculata (endemic) and *Quercus macranthera, Hesperis straussii, Rhus coriaria, Zeugandra iranica,* and *Juncus effuses* (nationally rare species)

B1. Site is a particularly species-rich example of defined habitat type

Mountain Forest Vegetation-Oak Forest-Medium Zone & Highest Zone habitat type and Mountain Forest Vegetation–Mountain Riverine Forest habitat type.

Additional Important Bird Observations: During the surveys, 37 bird species were observed. Four Mediterranean and one Eurasian High-Montane biome-restricted species were recorded but at levels that did not trigger inclusion under A3 criterion. The European Roller *Coracias garrulus* and Eastern Cinereous Bunting *Emberiza semenowi* (the latter mentioned above) are Near Threatened.

Other Important Fauna: According to local reports Wild Goat *Capra aegagrus* (Vulnerable) is present in small numbers on the mountain and appears to have decreased in number as a result of extensive hunting. During the interview survey, locals reported that Grey Wolves *Canis lupus* attacks their herds every year, and they also reported a sighting of Brown Bear *Ursus arctos* in 2009. No fish surveys were undertaken but electro-fishing was observed during a trip on Lesser Zab in 2011 and fish stocks in the river were reported as good.

Additional Plant & Habitat Information: This site contains a good population of P*istacia eurycarpa*, P. *khinjuk*, *Rheum ribes*, *Morus alba*, *Morus nigra*, and *Crataegus azarolus*, which are important both economically and for local heritage.

Conservation Issues: As the Lesser Zab River leaves its gorge and approaches Rania, the land opens up and gravel mining becomes a significant threat. In addition, a dam has been approved where the gorge ends, which will dramatically impact the biodiversity of the river and potentially the lower slopes of Assos Mountain. As this area attracts many hunters from nearby cities and towns, hunting pressure is a significant concern and electro-fishing appears to be common (three groups observed during one trip on the river were exclusively electro-fishing). It was reported that the forestry police will fine people for this and thus most people engaged in this activity do it within the gorge where access to the river is limited. Finally, there are also concerns with housing and road construction within the site.

Recommendations: If the dam project can be cancelled the natural beauty of the area and the river gorge lends this area well to eco-tourism development. The river, particularly in the gorge, is well suited to boating and a pilot program conducted with recently formed kayak and rafting companies in the region could be conducted to train local guides and encourage river tourism in the area. This would demonstrate the value of keeping this section of the river wild. Mining and resource extraction rules should be tightened and/or strengthened to limit the damaging effects of currently gravel mining on Iraqi rivers. Efforts need to be made to decrease the impacts of erosion and restore old mining sites to pre-mining, natural conditions. Forestry police require more access to and time on the river in order to discourage electro-fishing and overall need additional support to oversee this large area.

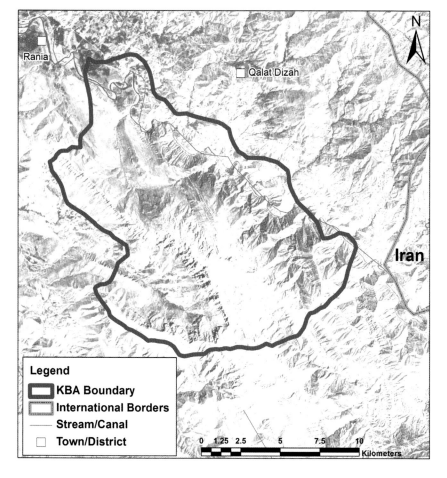

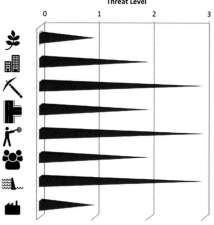

Taq Taq (IQ025)

Erbil – 35.892778°N 44.623333°E

IBA Criteria: A1 and A2
IPA Criteria: B1

Area: **1845 ha** - Altitude: **330-430 m**
Ecoregion: **Zagros Mountains Forest Steppe (PA0446)**
Status: **Unprotected**

©2012 N MUSA/NI

Site Description: This area is dominated by the Lesser Zab River, which has banks made of clay and silt and fluctuating speed and flow rates due to the opening and closing of the Dukan Dam upstream. The main part of the town of Taq Taq lies on the northern bank and the river on both sides is used for agriculture and grazing. There is also oil development in the area. The main habitat examined here is river riparian vegetation along the Lesser Zab River. The site is located in the Zagros Range and the geology is limestone and marls, with a soil type of sandy clay.

Important Bird Area Criteria	Observations made 2007-2009.	
A1. Globally threatened species	**Breeding**	**Wintering/ Passage**
Egyptian Vulture *Neophron percnopterus* (Summer visitor)	2 pairs (count 2008)	
A2. Restricted-range species	**Breeding**	**Wintering/ Passage**
Iraq Babbler *Turdoides altirostris* (Resident)	16 pairs (2007-2008)	
Important Plant Area Criteria		
B1. Site is a particularly species-rich example of defined habitat type		
Mountain Forest Vegetation – Riverine Forest habitat type		

Additional Important Bird Observations: During the surveys a total of 40 species were recorded. The site also held breeding populations of three Irano-Turanian, one Mediterranean, and three Sahara-Sindian Desert biome-restricted species, but this did not trigger inclusion under criterion A3. Eastern Cinereous Bunting *Emberiza semenowi* (Near Threatened) and Iraq Babbler *Turdoides altirostris* (Near Endemic) were breeding at the site.

Other Important Fauna: Mammal data were collected in 2007. Observations were made of Golden Jackal *Canis aureus* and tracks and signs of Wild Boar were seen in the riparian area (one dead female boar shot by locals as a pest species was found on an island in the river during the summer of 2007). One otter pelt was examined from a local hunter in the Taq Taq area and samples were sent to experts from King Khaled Wildlife Research Centre for DNA tests. The specimen was identified as Smooth-coated Otter *Lutrogale perspicillata*, a Vulnerable species, and the Taq Taq specimen indicates a range extension northward of more than 500 km for this species (Omer *et al.* 2012).

Fish: Data were collected in 2007 and 2008, during which 10 species were identified. According to Coad (2010) significant species were: *Alburnus mossulensis, Arabibarbus grypus* (Vulnerable), *Cyprinus carpio* (Vulnerable), *Cyprinion macrostomum, Leuciscus vorax, Luciobarbus esocinus* (Vulnerable), *L. xanthopterus* (Vulnerable), and *Silurus triostegus.*

Conservation Issues: Gravel mining is extensive along the whole river. It has affected the area and damaged instream and riparian habitats. The riparian area along the Lesser Zab River has also been heavily utilized for agriculture resulting in widespread habitat disturbance and possibly causing some pollution to the river. There is some moderate housing devlopment in the area.

Little is known about oil development impacts on the area, but they are likely to be extensive. In 2005, Addax Petroleum of Canada in a sharing agreement with Genel Enerji of Turkey and the Kurdistan Regional Government (KRG) entered this area for the purpose of oil production and in 2006 with agreement of the KRG expanded the geographic scope of the original agreement to include the Kewa Chirmila prospect. According to the information on their website, the license area covers approximately 951 km², which includes the Taq

Taq field and the Kewa Chirmila prospect. In 2007, the Taq Taq Operating Company "acquired 292 km² of 3D seismic covering the Taq Taq field and 218 km² of 2D seismic covering the remainder of the license area including the Kewa Chirmila prospect" (Addax Petroleum, 2012). Experts quoted in a Forbes article (Helmen, 2007) believe that Taq Taq holds in excess of 2.7 billion barrels of crude oil and could produce up to 200,000 bpd for up to ten years. The article also noted that at that time, due to lack of infrastructure, oil produced at the opening of the well had to be burned off.

Recommendations: Information on the nature and extent of oil development projects and the results of environmental impact assessments need to be made available to the general public for comment. Gravel mining needs better limitations and in-stream mining should be stopped altogether as it significantly impacts the ecological condition of the river. Many areas along the river are in urgent need of restoration due to years of this activity. It is also necessary to limit the agricultural practices (both land clearing and chemical use) by creating buffer zones along waterways.

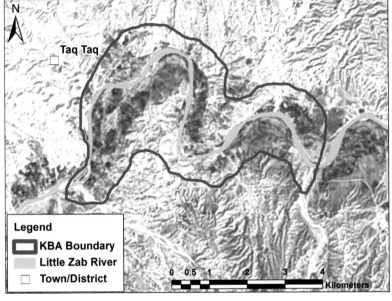

Legend
KBA Boundary
Little Zab River
Town/District

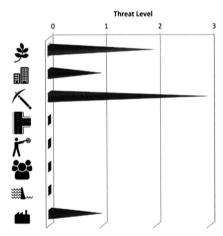

Chami Razan (IQ026)

Sulaimani – 35.809167°N 45.020556°E

IBA Criteria: **A1 and A3**
IPA Criteria: **A4 and B1**

Area: **4530 ha** - Altitude: **529-1228 m**
Ecoregion: **Zagros Mountains Forest Steppe (PA0446)**
Status: **Unprotected**

Site Description: The Tabin Stream flows from the east through a long valley into a narrow winding gorge, eventually joining the Lesser (Little) Zab River. It is an area of hills and rocky ridges with sparse mountain forest vegetation (primarily oak forests) and mountain riverine forest. The steep cliffs especially those with less exposure to the sunlight may provide a refuge area for plants. The site is located in the Zagros Range,

where the geology is a sedimentary limestone, and the soil is sandy clay.

Several villages are situated in and around the area. There is a spring from an intersecting valley that flows into the Tabin as well as a spring-fed waterfall (Bafil) that has been developed into a tourist site. Archaeological ruins and several important caves lie within the area, including Qizqapan cave (actually a

Midean (728 BC–549 BC) tomb carved in the cliff face) that is located west of the village of Zarzi; the remains of an Islamic-era fort further downstream in the gorge; and the Zarzi and Sara caves. According to Rosti (2011) traces of Neanderthals have been found in Zarzi and Sara, although the caves have not been fully explored. Some of these caves were also used by the Peshmerga in the 1980s and the area was chemically bombed during that period. The valley is used for grazing and growing vegetables, and there are a number of hoop-greenhouses upstream along the Tabin. The narrow gorge areas and table lands above are used primarily for grazing. Much of the area, especially near the caves, spring and streams, is popular for picnicking in spring.

Important Bird Area Criteria	Observations made 2007-2011.	
A1. Globally threatened species	**Breeding**	**Wintering/ Passage**
Egyptian Vulture *Neophron percnopterus* (Summer visitor)	2-10 pairs (counts 2007-2010)	
A3. Biome-restricted species		
Irano-Turanian biome	**Breeding**	**Wintering/ Passage**
See-see Partridge *Ammoperdix griseogularis* (Resident)	20 pairs (2008, 2010)	
Menetries's Warbler *Sylvia mystacea* (Summer visitor)	60 pairs (2009-2010)	
Eastern Rock Nuthatch *Sitta tephronota* (Resident)	80 pairs (2008-2010)	
White-throated Robin *Irania gutturalis* (Summer visitor)	3 pairs (counts, 2009)	
Finsch's Wheatear *Oenanthe finschii* (Resident)	10 pairs (2008-2010)	
Eastern Cinereous Bunting *Emberiza semenowi* (Summer visitor)	1-2 pairs (counts 2009)	
Mediterranean biome	**Breeding**	**Wintering/ Passage**
Masked Shrike *Lanius nubicus* (Summer visitor)	3 pairs (counts 2008-2010)	
Sombre Tit *Poecile lugubris* (Resident)	20 pairs (2008-2010)	
Western Rock Nuthatch *Sitta neumayer* (Resident)	80 pairs (2008-2010)	
Eastern Black-eared Wheatear *Oenanthe melanoleuca* (Summer visitor)	40 pairs (2008-2010)	
Black-headed Bunting *Emberiza melanocephala* (Summer visitor)	150 pairs (2008-2010)	
Important Plant Area Criteria		
A4. Site contains national endemic, near endemic, regional endemic and/or regional range-restricted species or infraspecific taxa		
Onosma albo-roseum var. *macrocalycinum*		
B1. Site is a particularly species-rich example of defined habitat type		
Mountain Forest Vegetation-Oak Forest-lowest and medium zones habitat type and Mountain Forest-Mountain Riverine Forest habitat type		

Additional Important Bird Observations: During the surveys 65 species were seen. The following Near Threatened species were also observed: Pallid Harrier *Circus macrourus* (passage), European Roller *Coracias garrulus* (breeding) and Semi-collared Flycatcher *Ficedula semitorquata* (passage). The Sahara-Sindian Desert biome-restricted White-eared Bulbul *Pycnonotus leucotis* was breeding. Note that Eastern Cinereous Bunting *Emberiza semenowi* mentioned in the table above is a Near Threatened species.

Other Important Fauna: Spur-thighed Tortoise Testudo graeca are often seen here in pairs, and could likely meet KBA Vulnerability criteria with more study. Fish exist in the Tabin Stream but have not been assessed. In the winter of 2010, people were observed electro-fishing in the stream.

Additional Plant & Habitat Information: This site contains a good population of pistachios *Pistacia eurycarpa* and *P. khinjuk*, which are economically and culturally important.

Conservation Issues: In 2012, construction began on the Surqawshan Dam downstream on the Tabin just upstream of its confluence with the Lesser Zab River (the western part of the delineated site). The lower riparian zone on the stream has been severely damaged from gravel mining operations created for the dam construction (this appears to be complete now and the area is now being planned for tourism development). These activities led to the release of high levels of sediment into the Lesser Zab River and significant erosion is likely to continue. The reservoir that will fill the gorge upstream of the Surqawshan Dam will submerge much of the lower valley in the delineated area causing extensive ecological change and biodiversity loss. There may also be possible flooding of the ancient Islamic fort (though this is not clear) and destabilization of slopes within the narrow gorge. There are also no plans for fish passages included in the dam.

Tourism use of the area is heavily impacting Chami Razan, particularly during the Nawruz holiday in spring. The tourism is also causing the accumulation of garbage and other wastes left behind by picnickers and along with tourism development, hunting and animal grazing, was considered a high threat. The electro-fishing observed in the Tabin is likely an infrequent practice due to the streams relatively small size. In 2010, development around the Qizqapan Cave also caused localized erosion.

Recommendations: The areas popular for tourist needs to be cleared of garbage and regular garbage collection

implemented during the peak periods for visits (spring). The site is close to the city of Sulaimani and draws people because of its ancient and modern history as well as its natural beauty. An education program aimed at raising the awareness of visitors during Nawruz would aid in improving the site and changing behaviors. There is no indication that an environmental impact assessment was conducted for the Surqawshan Dam despite the fact that his project has clear and negative impacts. In this case the resulting reservoir is to be used for irrigation and recreation, though the reservoir will flood the agricultural land behind the dam and the recreational value of the existing valley gorge could have been equally developed without the dam.

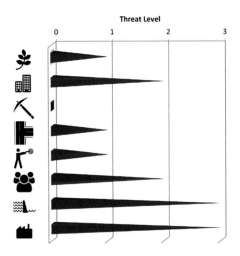

Legend
- KBA Boundary
- Stream/Canal
- Loose /Unpaved Road
- Town/District

Peramagroon Mountain (IQ027)

Sulaimani–35.76°N 45.241389°E

KBA Criteria: **V**
IBA Criteria: **A1 and A3**
IPA Criteria: **A4, B1and B2b**

Area: **16738 ha** - Altitude: **790-2613 m**
Ecoregion: **Zagros Mountains Forest Steppe (PA0446)**
Status: **Unprotected**

Site Description: This is the highest peak in Sulaimani Governorate, reaching 2613 m. The preliminary delineated area includes the mountain ridge and most of the Mergapan valley to the northeast, encompassing the valleys near the villages of Homer Qawm and Shadala and a small gorge between the villages of Shadala and Kani Shuk. Peramagroon Valley (which includes the small village of Zewe) lies within a wide bowl on its southeastern side below the highest peak on the mountain. Major settlements around the mountain include the town of Peramagroon and the villages of Kani Shuk, Sardasht, and Shadala.

An area above Zewe, from which a seasonal spring issues, is a popular picnicking site as are many areas along Chami Mergapan stream and the springs above Kani Shuk. This stream runs to the northwest through the Mergapan valley and bisects Peramagroon at its northern extent in a dramatic gorge; it becomes the Tabin stream that flows through the Chami Razan KBA site (IQ026) and eventually joins the Lesser Zab River.

Peramagroon Mountain is a botanically rich site and contains mountain riverine forests, oak woodlands and thorn-cushion vegetation. There are also extensive cliff habitats. The site is part of the Zagros Range, where the geology consists of sedimentary limestone and the soil is sandy clay. There is agriculture on the lower slopes (primarily grapes) and grazing throughout the area, though this is now rare on the higher slopes. Most of the mountain is free of mines but mines are reported on the top and in some valleys at the northern extent of the mountain ridge. Zewe and other villages were subject to chemical attacks in the past. Two sub-sites were surveyed at Peramagroon.

©2010 N ABDULHASAN/NI

Key Biodiversity Area Criteria	Notes	
V. Vulnerability Criteria: *Presence of Critically Endangeredand Endangeredspecies – presence of a single individual or Vulnerable species– 30 individuals or 10 pairs.*		
Capra aegagrus	Wild Goats *Capra aegagrus* were observed in the summer of 2011 and winter 2012.	
Testudo graeca	Spur-thighed Tortoise *Testudo graeca* has been observed and photographed.	
Important Bird Area Criteria	**Observations made 2007-2010.**	
A1. Globally threatened species	**Breeding**	**Wintering/ Passage**
Egyptian Vulture *Neophron percnopterus*	3-6 pairs (counts)	
A3. Biome-restricted species		
Irano-Turanian biome	**Breeding**	**Wintering/ Passage**
See-see Partridge *Ammoperdix griseogularis* (Resident)	150 pairs	
Plain Leaf Warbler *Phylloscopus neglectus* (Summer visitor)	8 pairs (counts 2011, 2012)	

Upcher's Warbler *Hippolais languida* (Summer visitor)	45 pairs (2011, 2012)	
Menetries's Warbler *Sylvia mystacea* (Summer visitor)	20 pairs (2008-2010)	
Eastern Rock Nuthatch *Sitta tephronota* (Resident)	120 pairs	
White-throated Robin *Irania gutturalis* (Resident)	80 pairs (2009-2010)	
Kurdistan Wheatear *Oenanthe xanthoprymna* (Resident)	110 pairs	
Finsch's Wheatear *Oenanthe finschii* (Resident)	170 pairs	
Grey-necked Bunting *Emberiza buchanani* (Summer visitor)	22 pairs (2012)	
Eastern Cinereous Bunting *Emberiza semenowi* (Summer visitor)	65 pairs (2009-2010, 2012)	
Pale Rock Sparrow *Carpospiza brachydactyla* (Summer visitor)	18 pairs (2012)	
Mediterranean biome	**Breeding**	**Wintering/ Passage**
Masked Shrike *Lanius nubicus* (Summer visitor)	30 pairs (2009-2010)	
Sombre Tit *Poecile lugubris* (Resident)	40 pairs (2008-2010)	
Western Rock Nuthatch *Sitta neumayer* (Resident)	200 pairs	
Eastern Black-eared Wheatear *Oenanthe melanoleuca* (Summer visitor)	150 pairs	
Black-headed Bunting *Emberiza melanocephala* (Summer visitor)	400 pairs (2008-2010)	
Eurasian High-Montane biome	**Breeding**	**Wintering/ Passage**
Yellow-billed Chough *Pyrrhocorax graculus* (Resident)	10 pairs (2010)	
White-winged Snowfinch *Montifringilla nivalis* (Resident)	30 pairs (2011)	
Red-fronted Serin *Serinus pusillus* (Resident)	15 pairs (2011)	

Important Plant Area Criteria

A4. Site contains national endemic, near endemic, regional endemic and/or regional range restricted species or infraspecific taxa.
Note: *historically recorded; **historically recorded and seen on recent surveys

Endemics at this site: *Acantholimon petraeum, *Anthemis micrantha, *Astragalus lobophorus var. pilosus, *Astracantha peristerea, *Astracantha zoharyi , *Camelinopsis kurdica, *Centaurea gudrunensis, Cousinia inflate, *C. macrolepis, *C. masu-shirinensis, C. odontolepis, *Fritillaria crassifolia subsp. poluninii, Himantoglossum hircinum, *Leutea rechingeri, *Pimpinella hadacii, **Scrophularia kurdica subsp. kurdica, and Stachys kotschyii.
Near endemics at this site: *Acantholimon blackelockii, Acanthophyllum kurdicum, *Asperula insignis, *Astracantha crenophila, *Bunium cornigerum, *Campanula acutiloba, *Cousinia haussknechtii, *C. straussii, Dianthus bassianicus, *Echinops parviflorus, Erysimum kurdicum, *Stachys kotschyii, *Trigonosciadium viscidulum, *Verbascum alceoides, *Veronica macrostachya var. schizostegia, and *Ziziphora clinopodioides subsp. Kurdica.
Nationally rare species here were: Aristolochia paecilantha, Cousinia odontolepis, Lactuca hispida, Phelypaea coccinea, and Quercus macranthera

B1. Site is a particularly species-rich example of defined habitat type

Mountain Forest Vegetation-Oak Forest-Medium Zone and Highest Zone habitat type and Mountain Forest Vegetation-Thorn-Cushion Vegetation habitat type

B2b. The site is a refuge for: biogeographically and bioclimatically restricted plants to 'retreat to' in the face of global climate change.

This site represents a good example of the Thorn-Cushion Vegetation habitat type and the top of mountain can provide a refuge for plants associated with this habitat. Also some gorges/cliffs in the mountain can be refuge for the Oak forests and associated plants in the case of climate changes.

Additional Important Bird Observations: During the surveys 131 species were observed. Also the Endangered Saker Falcon *Falco cherrug* was observed in the winter of 2012. Also the following Near Threatened species were recorded but in sub-IBA threshold numbers: European Roller *Coracias garrulus* (breeding) and Semi-collared Flycatcher *Ficedula semitorquata* (passage); the site also held breeding populations of two Sahara-Sindian biome-restricted species but this did not trigger inclusion under criterion A3. Eastern Cinereous Bunting *Emberiza semenowi* listed in the table above is Near Threatened.

Other Important Fauna: Mammal surveys conducted in 2009 and 2010 found considerable number of Persian Squirrels *Sciurus anomalus*. This is a Least Concern species, but their population trend is decreasing and a Nature Iraq interview survey in 2010 found that a large number were hunted at this site and transported to other parts of the country for sale as pets. A summer and winter survey in 2011 documented the presense of Wild Goats on Peramagroon. According to local reports, Persian Leopards *Panthera pardus saxicolor* (Endangered) occurred historically but are thought to no longer be present. One significant reptiles observation was the Urmia Rock Lizard *Apathya cappadocia urmiana*.

Fish: Fish exist in Chami Mergapan but have not been assessed. Portions of this stream currently go dry in summer and historical information would be needed to determine if this has always been the case or if development in the area has led to decreased water resources overall.

Additional Plant & Habitat Information: Peramagroon contains a good population of pistachios, *Pistacia eurycarpa* and *P. khinjuk*, which are economically and culturally important. *Gundelia tournefortii* and *Rumex ribes*, which are economically important as a traditional food, are found at the site as well as a good population of *Aegilops crassa*, and *A. umbellulata, Bromus diandrus, Hordeum bulbosum, Poa bulbosa,* and *Vitis vinifera*, which are important as genetic resources.

Conservation Issues: As expected, areas of high human use, such as the valleys and lower slopes, are more impacted, but evidence of human impacts can also be seen at higher elevations, including garbage in camping areas on the peak. A major road was completed in 2010 connecting the city of Sulaimani to the Mergapan valley. This has also increased land clearance and residential, commercial and tourism development throughout the valley, which are signifigant threats to Peramagroon. Though urban development has reached areas around the mountain urban planning lags far behind with no visible management or control of solid and liquid wastes and other habitat impacts related to these developments. For example, high threats from human disturbance and transport corridors have also resulted. More

recently another road was being built up to the ridge top on the northwest side of the mountain not far from Kani Shuk. The purpose of this road is not clear and it is uncertain if it will be completed. As with other tourist destinations in Iraq, a great deal of garbage is left in the area because there appear to be few if any waste management facilities offered to the public. Most solid waste is dumped off convenient hillsides and later burned (including plastics and other hazardous materials). While there is no energy production or mining at Peramagroon currently, oil surveys have been conducted within and around the KBA site, so oil exploration and development may affect the site in the future.

Large mammals such as Wild Goat appear to have low populations due to hunting pressure. Expansion and intensification of agiculture have some significant impact. There are a high number of farms in the area and the use of agricultural chemicals such as pesticides has been seen in the fields, many of which are used for grape cultivation. Sediment run-off and direct livestock access to Chami Mergapan are likely affecting water quality. Grazing throughout the site should also receive closer scrutiny and control. While the environmental conditions improve with altitude, the area around the peak of Peramagroon is affected by the collection of Rhume *Rheum ribes*, an edible plant that grows naturally in the area.

Recommendations: Peramagroon Mountain is a globally important site for biodiversity and it should be given high priority for protection. The site offers opportunities for ecotourism, recreation and tourism if carefully planned and controlled. The Mergapan valley is still rapidly developing and this requires swift action to address urban and environmental planning needs. Road building, particularly to the top of the ridge, should be discouraged as it will increase negative impacts to the mountain. Overall management planning is needed to put in place controls on development, waste management, animal grazing, hunting, agricultural chemical use and tourism. Clear short & long-term zoning should be done to determine where and what kind of development is allowed to occur with a focus on ensuring sustainability. Land mine clearance is also an issue to address in parts of the site.

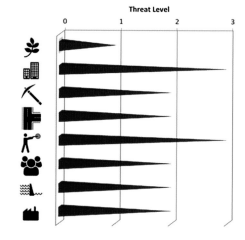

In 2013, a three-year conservation project funded by the Darwin Initaitive in the UK was started at this site. It includes extensive botanical surveys, development of modern conservation tools (photographic field guides that can be accessed through mobile phone applications), university and community environmental education programs, and socio-economic surveys of the area. It is recommended that this serve as a pilot project to encourage conservation planning at this and other sites in the region.

Sargalu (IQ028)

Sulaimani – 35.875278°N 45.165278°E

IBA Criteria: **A1**
IPA Criteria: **A4 and B1**

Area: **5235 ha** - Altitude: **936-1820 m**
Ecoregion: **Zagros Mountains Forest Steppe (PA0446)**
Status: **Unprotected**

©2010 K ARARAT/NI

Site Description: This is a mountainous area characterized by rocky ridges and valleys located just to the north of Peramagroon Mountain (IQ027). The main habitat types were oak woodlands and mountain riverine forest. Daban Mountain sits to the south of the area, Halaj Mountain to the north and the villages of Sargalu and Bargalu lie in the valley between them. Many small springs originating in the mountains join to form the stream that flows through Sargalu. The area is used for agriculture (primarily vineyards) and draws many visitors in the spring and summer months. Poison chemical attacks occurred here in the 1980s as an important Peshmerga base of operations was located in Sargalu. There is heavy grazing in the area and extensive road construction is in progress, especially access roads built by farmers in the foothills. Construction, particularly of summer houses was also observed. The area contains some culturally important caves.

Important Bird Area Criteria	Observations made 2007-2011.	
A1. Globally threatened species	**Breeding**	**Wintering/ Passage**
Egyptian Vulture *Neophron percnopterus* (Summer visitor)	1-3 pairs (counts 2007 & 2010-2011)	
Important Plant Area Criteria		
A4. Site contains national endemic, near endemic, regional endemic and/or regional range-restricted species or infraspecific taxa		
One endemic species was *Bellavalia kurdistanica*		
B1. Site is a particularly species-rich example of defined habitat type		
Mountain Forest-Mountain Riverine Forest ; Mountain Forest Vegetation-Oak Forest- Medium and High Zones and Mountain Forest Vegetation-Thorn- Cushion Vegetation habitat type		

Additional Important Bird Observations: During the surveys a total of 68 species were recorded. The site also held breeding populations of four Mediterranean, four Irano-Turanian and one Sahara-Sindian Desert biome-restricted species, but these did not trigger inclusion under criterion A3. European Roller *Coracias garrulus* and Eastern Cinereous Bunting *Emberiza semenowi* (both Near Threatened) were also breeding.

Other Important Fauna: The only non-avian observations were for fish, which were surveyed in 2007 during which four species were reported. According to Coad's (2010) criteria, significant species were: *Alburnus mossulensis, Cyprinion macrostomum,* and *Garra rufa.*

Additional Plant & Habitat Information: This site contains a good population of *Crataegus azarolus* and *Pistacia eurycarpa,* which are economically and culturally important and *Aegilops triuncialis, Avena fatua, Poa* sp. and *Vitis vinifera,* which are important as genetic resources.

Conservation Issues: Residential and transportation development continue to expand in Sargalu, and the site is a popular tourist area, all of which were assessed as very high threats. Since the site was subject to chemical weapons attack in the 1980s overall human intrusion and war impacts have been

great. There is a recent oil block dedicated for future drilling in the area but nothing further is known and no activities were seen during the survey so this was rated a medium threat but could have a larger impact in the future. Pollution (noise pollution, garbage from visiting tourists, as well as sewage and garbage from the villages) has a high impact on the site as did hunting. Grazing is impacting the site as well. Agricultural overall was considered a medium threat to the site.

Recommendations: Although many of the materials used in the chemical bombings of the 1980s have dissipated long ago, there is still a need for further investigation of their long-term effects on humans and the environment. But the main issues currently facing Sargalu have to do with tourism and residential housing construction as well as road building, with hunting, grazing and waste management being important as well. Recreational use of the site would benefit from having more clearly defined picnic sites established that have adequate waste collection services. Waste management systems are also needed for the village as well. Road construction needs to be better planned and controls to avoid habitat destruction. To prevent over-grazing, controls on the extent of the area utilized and/or the timing when livestock are grazed should be considered but this will necessitate local stakeholder discussions.

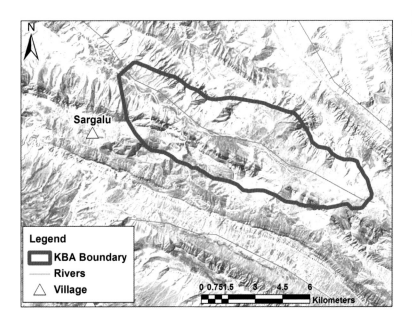

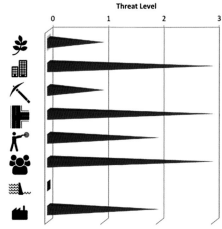

Mawat Area (IQ029)

Sulaimani – 35.93917°N 45.3825°E

KBA Criteria: **V**
IPA Criteria: **B1**

Area: **2699 ha** - Altitude: **680-1343 m**
Ecoregion: **Zagros Mountains Forest Steppe (PA0446)**
Status: **Unprotected**

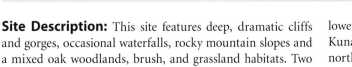

Site Description: This site features deep, dramatic cliffs and gorges, occasional waterfalls, rocky mountain slopes and a mixed oak woodlands, brush, and grassland habitats. Two sub-sites were surveyed in Mawat. An important river, the

lower Chami Chwarta, (also known by other names such as Kuna Masi River, Chami Qellachiwalan or Nahr Siwayl) flows north through a deep, narrow gorge west of the town of Mawat, eventually joining the Lesser Zab River at an area called Du

Choman (Two Rivers), where the Lesser Zab forms the border with Iran. The preliminary delineation includes just Chami Chwarta gorge but future surveys may indicate that refinement to this boundary should extending northward to include the Lesser Zab Valley. During the one botanical survey to one of the sub-sites (sub-site S36) 55 plant species were identified within the Mountain Forest Vegetation-Mountain Riverine Forest habitat, which is generally rich and relatively protected in the gorge. The common trees were *Salix* spp., *Platanus orientalis* and *Ficus carica* and the common shrub was *Rubus sanctus*.

The area was heavily mined (though some mine clearance has taken place) and saw periods of fighting during the Iran-Iraq war. There are agricultural fields and orchards but the economy is largely dependent on smuggling fuel and other items back and forth across the border, which has had damaging effects upon the Lesser Zab River. The area contains many small villages and the larger town of Mawat. Increased road construction was observed in the year of 2012 and rural electrification took place during the final years of the survey.

©2009 A BACHMANN/NI

Key Biodiversity Area Criteria	Notes
V. Vulnerability Criteria: *Presence of Critically Endangered and Endangered species – presence of a single individual or Vulnerable species – 30 individuals or 10 pairs.*	
Neurergus derjugini	Observed and photographed in 2007 and 2012.
Important Plant Area Criteria	
B1. Site is a particularly species-rich example of defined habitat type	
Mountain Forest- Mountain Riverine Forest	

Additional Important Bird Observations: During the surveys, a total of 33 species were recorded, including European Roller *Coracias garrulus* (Near Threatened). Though this site did not meet IBA criteria this was likely due to limited bird surveys in this area.

Other Important Fauna: Data were not collected for other fauna except the Critically Endangered Kurdistan Newt *Neurergus derjugini*, which was photographed at the site in 2007. During cooperative work between Nature Iraq and a PhD student, Elnaz Najafi Majd, in May of 2012 this species was found again (as egg masses and in the larval-stage).

Conservation Issues: Large areas were heavily mined in the Iran/Iraq War in the 1980s and mine clearance activities are conducted in this region. Significant threats to the site are from smuggling and continued border insecurity. The degree and level of smuggling as well as the items being smuggled is largely dependent on price differentials between Iraq and Iran. For a period of time fuel smuggling was causing extensive damage to the river and water quality but this issue fluctuates when pricing changes. Smuggling causes disturbance to

riparian areas, as well as accumulation of garbage and other pollutants along the banks and in the Lesser Zab River (smuggling has little impact on the lower Chami Chwarta but some garbage and wastes from upstream towns and villages collects here as well). For these reasons human intrusion is a very high threat and pollution a high threat. Hunting is also carried out in the area and is considered a high threat as is the widening and paving of existing roads, which has caused erosion due to poor methods. Though road construction was originally considered a medium threat, more extensive work on roads occurred in the area in 2012 and during the survey for the Kurdistan Newt, it was found that road construction had damaged some of the habitat for this species. Agriculture and grazing were considered a moderate threat. There is a forestry police station in the town of Mawat but there are not enough staff and resources to manage the large scope of territory for which they are responsible.

Recommendations: Efforts to stop smuggling outright will likely cause economic hardship to many residents involved in the activity and will, regardless, require the normalizing of relations on the border with Iran. Opening an official border crossing could possibly lead to legalization and control of the border trade and decrease border tensions. The entire region offers stunning opportunities for sustainable development through ecotourism initiatives, for example smugglers could be turned into local guides using their horses and knowledge

of the area. Continued mine clearance will remain a priority here and hunting, agricultural and grazing activities should be investigated to determine the best management practices. More resources, including staff and training, are clearly needed for the local Forestry Police. The Mawat Area is also close to several other KBA sites including the Kuradawe & Waraz (IQ030) and Sharbazher area (IQ031) on the east and Assos Mountain (IQ024) to the northwest so integrated planning to protect all of these sites is recommended.

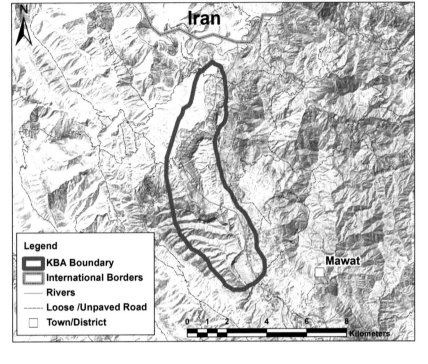

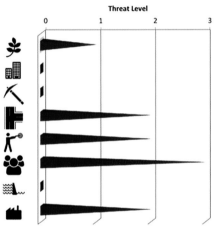

Kuradawe & Waraz (IQ030)

**Sulaimani – 35.840833°N
45.499444°E**

IBA Criteria: **A1**
IPA Criteria: **B1**

Area: **7272 ha** - Altitude: **894-2173 m**
Ecoregion: **Zagros Mountains Forest Steppe (PA0446)**
Status: **Unprotected**

Site Description: Waraz is a valley surrounded by mountains and villages, including Kanarwe, Ballikhy, and Betwate. To the south the Kanarwe River flows from the Penjween Area (including waters from IQ032) towards Kuna Masi River (also called Chami Chwarta River), a tributary of the Lesser Zab. The main habitat types were oak woodlands and mountain riverine forest. The site is located in the Zagros Range, where the geology is basic igneous rock, radiolarian chert, siliceous and calcareous shale, and metamorphic schist and limestone of unknown age. The soil type is often serpentine and in fact this site is one of the most important serpentine areas in the Kurdistan region. *Platanus orientalis* and walnut trees dominate most of the valley but agricultural fields cover parts as well and there is also gravel mining that affects the area in and around the rivers. The area is also popular for picnicking.

Further northwest is the forested mountainous area known as Kuradawe (named for the largest village in the area). Kuna Masi stream flows north to the west of the site and there area other streams that flow to join it through the area. This part of the site is surrounded by the villages of Amaden, Dashty Tile and Pirahmad, Bewre, Gomle, Basne, and Saraw and extends less than 1 km north of Kuradawe Village. There are farmlands, orchards, and vineyards, which mainly produce grapes, walnuts, pears and almonds. Two sub-sites visited here.

Additional Important Bird Observations: A total of 40 species was seen. Breeding populations of four Irano-Turanian, three Mediterranean, one Eurasian High-Montane and one Sahara-Sindian Desert biome-restricted species were found but did not trigger inclusion under criterion A3. No other significant observation were made for non-avian fauna.

©2009 S. ABDULRAHMAN/NI

Important Bird Area Criteria	Observations made in 2009.	
A1. Globally threatened species	**Breeding**	**Wintering/ Passage**
Egyptian Vulture *Neophron percnopterus* (Summer visitor)	1 pair	
Important Plant Area Criteria		
B1. Site is a particularly species-rich example of defined habitat type		
Mountain Forest-Mountain Riverine Forest habitat type and Mountain Forest Vegetation-Oak Forest-Lowest and Medium Sub-zones habitat type.		

Conservation Issues: Livestock production and grazing as well as farming and human intrusion related to picnicking activities particularly near Kuradawe village were considered very high threats at this site. Garbage dumped, mostly related to the latter activity, in and around the streams is a moderate threat. Hunting pressure is high in the area. Gravel mines are impacting the stream and the land around it but are somewhat limited in scope. Road construction and residential development are under way (especially close to the villages), which were judged as medium threats but may become more urgent if not controlled properly.

Recommendations: As with other sites, this area will benefit from increased and improved natural resource management planning. The

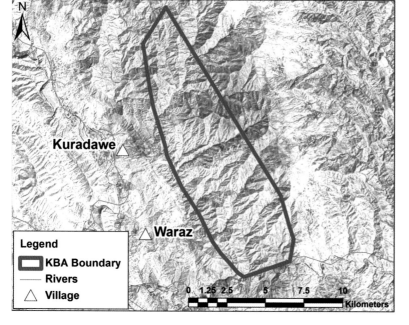

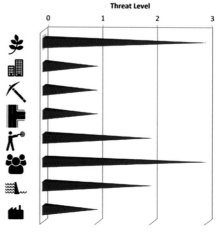

focus should be on agricultural practices, but hunting, general development, recreational uses, and sanitary services should not be overlooked. This site is close to several other KBA sites including the Mawat Area (IQ029) and Sharbazher area (IQ031) so integrated planning to protect all of these sites is recommended.

Sharbazher Area (IQ031)

Sulaimani - 35.950281°N 45.563611°E

IBA Criteria: **A3**
IPA Criteria: **A4 and B1**

Area: **2322 ha** - Altitude: **931-12,325 m**
Ecoregion: **Zagros Mountains Forest Steppe (PA0446)**
Status: **Unprotected**

Site Description: This is a mountainous area near the border with Iran where the main habitat types were oak woodlands and mountain riverine forest. The geology here is metamorphic schist and limestone of unknown age and the soil types are sand and sandy clay. Gmo Mountain, at over 2300 meters, is the largest peak in this site located on the western edge of the site. Farmland with field crops and orchards (grape, walnut, almond, and pear) are also present. Villages located inside the delineated area are Derey Saru and Derey Khwaru. Tributaries of the Lesser Zab River pass through the site and the Lesser Zab itself (upstream of the town of Mawat) is located north-northeast of the delineated area. Dirt access roads to farms are being constructed and roads to popular areas were being upgraded during the survey period. Two sub-sites were surveyed in this area.

Important Bird Area Criteria	Observations made 2010.	
A3. Biome-restricted species		
Irano-Turanian biome	**Breeding**	**Wintering/ Passage**
See-see Partridge *Ammoperdix griseogularis* (Resident)	12 pairs	
Menetries's Warbler *Sylvia mystacea* (Summer visitor)	4 pairs	
Eastern Rock Nuthatch *Sitta tephronota* (Resident)	15 pairs	
Finsch's Wheatear *Oenanthe finschii* (Resident)	25 pairs	
White-throated Robin *Irania gutturalis* (Summer visitor)	4 pairs	
Important Plant Area Criteria		
A4. Site contains national endemic, near endemic, regional endemic and/or regional range-restricted species or infraspecific taxa		
Two nationally rare species found at the site were: *Equisetum arvense* and *Juncus effuses*		
B1. Site is a particularly species-rich example of defined habitat type		
Mountain Forest-Mountain Riverine Forest habitat type; Mountain Forest Vegetation-Oak Forests (Medium Sub-Zone) habitat type and Mountain Forest – Thorn Cushion (subalpine) Vegetation habitat type		

Additional Important Bird Observations: During the surveys a total of 50 species were recorded. Three Mediterranean biome-restricted breeding species, as well as European Roller *Coracias garrulus* (Near Threatened) were recorded in the breeding season.

Other Important Fauna: Few other observations were made at the site, but during the 2008 survey locals reported wolves attacking a herd of sheep. Other carnivores such as the Golden Jackal *Canis aureus* and Red Fox *Vulpes vulpes* have been seen, which may indicate a healthy diversity of prey species at the site. There are streams and rivers in the area but no fish survey was conducted.

Additional Plant & Habitat Information: *Pistacia eurycarpa* and *P. khinjuk* are economically and culturally important. Also a good population of *Rheum ribes* is present which is economically important. The site also contains a good population of *Hordeum bulbosum*, *H. vulgare* ssp. *spontaneium*, and *Vitis vinifera*, which are important as genetic resources. Additionally, this site contains a good population of *Anchusa italica*, which is important as a traditional food.

Conservation Issues: Gravel mining is a significant threat in Sharbazher, which occur along streams with both riparian and in-stream habitats are often heavily impacted. This site appears to be is less impacted by overgrazing than most sites visited, but agriculture, in particular orchards and vineyards, may pose potentially high threats. Sharbazher still contains large minefields though some clearance operations are ongoing there. Regardless, the area, particularly around the streams, is popularly used for picnicking. Visitors from Sulaimani and other locations leave garbage behind and cause pollution to the land and water resources. Also extensive road works and new road construction allowing greater ease of access to both farms and picnicking areas is also impacting the area.

Recommendations: Landmine clearance is a priority for this area especially in areas used by locals and those suitable for further ecological studies. Road construction taking place, particularly on steep hillsides, should always use techniques that minimize erosion. Here as elsewhere gravel mining needs stricter controls and restriction. Habitat protection and restoration of both riparian and in-stream habitats should be the primary goal of such actions. This site is near two other KBA sites (Kuradawe & Waraz (IQ030) and Mawat (IQ031)) so integrated planning that addresses the many threats to all of these sites is recommended.

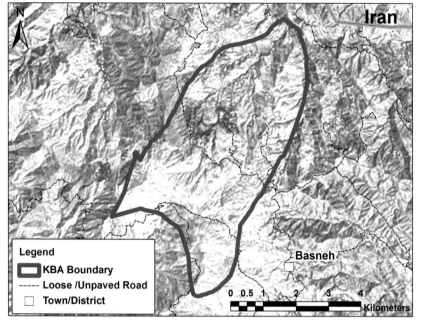

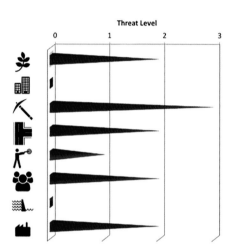

©2008 K ARARAT/NI

Penjween (IQ032)

Sulaimani – 35.755556°N 45.943889°E

IBA Criteria: **A1**
IPA Criteria: **A4 and B1**

Area: **4035 ha** - Altitude: **1178 - 1480 m**
Ecoregion: **Zagros Mountains Forest Steppe (PA0446)**
Status: **Unprotected**

Site Description: This is an open area that includes highlands and foothills not far from the town of Penjween. A small river (Chami Mishiaw) flows through the site and eventually joins with other rivers and streams to reach the Lesser Zab River south of the town of Mawat. Land is used primarily for farming and there are some orchards along the stream. The habitat investigated here is mountain riverine forest. The site is located in the Zagros mountain range, where the geology is limestone and soil type is sandy clay.

Important Bird Area Criteria	Observations made 2007-2009.	
A1. Globally threatened species	**Breeding**	**Wintering/ Passage**
Egyptian Vulture *Neophron percnopterus* (Summer visitor)	2 pairs (counts 2007 & 2009)	
Important Plant Area Criteria		
A4. Site contains national endemic, near endemic, regional endemic and/or regional range-restricted species or infraspecific taxa		
Three nationally rare species *Juncus effusus, Ranunculus sphaerospermus* and *Tamarix brachystachys* were found in the area.		
B1. Site is a particularly species-rich example of defined habitat type		
Mountain Forest-Mountain Riverine Forest habitat type		

Additional Important Bird Observations: During the survey period 79 bird species were seen. European Roller *Coracias garrulus* (Near Threatened) and Iraq Babbler *Turdoides altirostris* (Sahara-Sindian Desert biome-restricted and restricted-range species) were found at the site but did not meet IBA criteria levels. In addition the site held breeding populations of two Irano-Turanian and two Mediterranean biome-restricted species, but these did not trigger inclusion under criterion A3.

Other Important Fauna: Little data for other fauna was collected but Nature Iraq assisted in the publishing of an account of a Syrian Brown Bear *(Ursus arctos syriacus)* the smallest of the Brown Bear subspecies, that had been killed by local hunters in the Penjween Area in 2011 (Garshelis & McLellan, 2011). A small population of this subspecies exists in northern Iraq but may be nationally threatened due to hunting. No significant reptiles and amphibians were observed. In 2012, near this area, the Critically Endangered Kurdistan Newt *Neurergus derjugini* was found (Elnaz Najafi Majd, pers. Communication 2013).

Fish: Data were collected in 2007, 2009 and a separate 2012 survey done by Dr. Jörg Freyhof from the Leibniz Institute of Freshwater Ecology and Inland Fisheries. A total of 11 species were observed. Important species were: *Alburnus mossulensis, Garra rufa, Hypophthalmichthys molitrix, Oxynoemacheilus* sp., *Squalius cephalus*. Also species that were found by Dr. Freyhof and not mentioned in Coad were: *Oxynoemacheilus cf. bergianus, Oxynoemacheilus frenatus,* and *Turcinoemacheilus kosswigi*. Some of these are undescribed species that are very likely to be endemic to the drainage basin.

©2010 S ABDULRAHMAN/NI

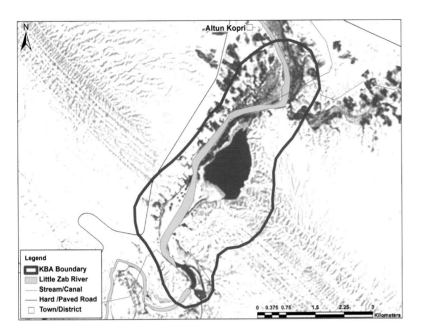

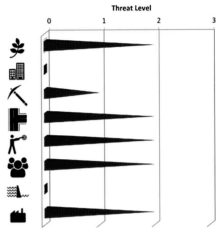

Parazan (IQ034)

Sulaimani – 35.626944°N 45.738611°E

IBA Criteria: **A1**
IPA Criteria: **A4 and B1**

Area: **3617 ha** - Altitude: **966-1480 m**
Ecoregion: **Zagros Mountains Forest Steppe (PA0446)**
Status: **Unprotected**

©2010 H RAZA/NI

Site Description: The main habitat here is open oak woodlands with many large oak trees. The geology of the area is sedimentary and the soil type is sandy clay. Kura Kazhaw Mountain, where there is a spring, is located in the south. A small stream flows through Nalparez District. The villages of Haji Mamand, Gweze Resh, Dolle Pemu, Deremyaneh and Sarkhwar surround the site. The area is a popular picnic site despite many minefields on the surrounding mountains.

Important Bird Area Criteria	Observations made 2007-2010.	
A1. Globally threatened species	Breeding	Wintering/ Passage
Egyptian Vulture *Neophron percnopterus* (Summer visitor)	1-2 pairs (counts)	
Important Plant Area Criteria		
A4. Site contains national endemic, near endemic, regional endemic and/or regional range-restricted species or infraspecific taxa		
Three nationally rare species were found: *Juncus effusus*, *Ranunculus sphaerospermus*, and *Tamarix brachystachys*		
B1. Site is a particularly species-rich example of defined habitat type		
Mountain Forest Vegetation-Oak Forest-Medium Sub-Zone habitat type		

Additional Important Bird Observations: During the surveys, 53 species were observed. The site held breeding Levant Sparrowhawk *Accipiter brevipes* (Eurasian Steppe and Desert biome-restricted species) as well as two Irano-Turanian and four Mediterranean biome-restricted species but these did not trigger inclusion under criterion A3. Eastern Cinereous Bunting *Emberiza semenowi* (Near Threatened) was breeding.

Other Important Fauna: Data for mammals were collected in 2010 only. Locals reported the presence of Jungle Cat *Felis chaus* and recent wolf attacks on livestock. Also, in 2012 there was a report of a leopard kill on Kartoo Mountain near Mame Khelan village approximately 5 km north of Parazan. Direct observations were made of Persian Squirrel *Sciurus anomalus*, which is a Least Concern species but has a declining population trend overall and is heavily persecuted for the pet trade in Iraq. Fish data were not collected for this site.

Additional Plant & Habitat Information: There is a good population of *Anchusa italica*, *Gundelia tournefortii*, *Morus alba*, and *M. nigra*, which are important as traditional foods and a vinyard containing *Vitis venefira* is also present, which is important as a genetic resource.

Conservation Issues: The salient threats are human intrusion from tourism as well as minefields remaining from the Iran/Iraq war (rated very high). High threats were from livestock production/grazing and agriculture as well as municipal and residential development including the construction of a new health center and renovation of some older houses in the nearby village. However, the upper part of area was in better ecological condition and it is considered one of the best areas for oak forests in the Kurdistan region. Pollution due to garbage from picnickers is also a big problem at the site, especially in the spring, as is waste management in local villages.

Recommendations: Minefield clearance remains a priority and strong controls on development are recommended to prevent damage to the oak forest. Grazing should be monitored and controlled more closely to prevent damage to natural habitats. Educational campaigns are crucial to raise the environmental awareness of the people who visit the site for picnicking in the spring.

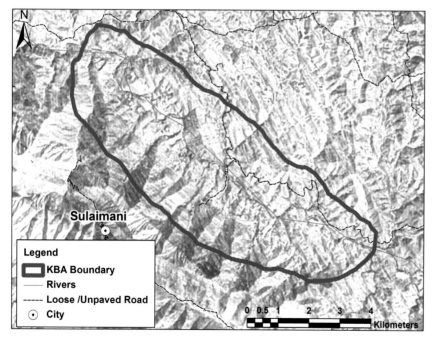

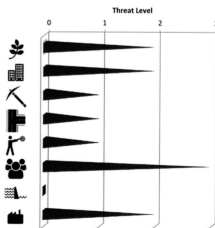

Chamchamal (IQ035)

Sulaimani–**35.421389°N 44.618333°E**

IBA Criteria: **A2**
IPA Criteria: **A4 and B1**

Area: **24745ha** - Altitude: **420-900 m**
Ecoregion: **Middle East Steppe (PA0812)**
Status: **Unprotected**

©2005 S ABDULRAHMAN/NI

Site Description: This area includes the foothills located approximately 12 km southwest of the town of Chamchamal. Main habitat is steppe but the botanical survey evaluated an area of the Riverine Forest of the Plains (Al-Ahrash) along the Khra Azizy, which is a small seasonal stream that runs southwards through the area. There are also small drainages running between the hills made by seasonal rainfall. The site is located in the Moist Steppe Zone. The geology consists of sandstone, clay and sandy gravel and the soil type is sandy clay. The non-vegetated area covered about 50% of the site.

Important Bird Area Criteria	Observations made 2007-2009.	
A2. Restricted-range species	**Breeding**	**Wintering/ Passage**
Iraq Babbler *Turdoides altirostris* (Resident)	90 pairs (2007-2008)	
Important Plant Area Criteria		
A4. Site contains national endemic, near endemic, regional endemic and/or regional range-restricted species or infraspecific taxa		
Endemics historically recorded here include: *Anthemis plebeia*, *Centaurea rigida* var. *schizophylla*, and *Thymus neurophyllus*; a near endemic historically recorded was: *Echinops mosulensis*, and nationally rare species were *Juncus effusus* and *Glaucium corniculatum*		
B1. Site is particularly species-rich example of defined habitat type		
Riverine Forest of the Plains (Al-Ahrash) habitat type		

Additional Important Bird Observations: A total of 64 bird species were seen during the survey period. Eastern Imperial Eagle *Aquila heliaca* (Vulnerable) was observed in winter; the site held breeding populations of two Mediterranean, three Irano-Turanian and three Sahara-Sindian Desert biome-restricted species but these did not trigger inclusion under criterion A3.

Other Important Fauna: There have been reports of Goitered Gazelle *Gazella subgutturosa* (Vulnerable) being hunted in the area. Not enough data were collected to assess this population.

Fish: This site has a small, shallow stream. Data were collected in 2007 and 2008, and three significant species were observed: *Acanthobrama marmaid*, *Alburnus mossulensis*, and *Garra rufa*.

Additional Plant & Habitat Information: This site contains some plants, which are important as genetic resources including *Triticum aestivum*, *Triticum durum*, and *Hordeum vulgare* ssp. *Spontaneum*.

Conservation Issues: This is an important plain areas in Iraq and is under serious threat from agriculture and hunting. Field crops and vegetable fields were observed throughout, and livestock grazing had a noticable impact on vegetation at the site. The site is also affected by picnicking (people were seen cooking and throwing garbage); related human intrusion, pollution but also transportation threats were considered high.

Recommendations: Hunting needs to be better regulated and agriculture, including grazing activities should be more closely examined and controlled. Agreements with local

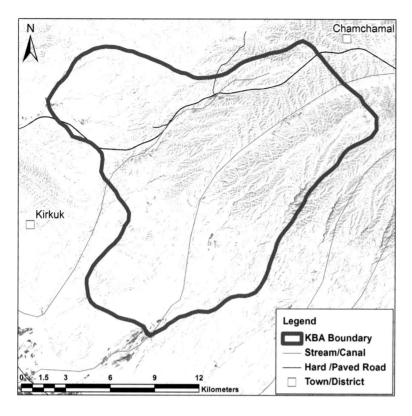

stakeholders utilizing the site could result in a more sustainable schedule for grazing to prevent habitat degradation. Springtime is the period of time when the area gets the highest recreational use, so this should be the period when focused environmental education and awareness programs are implemented.

De Lezha (IQ036)

Sulaimani – 35.460278°N 45.19444°E

KBA Criteria: **V**
IBA Criteria: **A1 and A3**
IPA Criteria: **A4 and B1**

Area: **8110 ha** - Altitude: **630-1310m**
Ecoregion: **Zagros Mountains Forest Steppe (PA0446).**
Status: **Unprotected**

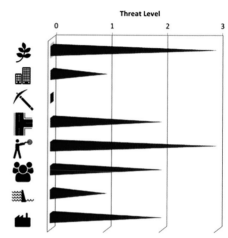

Site Description: De Lezha is part of same ridge that extends northwest from the Qara Dagh ridge and could be effectively considered as a part of the Qara Dagh KBA (IQ039). The main habitat types here were mountain forest vegetation (primarily oak forests) and mountain riverine forest. There are also dense grassland and scattered bushes cut through by a stream, which is densely wooded on both banks and crossed by a bridge. A line of old willows stands along parts of the local river. The site is located in the Zagros mountain range, where the geology is limestone and the soil type is sandy clay.

Some cultivated areas and three small villages (De Lezha, Solai Kabinakan and Solai Shekh Qadir) lie nearby. It is a popular picnic site. A ridge of mountains to the south was not surveyed but is primarily used for grazing. East of De Lezha, a residential development has been built that will house employees for a future dam that is proposed to be constructed on the stream.

Key Biodiversity Area Criteria	Notes	
V. Vulnerability Criteria: _Presence of Critically Endangered and Endangered species – presence of a single individual or Vulnerable species – 30 individuals or 10 pairs._		
Testudo graeca	Two individuals of Spur-thighed Tortoise were observed at the site.	
Important Bird Area Criteria	**Observations made 2009-2010.**	
A1. Globally threatened species	**Breeding**	**Wintering/ Passage**
Egyptian Vulture _Neophron percnopterus_ (Summer visitor)	4-8 pairs (counts)	

A3. Biome-restricted species

Irano-Turanian biome	Breeding	Wintering/ Passage
See-see Partridge *Ammoperdix griseogularis* (Resident)	40 pairs	
Menetries's Warbler *Sylvia mystacea* (Summer visitor)	80 pairs	
Eastern Rock Nuthatch *Sitta tephronotus* (Resident)	300 pairs	
White-throated Robin *Irania gutturalis* (Summer visitor)	5 pairs (2010)	
Finsch's Wheatear *Oenanthe finschii* (Resident)	5 pairs (2010)	
Pale Rockfinch *Carpospiza brachydactyla* (Summer visitor)	3 pairs (2010-2012)	

Important Plant Area Criteria

A4. Site contains national endemic, near endemic, regional endemic and/or regional range-restricted species

Aristolochia paecilantha and *Briza minor* are nationally rare species

B1. Site is a particularly species-rich example of defined habitat type

Mountain Forest-Mountain Riverine Forest habitat type and Mountain Forest Vegetation-Oak Forest-Lowest habitat type

©2009 S ABDULRAHMAN/NI

Additional Important Bird Observations: During the survey period a total of 79 species were seen. European Roller *Coracias garrulus* (breeding) and Semi-collared Flycatcher *Ficedula semitorquata* (migrant), both Near Threatened, were also observed, but at sub IBA-threshold numbers. Four Mediterranean and two Sahara-Sindian Desert biome-restricted species were found breeding.

Other Important Fauna: Little data were obtained on non-avian fuana under the rapid assessment. Only one Indian Grey Mongoose *Herpestes edwardsii* was observed. The stream supports fish but no surveys were conducted.

Additional Plant & Habitat Information: This site contains a good population of *Morus alba, Morus nigra, Pistacia eurycarpa* and *P. khinjuk*, which are economically and culturally important.

Conservation Issues: A proposed dam construction project that included the construction of worker housing is a major threat here as is hunting, and human intrusion. Pollution (from village waste and also picnicking waste) and transportation were rated as high threats and livestock production and grazing was assessed as a medium concern.

Recommendations: Tourism impacts need to be monitored and controlled. Also grazing examined to determine its effects and perhaps confine it to a smaller area. Authorities should make sure that all dam projects have adequate environmental impact assessments, cost-benefit analysis and protection for biodiversity.

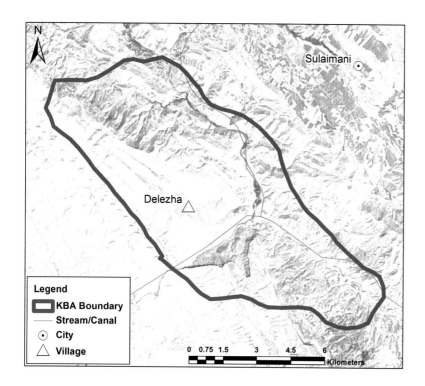

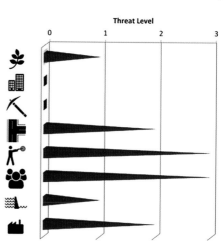

Threat Level

Hazarmerd (IQ037)

Sulaimani – 35.4975°N 45.311663°E

IBA Criteria: A1

Area: 607 ha - Altitude: **912-1260 m**
Ecoregion: **Zagros Mountains Forest Steppe (PA0446)**
Status: **Unprotected**

©2010 R PORTER/NI

Site Description: This is an agricultural area dominated by foothills and Baranan Mountain ridge that runs roughly east-west and includes the Hazarmerd Cave archaeological site. Close by are the villages of Hazarmerd and Qorkhu, and a stream that is polluted with sewage and garbage from the local communities. A rough gravel road switchbacks up close to the cave, which overlooks the city of Sulaimani. This site, with its captivating scenery is used as a hiking and picnic area by many people due to its close proximity to the city.

Important Bird Area Criteria	Observations made in 2010.	
A2. Globally threatened species	Breeding	Wintering/ Passage
Egyptian Vulture *Neophron percnopterus* (Summer visitor)	5 pairs	

Additional Important Bird Observations: During the surveys 34 species were recorded. Additionally there were four breeding Irano-Turanian and two breeding Mediterranean biome-restricted species but these did not trigger inclusion under criterion A3. European Roller *Coracias garrulous* (Near Threatened) was also breeding.

Conservation Issues: This site is given a very high threat rating due to its popularity as a tourist area especially after the fall of Saddam in 2003. This is causing pollution and disturbance to breeding birds. Agriculture and grazing are practised in the fields below the cliffs. There is a paved road leading partially up the mountain and it is expected that there will be more road construction if the site continues to be of interest to visitors. Housing development is undertaken near the base of the mountain and even close to the cliff. Hunting and energy production are moderately impacting the area.

Recommendations: Recognizing the popularity of this site, education and awareness programmes are the priority actions recommented to regulate the use and the protection of this site.

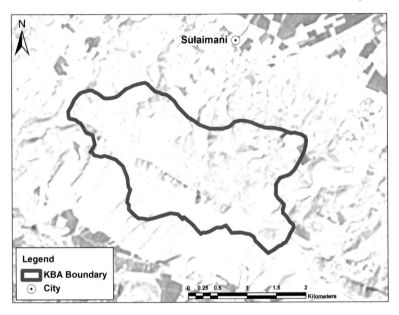

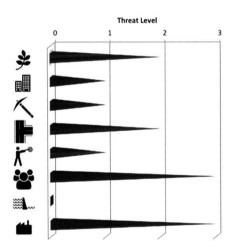

Sangaw Area (IQ038)

Sulaimani – 35.290278°N 45.218889°E

KBA Criteria: **V**
IPA Criteria: **A4 and B1**

Area: **1873 ha** - Altitude: **810-1030 m**
Ecoregion: **Middle East Steppe (PA0812)**
Status: **Unprotected**

Site Description: Sangaw is in the moist steppe. The geology is limestone and marls, and the soil type is sandy clay. This site consists mainly of flat agricultural land (wheat, barley, chickpeas, and lentils) and approximately 55% of the area was unvegetated during the survey visits. The Qara Dagh Range (IQ039) is located to the east of the site. The foothills extend from the middle of the site to the west and a valley also features a small water body formed by springs. Some of these are seasonal and water disappears in the summer. There are some villages in the area and the nearest town is Sangaw to the west.

Key Biodiversity Area Criteria	Notes
V. Vulnerability Criteria: *Presence of Critically Endangered and Endangered species – presence of a single individual or Vulnerable species – 30 individuals or 10 pairs.*	
Gazella subgutturosa	No observation was made during the surveys, but reliable local reports indicate the presence of the globally Vulnerable Goitered Gazelle.
Important Plant Area Criteria	
A4. Site contains national endemic, near endemic, regional endemic and/or regional range-restricted species or infraspecific taxa	
One nationally rare species (*Dianthus floribundus*) and one new record for Iraq (*Outreya carduiformis*) were found at this site.	
B1. Site is a particularly species-rich example of defined habitat type	
Moist Steppe Zone- Steppe habitat type	

Additional Important Bird Observations: During the surveys 22 species were recorded but the site does not qualify as an IBA. For non-avian fauna, there were no other noteworthy observations.

Additional Plant & Habitat Information: This site contains a good population of *Gundelia tournefortii*, which is important as a traditional food. *Hordeum vulgare* ssp. *spontaneum*, *Triticum aestivum*, and *T. durum* are important as genetic resources.

Conservation Issues: An oil exploration area covering approximately 492 km² is located in the area but it is unclear to what degree this coincides with the delineated KBA site. It is also not clear if an Environmental Impact Assessment (EIA) has been conducted for these activities, which presents the one of the highest potential threats to the biodiversity of the area.

Hunting is also a very high threat, as it especially targets the globally Vulnerable Goitered Gazelle *Gazella subgutturosa*, for which the site is important. Additionally this is an agricultural area and grazing presents a high threat. Road

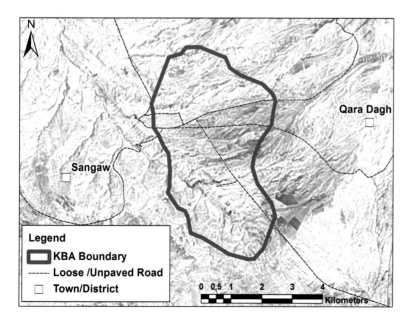

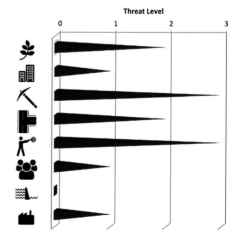

construction is also considered an issue because more roads are being built to access the agricultural fields and possibly for oil development. Residential and commercial development, human intrusion and pollution are considered medium threats but may have increasingly negative impacts in the future if not controlled.

Recommendations: Details should be made more available on the nature and extent of oil development projects and access to environmental impact assessments need to be made available to the general public for comment. Hunting can only be controlled through awareness raising programs as well as stricter implementation of existing hunting laws. Agriculture and particularly grazing activities should be more closely examined to better understand their impacts. It is likely that these threats may be higher in this area that was determined by the team. New road construction should also be the subject of EIAs and construction standards should focus on methods that minimize habitat destruction and erosion.

Qara Dagh (IQ039)

Sulaimani–35.331111°N 45.290278°E

KBA Criteria: **V**
IBA Criteria: **A1 and A3**
IPA Criteria: **A4, B1 and B2b**

Area: **25345 ha** - Altitude: **700-1719 m**
Ecoregion: **Zagros Mountains Forest Steppe (PA0446)**
Status: **Unprotected**

©2009 K ARARAT/NI

Site Description: This area consists of a mountain ridge with rocky slopes and many gorges and valleys. It contains mountain riverine forest and relatively dense oak woodlands. The geology of the area was sedimentary and the soil type was clay. There are farmland and villages nearby (primarily around the base of the ridge but there are some houses on the ridge itself.

The ridge is over 55 km long and runs from northwest to southeast, but the survey area encompasses only about half of the ridge, primarily the southeastern part. A paved road traverses the ridge. There is also at least one deep cave system, Kuna Ba, within the Qara Dagh range. An important archeological site is also located in Qara Dagh, which consists of an ancient rock carving along a stream, which is now part of a water diversion project related to oil development.

A water quality sampling point was located at the base of the ridge in another stream coming from Kani Bajga, which is spring fed. Past water sampling has shown that the water flow in the valley below can be very slow or stagnant and the site was dry in winter 2009 at the end of a period of drought. The stream riparian is well vegetated but in general the area is rocky. Plant decomposition and trash were observed in and around the water. The entire area attracts many picnickers, particularly in spring.

Additional Important Bird Observations: During the surveys a total of 87 species was recorded. The site also held breeding White-eared Bulbul *Pycnonotus leucotis*, a Sahara-Sindian Desert biome-restricted species. The Eastern Cinereous Bunting *Emberiza semenowi* mentioned in the table above is Near Threatened.

Key Biodiversity Area Criteria	Notes
V. Vulnerability Criteria: *Presence of Critically Endangered and Endangered species – presence of a single individual or Vulnerable species – 30 individuals or 10 pairs.*	
Capra aegagrus	23 Wild Goats were seen in summer and 26 in winter of 2011 and goats are regularly reported here; available habitat suggests that the population would meet the KBA requirements.
Panthera pardus saxicolor	A camera trap photographed one male Persian Leopard in October 2011 and another camera trap photo, possibly of the same individual, was taken in February 2012.
Testudo graeca	Spur-thighed Tortoise was observed at the site frequently and available habitat suggests that the population would meet the KBA requirements

Important Bird Area Criteria	Observations made 2007-2010.	
A1. Globally threatened species	**Breeding**	**Wintering/ Passage**
Egyptian Vulture *Neophron percnopterus* (Summer visitor)	6-11 pairs (counts)	
A3. Biome-restricted species		
Irano-Turanian biome	**Breeding**	**Wintering/ Passage**
See-see Partridge *Ammoperdix griseogularis* (Resident)	150 pairs (2008-2011)	
Upcher's Warbler *Hippolais languida* (Summer visitor)	15 pairs (count 2010)	
Menetries's Warbler *Sylvia mystacea* (Resident)	50 pairs (2008-2010)	
Eastern Rock Nuthatch *Sitta tephronota* (Resident)	600 pairs	
White-throated Robin *Irania gutturalis* (Summer visitor)	40 pairs (2009-2010)	
Finsch's Wheatear *Oenanthe finschii* (Resident)	140 pairs	
Eastern Cinereous Bunting *Emberiza semenowi* (Summer visitor)	70 pairs (2007, 2009 and 2010)	
Mediterranean biome	**Breeding**	**Wintering/ Passage**
Masked Shrike *Lanius nubicus* (Summer visitor)	200 pairs (2008, 2010)	
Sombre Tit *Poecile lugubris* (Resident)	150 pairs (2007, 2008 and 2010)	
Western Rock Nuthatch *Sitta neumayer* (Resident)	150 pairs (2009-2010)	
Eastern Black-eared Wheatear *Oenanthe melanoleuca* (Summer visitor)	2000 pairs (2009-2010)	

Important Plant Area Criteria
A4. Site contains national endemic, near endemic, regional endemic and/or regional range restricted species or infraspecific taxa   *Note: *historically recorded; **historically recorded and seen on recent surveys*
Historical endemics at this site include:*Alyssum penjwinense, *Asperula friabilis, *Erysimum boissieri , *Galium hainesii, **G. qaradaghense, *Onosma albo-roseum var. macrocalycinum, *Ornithogalum kurdicum, and *Scilla kurdistanica; Historical near endemics at this site include: *Acantholimon blackelockii, *Bunium cornigerum, *Campanula acutiloba, *Centaurea gigantea, C. irritans, **Cousinia kopi-karadaghensis, *Echinops inermis, *Ferulago bracteata, *Korshinskia assyriaca, *Malabaila secacul subsp. aucheri, *Serratula grandifolia, *Stachys kotschyii, and *Verbascum alceoides, and A nationally rare species was Bromus brachstachys
B1. Site is a particularly species-rich example of defined habitat type
Mountain Forest Vegetation-Mountain Riverine Forest habitat type and Mountain Forest Vegetation- Oak Forest-Medium Zone and Highest Zone habitat type
B2b.The site is a refuge for: biogeographically and bioclimatically restricted plants to 'retreat to' in the face of global climate change.
This site represents a good example of Mountain Forest Vegetation- Oak Forest-Medium Zone and Highest Zone habitat types. The mountain top and some gorges/cliffs can provide refuge for these Oak forests and associated plants under possible future climate change.

Other Important Fauna: Data collected on 2007-2010 surveys found field signs of a considerable number of the Muridae family that remain unidentified. Wild Goats *Capra aegagrus* (Vulnerable) were the subject of a specific study in 2011. A Persian Leopard *Panthera pardus saxicolor* (Endangered) was camera-trapped during the survey, which was the first photographic record of the species in Iraq. Wild Cat *Felis silvestris*, Golden Jackal *Canis aureus*, Red Fox *Vulpes vulpes*, Indian Crested Porcupine *Hystrix indica*, Wild Boar *Sus scrofa*, and Persian Squirrel *Sciurus anomalus* were also camera-trapped. Grey Wolf *Canis lupus*, Goitered Gazelle *Gazella subgutturosa* (Vulnerable), Indian Grey Mongoose *Herpestes edwardsii*, and Jungle Cat *Felis chaus* are likely present. It is also likely that a variety of bat species inhabit the Kuna Ba cave system and other caves. Iraq's mammal checklist includes several bats of conservation concern but these species are poorly studied.

Fish: Surveys were only conducted in 2007 and 2008 (summer) and two significant species according to Coad (2010) were: *Alburnus mossulensis* and *Garra rufa*.

Additional Plant & Habitat Information: Qara Dagh contains a good population of *Pistacia eurycarpa* and *P. khinjuk*, which are culturally and economically important, as well as *Arum conophalloides, Anchusa italica, Crataegus azarolus, Gundelia tournefortii,* and *Rumex ribes*, which are economically important as traditional food plants. Also *Aegilops crassa, A. columnaris, A. umbellulata, Bromus danthoniae, B. brachystachys, Hordeum bulbosum,* and *Triticum aestivum* are important genetic resources.

Conservation Issues: Industrial and oil development present the most serious threats to Qara Dagh. In 2010 oil drilling began, which has led to increased utilization of ground and surface water. A representative of the company indicated that the entire range would be dedicated to oil development. This poses the most serious threat, though with proper planning and oversight, oil exploration can coincide with development of the site as a protected area. Another resource extraction threat is limestone quarrying with related road construction that is causing erosion and

threatens the Kuna Ba Cave system. Finally a dam is under construction just east of Kuna Ba Cave that will affect the stream flowing along the northeast base of Qara Dagh and may negatively impact biodiversity.

In 2008 and 2009 the area experienced a period of drought, which led to low water conditions. In 2008, the target area showed heavy metal contamination in the sediments, particularly cadmium, zinc and nickel during water quality and sediment monitoring. In 2009, also copper and manganese were found to be elevated above Iraqi standards.

Qara Dagh is also seriously impacted by illegal hunting, which is widely practiced. According to locals, hunting has led to a decline in the status of wildlife, particularly large mammals, such as the Wild Goat *Capra aegagrus*. This area also has overgrazing issues as well as road construction, extension of the electrical grid along the road, and trash from tourists, all of which have caused some level of damage and/or fragmentation of the landscape. However, Qara Dagh is presently still in overall good condition and there are many endemic and rare plants in addition to important bird species.

Recommendations: This area has previously been recommended as a site that should receive additional protection (Nature Iraq and Iraqi Ministry of Environment, 2010). If this site is identified as a future protected area, funds should be made available for a more comprehensive and targeted survey to assess the site in more detail and build a comprehensive inventory of its resources (both natural and social-economic). It is unclear what the sources of heavy metal contamination are but this requires more study. An adjacent site, De Lezha (IQ036) could potentially be incorporated in any protected area developed here. Also as previous attempts to fence part of the site were not successful, protection of the site will require extensive local stakeholder input. Other issues such as cultural sites and land ownership will need to be addressed, as well as oil development within the site, which should not stop the potential designation of the site as a protected area.

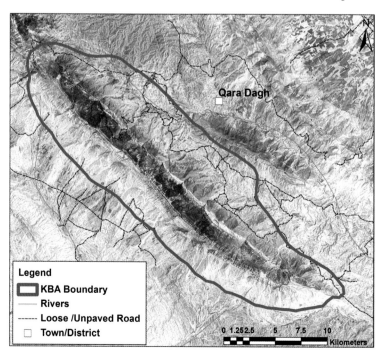

Legend
KBA Boundary
Rivers
Loose /Unpaved Road
Town/District

Qara Dagh

0 1.25 2.5   5   7.5   10
Kilometers

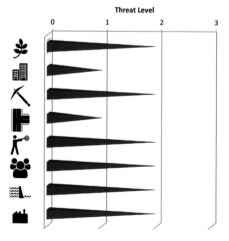

Threat Level

Darbandikhan Lake (IQ040)

Sulaimani – 35.144722°N 45.755°E

KBA Criteria: **V and Ia**
IBA Criteria: **A1 and A4i**
IPA Criteria: **A4 and B1**

Area: **36842 ha** - Altitude: **496-1350 m**
Ecoregion: **Zagros Mountains Forest Steppe (PA0446)**
Status: **Unprotected**

©2012 S ABDULRAHMAN/NI

Site Description: Darbandikhan is a large freshwater reservoir created by the Darbandikhan Dam that is fed by two main rivers, the Tanjero in the north and the Sirwan in the east, and covers (depending on the time of year) approximately 7,500 ha. Evans (1994) listed Darbandikhan, along with Dukan and Bakhma, as an Important Bird Area (IBA004). The main habitat types investigated here were mountain forest vegetation (primarily oak forests) and mountain riverine forest. The site is located in the Zagros Range and the geology is limestone with soil types of clay and sandy clay. The lake is surrounded by hills covered with grass and small shrubs and mountains (including Bashari, Zmnako and Zawaly) that are covered in oak forests.

The area as a whole supports a significant amount of bird life. The rock-filled embankment dam was constructed between 1956 and 1961 for irrigation, flood control and power generation. The lake also supports recreational uses and a fishery. Due to problems after construction there have been several slope failures upstream and repairs required of the dam. Water levels decline in summer after the spring melt due to dam release and rise again when winter rains return in the late fall. The agricultural lands around the lake are used for field crops especially wheat and barley.

Key Biodiversity Area Criteria	Notes	
V. Vulnerability Criteria: *Presence of Critically Endangered and Endangered species – presence of a single individual or Vulnerable species – 30 individuals or 10 pairs.*		
Luciobarbus subquincunciatus	Critically Endangered fish species	
Capra aegagrus	No direct observation, but reliably reported by locals.	
Panthera pardus saxicolor	Reported by locals and one Persian Leopard was killed by a landmine near the village of Mortka in October 2008.	
Rafetus euphraticus	One individual was photographed in May 2009.	
Ia. Irreplaceability Sub-criterion: Restricted-range species based on global range		
Rafetus euphraticus	See above	
Important Bird Area Criteria	**Observations made 2007-2009.**	
A1. Globally threatened species	**Breeding**	**Wintering/ Passage**
Egyptian Vulture *Neophron percnopterus* (Summer visitor)	2-3 pairs (counts 2007-2009)	
A4i. 1% or more of biogeographical population of a congregatory waterbird species		
Slender-billed Gull *Chroicocephalus genei* (Summer visitor)	3200 pairs (count 2009)	
Important Plant Area Criteria		
A4. Site contains national endemic, near endemic, regional endemic and/or regional range-restricted species or infra-specific taxa		

Historically recorded endemics here are: *Onosma albo-roseum* var. *macrocalycinum.* and *Ornithogalum iraqense*; one historically recorded near endemic here is: *Cousinia kopi-karadaghensis* and the nationally rare species here are: *Alcea sulphurea, Alkanna orientalis*, and *Muscari tenuiflorum*.

B1. Site is a particularly species-rich example of defined habitat type

Mountain Forest Vegetation-Mountain Riverine Forest habitat type and Mountain Forest Vegetation-Oak Forest-Medium habitat type

Additional Important Bird Observations: During the surveys 56 species were recorded. The site also held breeding populations of three Irano-Turanian, one Mediterranean and one Sahara-Sindian Desert biome-restricted species but these did not trigger inclusion under criterion A3. European Roller *Coracias garrulous*, a Near Threatened species was also recorded.

Other Important Fauna: Mammal surveys were conducted in 2007 and 2010. In addition to those listed in the table, important species present at the site include the Near Threatened Eurasian otter *Lutra lutra*, Golden Jackal *Canis aureus* and one local fisherman reported seeing a Eurasian Lynx *Lynx lynx* in 2006. No significant reptiles or amphibians were observed on the survey.

Fish: Data were collected in 2007 and 2008, when 26 species were observed. In addition to the species listed in the table the following significant species according to Coad (2010) were: *Acanthobrama marmid, Alburnus mossulensis, Arabibarbus grypus* (Vulnerable), *Carassius auratus, Cyprinus carpio* (Vulnerable), *Cyprinion macrostomum, C.kais, Heteropneustes fossilis, Hypophthalmichthys molitrix, H. nobilis, Leuciscus vorax, Luciobarbus esocinus (Vulnerable), L. xanthopterus* (Vulnerable), and *Silurus triostegus.*

Additional Plant & Habitat Information: This site contains a good population of pistachio Pistacia eurycarpa, an economically and culturally important species. In addition, some species important for traditional foods included: and *Anchusa italica, Arum conophalloides*, and *Gundelia tournefortii*. Also, species important as genetic resources included: *Hordeum vulgare ssp. spontaneum, H. bulbosum, Lens culinaris* and *Triticum aestivum*.

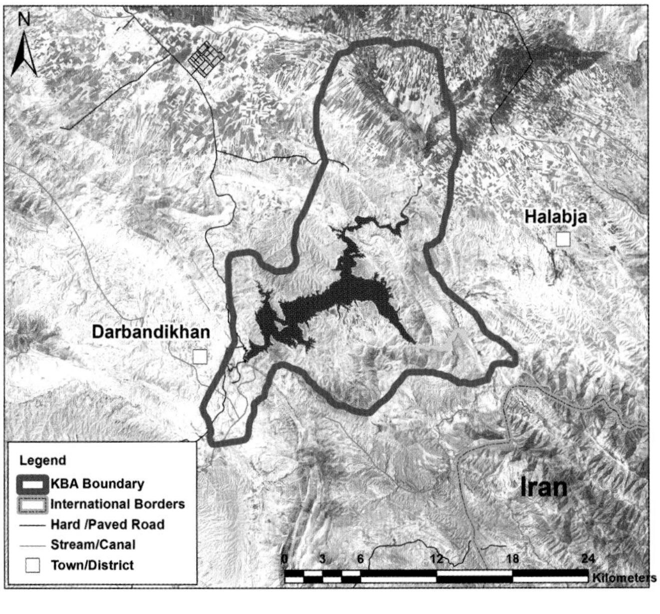

Conservation Issues: Very high threats include livestock production/grazing, hunting and fishing, tourism (mainly along the Sirwan River, which people from Sulaimani City, Darbandikhan and Kalar District visit for picnicking), and pollution. Though the presence of the dam attracts birds and other fauna, natural systems modification was considered a very high threat because of how water in the reservoir is being managed (with environmental impacts downstream and the widely fluctating level of the reservoir from year to year leave a wide belt of unvegetated land above the waters' edge). Pollution due to sewage and garbage were assessed as very high threats to the area. Parts of the lake see regular fish kills (several large enough to receive attention in the local media in 2008, 2009 & 2011, but locals report that they are a yearly occurrence in summer). These are most likely caused by untreated sewage from Suliamani and other towns polluting the Tanjero River.

Also there are gravel mines throughout the Tanjero basin and along the Sirwan River that have a very high impact. Another critical threat is the Iranian state-built dam on the Sirwan River, which restricts water feeding into the lake. This has lowered the water level and impacted the biodiversity of the lake and river. Previous water quality surveys (Mahir et al 2009) indicate that there is heavy metal contamination in the lake with lead, zinc, cadmium and nickel often above Iraqi standards for many survey areas within the sites.

Recommendations: Several ecologically important plants and birds occur here, highlighing the urgent need to both raise awareness of and address the conservation needs of this area. Further work should be done in the Darbandikhan Basin in general to follow the track of pollution through the basin and pinpoint all contamination sources. These issues were examined further in additional Nature Iraq survey efforts following a fish kill that occurred on the lake at the end of July 2008, culminating in a State of the Environment Report on the basin (Nature Iraq & Twin Rivers Institute, 2009). The report outlined a roadmap to fill gaps in information and take steps to address a wide-range of environmental issues that threaten the basin. To date, no action has been taken by local, regional or national authorities on the recommendations of this report.

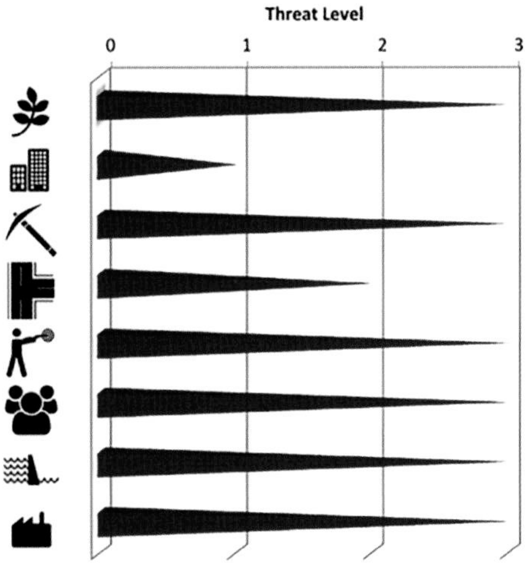

Zalm (IQ041)

Sulaimani – **35.306111°N**
45.970003°E

IPA Criteria: **A4**, **B1**, and **C**

Area: **550 ha** - Altitude: **477-513 m**
Ecoregion: **Middle East Steppe**
(PA0812)
Status: **Unprotected**

©2013 A KEMMAN

Site Description: The main habitat here includes the stream, riparian zone and steppe or grassland habitats (the latter has been largely displaced by agricultural fields), but the main habitat type examined in the KBA surveys was the Riverine Forest of the Plains. The site is located in the Moist Steppe Zone and the geology is clay and sandy gravel with a soil type of clay and sandy clay.

Water levels at the site are affected by Darbandikhan Dam and, as with the rest of the lake, are extremely low in winter. A series of mountains, including Suren, lie to the north of the site and the Zalm Stream flows west-southwest from its headwaters near Ahmed Awa (IQ042) to join the waters of the Tanjero River entering Darbandikhan Lake (IQ040). The Hawraman Area (IQ043) lies to the southeast of the site.

Important Plant Area Criteria
A4. Site contains national endemic, near endemic, regional endemic and/or regional range-restricted species or infraspecific taxa
Nationally rare species are: *Alkanna orientalis*, *Dipsacus laciniatus*, *Juncus effusus*, *Sparganium erectum*, and *Teucrium melissoides*
B1. Site is a particularly species-rich example of defined habitat type
Riverine Forest of the Plains (Al-Ahrash) habitat type and Inland running water, rivers or canals- Submerged river or canal vegetation habitat type
C. The site is identified as an outstanding example of a globally or regionally threatened habitat type.
This site represents a good example of the Riverine Forest of the Plains (Al-Ahrash) habitat type, which is a globally important habitat and extends along the basins of Tigris- Euphrates Rivers. Also this site is under threats of gravel mining, agriculture and pollution from untreated sewage and other wastes.

Important Bird Observations: During the surveys a total of 47 species were recorded. The site held one Irano-Turanian and one Mediterranean biome-restricted species, but these did not trigger inclusion under A3 criteria. Additionally,

European Roller *Coracias garrulus* (Near Threatened) was recorded during the breeding season.

Other Important Fauna: Data on mammals were recorded in 2008, during which one Golden Jackal *Canis aureus* was observed.

Fish: Data were collected in 2008 and during a separate survey in 2012 done by Dr. Jörg Freyhof from the Leibniz Institute of Freshwater Ecology and Inland Fisheries. A total of 17 species were reported. *Alburnus mossulensis, Carassius auratus, Carasobarbus luteus, Cyprinus carpio* (Vulnerable), *Cyprinion macrostomum, Gambusia holbrooki, Garra rufa, Hemiculter leucisculus,* and *Squalius berak* were all significant species according to Coad (2010). *Alburnoides* sp., *Paracobitis* sp., and *Oxynoemacheilus* sp. were found by Dr. Freyhof and not mentioned in Coad. These are undescribed species, which are very likely to be endemic to the drainage basin.

Additional Plant & Habitat Information: This site contains a good population of *Cicer arietinum* ssp. *arietinum, Hordeum vulgare* ssp. *spontaneium, Linum usitatissimum, Triticum aestivum,* and *T. durum,* which are important genetic resources.

Conservation Issues: This site is heavily used by picnickers and car washers, which has resulted in disturbance to the riparian area and the accumulation of garbage in and pollution

to the stream. Particularly the area of the stream that is nearest the road is heavily impacted by car washing. Areas along the stream are also affected by agriculture (extending croplands into the riparian area, burning in the riparian area, and agricultural run-off). Animal grazing impacts the vegetation of the area and there is some hunting activity.

Recommendations: Car washing could be at least partially controlled by placing permanent barriers in areas where cars currently access the stream from the road but this should include the installation of signs that help raise awareness and explain the purpose of restrictions. It is also possible to develop public-private economic initiatives by building car washing facilities nearby that recycle/clean wastewater before releasing it into local streams. Public picnic areas here should be better designated, planned and managed, with waste services provided and environmental awareness (signage, etc.) programs provided. Additionally, stream buffers need to be established in cooperation with local stakeholders to protect the stream from agricultural impacts.

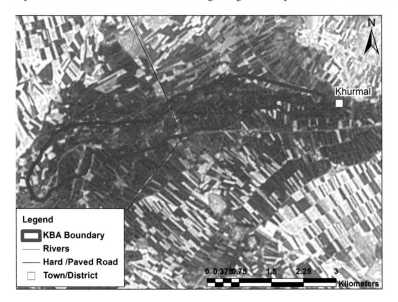
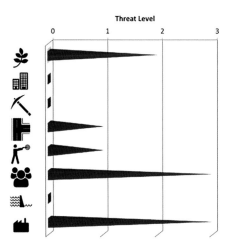

Ahmed Awa (IQ042)

Sulaimani –35.299722°N 46.078056°E

KBA Criteria: **V**
IBA Criteria: **A1 and A3**
IPA Criteria: **A4 and B1**

Area: **887 ha - 638-1610 m**
Ecoregion: **Zagros Mountains Forest Steppe (PA0446)**
Status: **Unprotected**

Site Description: This is a mountainous area in the Zagros Range located close to the Iranian border. It is dominated by rocky slopes and wooded valleys. The geology here is basic igneous rock, radiolarian chert, siliceous and calcareous shale, and metamorphic schist and limestone of unknown age, and the soil type is sandy clay. The two main habitats at the site are oak woodlands and mountain riverine forest.

A large spring and waterfall, joined by streams coming down from higher elevations, make up the headwaters of the Zalm Stream (IQ041), a large fast-moving waterway that flows towards the Tanjero River and into Darbandikhan Lake (IQ040). Several villages are situated in the narrow valley, the largest being Khurmal. While the area has been known for poor security in the past, it remains a popular tourist spot.

The waterfall and upper stream are reached by a narrow gravel road (currently a paved road is under construction), which begins at a large parking and popular picnic area further down the valley. During the spring and summer, many teashops and other shops line the upper stream below the waterfall. There are vineyards, pomegranate and walnut orchards (inside the delineated area) and wheat and barley farms where the valley opens up (outside the delineated area). The villages around the stream use the water for drinking and other domestic usage.

©2011 A BACHMAN/NI

Key Biodiversity Area Criteria	Notes	
V. Vulnerability Criteria: *Presence of Critically Endangered and Endangered species – presence of a single individual or Vulnerable species – 30 individuals or 10 pairs.*		
Capra aegagrus	No direct observation were made, but reliably reported by locals	
Neurergus derjugini	From direct observations.	
Important Bird Area Criteria	**Observation made 2007-2010.**	
A1. Globally threatened species	**Breeding**	**Wintering/ Passage**
Egyptian Vulture *Neophron percnopterus* (Summer visitor)	3-6 pairs (counts 2009-2010)	
A3. Biome-restricted specie		
Irano-Tauranian biome	**Breeding**	**Wintering/ Passage**
See-see Partridge *Ammoperdix griseogularis* (Resident)	10 pairs (2009-2013)	
Menetries's Warbler *Sylvia mystacea* (Summer visitor)	8 pairs (counts 2009-2013)	
Eastern Rock Nuthatch *Sitta tephronota* (Resident)	20 pairs (2007-2013)	
White-throated Robin *Irania gutturalis* (Resident)	5 pairs (count 2009-2013)	
Finsch's Wheatear *Oenanthe finschii* (Resident)	15 pairs (2009-2013)	
Pale Rockfinch *Carpospiza brachydactyla* (Summer visitor)	6 pairs (2008-2013)	
Mediterranean biome	**Breeding**	**Wintering/ Passage**
Masked Shrike *Lanius nubicus* (Summer visitor)	2 pairs (2009)	
Sombre Tit *Poecile lugubris* (Resident)	12 pairs (2008-2009)	
Western Rock Nuthatch *Sitta neumayer* (Resident)	30 pairs (2009)	
Mediterranean biome	**Breeding**	**Wintering/ Passage**
Eastern Black-eared Wheatear *Oenanthe melanoleuca* (Summer visitor)	20 pairs (2009-2010)	
Black-headed Bunting *Emberiza melanocephala* (Summer visitor)	45 pairs (2008-2010)	
Important Plant Area Criteria		
A4. Site contains national endemic, near endemic, regional endemic and/or regional range-restricted species or infraspecific taxa		

Endemics at this site found during recent surveys include: *Cousinia inflata, Ferula shehbaziana, Gypsophila sarbaghiae,* and *Onosma hawramanensis;* a near endemic found was *Onosma cardiostegium,* and nationally rare species here were: *Alcea arbelensis, Dionysia bornmuelleri, Onosma macrophyllum* var. *angustifolium, Phlomis kurdica, Silene araratica, S. avramana, Stachys kurdica,* and *Teucrium melissoides.*

B1. Site is a particularly species-rich example of defined habitat type

Mountain Forest-Mountain Riverine Forest habitat type and Mountain Forest Vegetation-Oak Forest- Lowest and Medium Zones habitat type.

Additional Important Bird Observations: During the survey period 67 species were recorded. In addition to those listed in the table above, European Roller *Coracias garrulus* (Near Threatened) was present in the breeding season.

Other Important Fauna: Two locals, who were interviewed separately, stated that a local hunter killed a large cat in 2002, which would most likely have been a Persian Leopard *Panthera pardus saxicolor* (Endangered). Hunting may be rather restricted in the region overall due to the presence of minefields. No fish data were collected due to high water flows.

Additional Plant & Habitat Information: This site contains good populations of pistachios *Pistacia eurycarpa* and *P. khinjuk*, as well as *Morus alba* and *M. nigra*, which are economically and culturally important. Species important as a traditional food were *Bongardia chrysogonum* and *Arum conophalloides*, and important genetic resource species included: *Hordeum vulgare* ssp. *spontaneum*, *H. bulbosum*, and *Lens culinaris*.

Conservation Issues: Ahmed Awa is very popular with tourists especially in spring and summer. Work to build new roads to the upper stream have damaged village buildings and caused extensive erosion. The local government is now developing these roads to accommodate more tourism of the site, which in turn will increase the threat from human intrusion, as well as erosion and habitat destruction. High threats come from the development of both houses and shops, hunting, channelization due to irrigation, and pollution generated from tourism activities. Water is also diverted from the falls to the settlements below (and there is also an smallscale hydropower station at the base of the falls that is currently not functional). Villages discharge sewage and garbage into the stream as well. Shops along the waterway have paved, trampled or otherwise destroyed much of the near-stream riparian areas. The upper part of the falls has been surrounded by chainlink fencing (possibly for safety reasons). The northeastern part of the site beyond the waterfall suffers less disturbance from tourism as it is close to the Iranian border and minefields are present.

Recommendations: Local tourism of this site and related threats such as waste management and development needs to be more strongly regulated to decrease impacts on the stream and surrounding slopes. Instead of increasing road building in the narrow valley, it is recommended that this be stopped and walking paths provided instead. Driving access should be limited to local residents and regulated drivers. For example, previously locals have generated some income by driving visitors up the valley closer to the falls in 4WD vehicles … this was far less damaging than building paved roads to the upper stream and potentially provides greater support to the local economy. Hunting as well as grazing should also be further examined. Overall, given the popularity of the site, recreational and tourism plans need to integrate more natural resource protection.

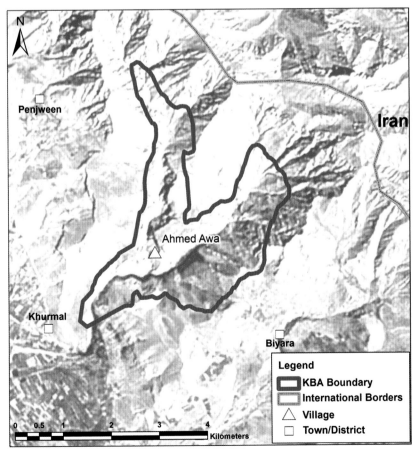

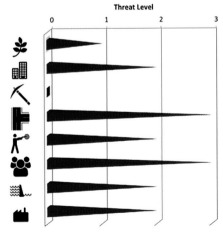

Hawraman Area (IQ043)

Sulaimani – 35.215833°N 46.186111°E

KBA Criteria: **V**
IBA Criteria: **A3**
IPA Criteria: **A4 and B1**

Area: **4463 ha** - Altitude: **1500-1925 m**
Ecoregion: **Zagros Mountains Forest Steppe (PA0446)**
Status: **Unprotected**

©2011 S ABDULRAHMAN/NI

Site Description: This site consists of a valley surrounded by foothills and highlands within the Zagros Range. Many springs are present and the habitats here are mountain forest vegetation with riverine forest and oak woodland vegetation. Vineyards cover most of the mountainsides and hills between Biyara and Awesar and the common cultivated species is *Vitis vinifera*. There are also a large number of aged walnut, mulberry, and fig treesas well as stands of poplars. The geology is basic igneous rocks, radiolarian chert, siliceous and calcareous shale, and metamorphic schist and limestone of unknown age, and the soil type is sandy clay.

In the area close to the town of Tawela much of the land is used for agriculture. Well-digging and newly constructed homes and roads were observed. The site also contains popular tourist areas during spring and summer. The border with Iran is close by and there is a border crossing but smuggling of fuel and other items across the border is also a common practice.

Key Biodiversity Area Criteria	Notes
V. Vulnerability Criteria: *Presence of Critically Endangered and Endangered species – presence of a single individual or Vulnerable species – 30 individuals or 10 pairs.*	
Capra aegagrus	Reported by locals. No direct observations were made.
Neurergus derjugini	From direct observations.
Important Bird Area Criteria	**Observations made 2007-2010.**
A2. Restricted-range species	

Mediterranean biome	Breeding	Wintering/ Passage
Masked Shrike *Lanius nubicus* (Summer visitor)	2 pairs (count 2010)	
Sombre Tit *Poecile lugubris* (Resident)	2 pairs (count 2009)	
See-see Partridge *Ammoperdix griseogularis*	8 pairs (count 2010)	
Western Rock Nuthatch *Sitta neumayer* (Resident)	1 pair (count 2009)	
Eastern Black-eared Wheatear *Oenanthe melanoleuca* (Summer visitor)	1 pair (count 2007)	
Black-headed Bunting *Emberiza melanocephala* (Summer visitor)	8 pairs (2008-2009)	

Important Plant Area Criteria
A4. Site contains national endemic, near endemic, regional endemic and/or regional range-restricted species or infraspecific taxa *Note: *historically recorded*

Endemics at the site include: *Alyssum penjwinense, Astragalus sarae, *Bunium avromanum, Cousinia inflata, *Delphinium micranthum,
Erysimum boissieri, Onosma albo-roseum var. *macrocalycinum, O. hawramanensis, *Scrophularia kurdica* subsp. *kurdica, Scrophularia sulaimanica, *Turgenia lisaeoides,* and *Vitis hissarica* subsp. *Rechingeri*

Near endemics at the site include: *Astragalus carduchorum, A. caryolobus, *A. dolius, A. globiflorus, A. gudrunensis, A. octopus, *Astracantha crenophila *Bunium cornigerum, *Centaurea imperialis, *C. koeieana ,*Cousinia inflata, *C. leptolepis, *Echinops inermis, *Hymenocrater longiflorus, Ferulago bracteata, *Malabaila secacul* subsp. *aucheri, **Onosma cardiostegium, *Picris strigosa* subsp. *kurdica, *Tragopogon bornmuelleri, *Verbascum phyllostachyum, *Veronica macrostachya* var. *schizostegia, *Ziziphora clinopodioides* subsp. *kurdica,* and **Zoegea crinita* subsp. *crinita*

Nationally rare species observed were: *Astragalus tawilicus, Buffonia calycina, Delphinium pallidiflorum, Fibigia suffruticosa, Galium kurdicum, Hesperis novakii, Hymenocrater longiflorus, Iris germanica, Minuartia sublineata, Orchis collina, O. tridentata, Phlomis kurdica, Silene araratica, S. avramana, Stachys kurdica, Stipa kurdistanica,* and *Teucrium melissoides*

B1. Site is a particularly species-rich example of defined habitat type

Mountain Forest Vegetation-Mountain Riverine Forest habitat type and Mountain Forest Vegetation-Oak Forests (medium and high) habitat type

Additional Important Bird Observations: During the survey 67 species were seen. Also breeding at the site but in sub-IBA threshold numbers were: European Roller *Coracias garrulus* and Eastern Cinereous Bunting *Emberiza semenowi* (both Near Threatened); in addition the site held breeding populations of four Irano-Turanian biome-restricted species, but these did not trigger inclusion under criteron A3.

Other Important Fauna: Mammal data were collected in 2007 and 2010 only. According to local reports, Wild Goats *Capra aegagrus* (Vulnerable) are present in the mountains in the border area. Residents state that the globally Near Threatened Eurasian Otter *Lutra lutra* is likely present. Grey Wolves *Canis lupus* were also reported. The Near Eastern fire salamander *Salamandra infraimmaculata* (Near Threatened) was also observed. No fish surveys were carried out because of the small size of the streams.

Additional Plant & Habitat Information: This site contains pistachio *Pistacia eurycarpa* as well as *Diosphyros*

kaki, Morus alba, and M. *nigra*, which are economically and culturally important. Riverine Forest habitatis commonly found alongside mountain valleys in Hawraman, especially between Biyara and Tawela and are dominated by *Juglans regia*.

Conservation Issues: The primary threat comes from residentional and tourism development. The local government has recently begun expanding a road to ease access for tourism and border crossing traffic and smuggling also have impacts. The streams near villages and popular picnicking areas are affected by sewage and garbage and the stream near Tawela has also been impacted due to modifications and clearing for irrigation purposes. Ecological conditions are noticeably healthier on the surrounding mountain slopes. This is in part because minefields restricts access to these areas.

Recommendations: Tourism areas and villages such as Tawela should be the targets for environmental education initiatives. This is a mountainous area and development and road expansion causes significant erosion and habitat distruction that requires better planning and mitigation activities. Stream modifications must be restricted to allow further study of the biodiversity and ensure the security of people, but with a recognition that plans to sustainably manage the local natural resources should be active once areas are opened up to human access.

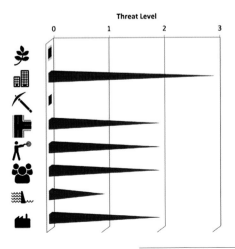

Kalar Area (IQ044)

Sulaimani - 34.555°N 45.285275°E

KBA Criteria: **V**
IBA Criteria: **A1**
IPA Criteria: **B1 and C**

Area: 1130 ha - Altitude: **177-195 m**
Ecoregion: **Mesopotamian Shrub Desert (PA1320)**
Status: **Unprotected**

©2009 K ARARAT/NI

Site Description: The Diyala River flows south-southwest from the outlet of Darbandikhan Reservoir and eventually enters the Tigris River south of Baghdad. This site is located on the Diyala River between two important bird areas along this corridor, Darbandikhan (IQ040) and Maidan Area (IQ045) to the north and Himreen Lake (IQ053) to the south. There are surrounding hills, some cultivated areas, and an extensive but often fragmented riparian zone.

The main habitat in the region is steppe, but the botanical survey evaluated the riverine forest. The site is located in the Moist Steppe Zone, where the geology is sandstone, clay, and sandy gravel, and the soil type is sandy clay.

The non-vegetated area covered approximately 50% of the terrain.

There is continuous construction of housing and commercial development, especially between Kalar and Kifri, northwest of the delineated area. Crops are grown (primarily wheat and barley) and hunting is practised especially in the foothills near Mil Qasm and Sar Qala (about 26 km northwest of the study site). Oil wells are being drilled in Mil Qasm and Sar Qala by Western Zagros Oil Company and the area northwest of the site is now targeted for future drilling by the same company. Gravel is mined along the Diyala River in many locations.

Important Bird Area Criteria	Notes	
V. Vulnerability Criteria: *Presence of Critically Endangered and Endangered species – presence of a single individual or Vulnerable species – 30 individuals or 10 pairs.*		
Gazella subgutturosa	Four individuals of Goitered Gazelle were seen over a two-day period (possibly the same group on both days). In addition, this species has been reliably reported by locals and in a separate 2011 survey near Mil Qasm and Sar Qala, eight were observed.	
Important Bird Area Criteria	**Observations made 2007-2009.**	
A1. Globally threatened species	**Breeding**	**Wintering/ Passage**
Egyptian Vulture *Neophron percnopterus* (Summer visitor)	2 pairs (count 2007)	
Important Plant Area Criteria		
B1. Site is a particularly species-rich example of a defined habitat type		
Riverine Forest of the Plains (Al-Ahrash) habitat type.		
C. The site is identified as an outstanding example of a globally or regionally threatened habitat type.		

This site represents a good example of the Riverine Forest of the Plains (Al-Ahrash) habitat type, which is a globally important habitat and extends along the basins of Tigris- Euphrates Rivers. Also this site is under threats of gravel mining, agriculture and pollution from untreated sewage from the city of Kalar as well as towns and villages upstream.

Additional Important Bird Observations: During the surveys, 60 species were recorded. The site also held one breeding Irano-Turanian, one Mediterranean and three Sahara-Sindian Desert biome-restricted species but this did not trigger inclusion under criterion A3. European Roller *Coracias garrulus* (Near Threatened) was recorded in the breeding season.

Other Important Fauna: Data for non-avian fauna were collected during 2007 and 2008 only, when 18 Wild Boars *Sus scrofa*, which are often hunted, and one Indian Grey Mongoose *Herpestes edwardsii* were observed.

Fish: Data were collected in 2007 and 2008 only, during which five significant species were reported: *Alburnus mossulensis*, *Carasobarbus luteus*, *Cyprinion macrostomum*, *Gambusia holbrooki*, and *Garra rufa*.

Conservation Issues: The area is highly impacted by gravel mining. Agriculture and pollution from untreated sewage

from the city of Kalar as well as towns and villages upstream that enter the Diyala River also represent a threat to the site. Residential developments such as housing complexes are increasing in the city of Kalar and have expanded into the areas between Kalar and Kifri, especially in the years after the fall of Saddam. Hunting is another high threat with a focus on Goitered Gazelle and See-See Partridge, despite the fact that hunting is forbidden in this area.

Recommendations: Gravel mining occurring up and down the Diyala River must be controlled to eliminate and/or mitigate its negative impacts on water quality, sediment transport, riparian forests and ultimately, river-dependent species. This is a common problem on most Iraqi rivers. Overall the Kalar Area would benefit greatly from increased environmental planning and better land management. Monitoring of impacts from and controls on agriculture, development and pollution, particularly as the Diyala River is an important water source for locals and downstream populations as well as wildlife, are a priority. Additional monitoring of birds and other wildlife along the entire Diyala River corridor is needed.

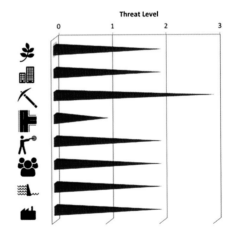

Maidan Area (IQ045)

Sulaimani – 34.655833°N 45.680278°E

KBA Criteria: **V**

Area: **53785 ha** - Altitude: **244-950 m**
Ecoregion: **Middle East Steppe (PA0812)**
Status: **Unprotected**

Site Description: This is an open area largely characterized by foothills and agricultural land. A string of mountains including Bamo and Sharwal Dra, run from the northeast to northwest. The Diyala River runs through the western part

of this area. There are several villages, including Kani Pamu, Razwar, Khurkhuri Khwaru and Khrkhuri Saru, most of which were abandoned during the Anfal campaign.

©2010 K ARARAT/NI

Key Biodiversity Area Criteria	Notes
V. Vulnerability Criteria: *Presence of Critically Endangered and Endangered species – presence of a single individual or Vulnerable species – 30 individuals or 10 pairs.*	
Gazella subgutterosa	Reported by local hunters. No direct observation.

Important Bird Observations: During the survey 37 bird species were observed. This site does not meet IBA criteria, but as it was only surveyed in winter, it could not be fully assessed and should be revisited during summer.

Other Important Fauna: Data for mammals were collected in 2010 only. Reports from hunters suggest the presence of Goitered Gazelle *Gazella subgutterosa* (Vulnerable) and Roe Deer *Capreolus capreolus*.

Fish: Data were not collected for this specific site. However, data were collected for the Diyala River in 2007 and 2008 in the Kalar Area (IQ044) located downriver of this area and so we have included these observations as they also likely apply to Maidan as well: *Alburnus mossulensis*, *Carasobarbus luteus*, *Cyprinion macrostomum*, *Gambusia holbrooki*, and *Garra rufa*.

Conservation Issues: This area is used for livestock grazing. Locals report that several gazelles were found dead in the area in 2009, possibly due to the 2008/2009 drought.

There are extensive gravel mines along the Diyala River causing damage to the river and river riparian habitat. There are minefields on the Iraq-Iran border located inside the delineated area, particularly near the villages of Rabayay Qabrsanaka, Pariawllay Khwarw, Awa Khwer, Zalkawaka, Rabayay Qara-Cham, Doli Dartel, Rabayaybari-Kalat, Quchalli, Grda Barz, Qalla Gawri, Kana Sor, and Makhfara Konaka. Most critically, over-exploitation of species was assessed as a very high threat, as gazelles and other wildlife face significant pressure from hunting. Residential and commercial development, transportation and service corridors, natural system modification and pollution are rated medium threats in this region.

Recommendations: Minefields should be cleared as they impact both humans and wildlife. The presence of gazelles warrants research to determine sustainable hunting limits and protective action through legislation, enforcement and education programs. Thought surveys of this area were limited it appears to be a rich site for biodiversity and warrants further surveys at different seasons of the year. As the area is important as livestock rangeland, the level and management of grazing should receive closer examination and management.

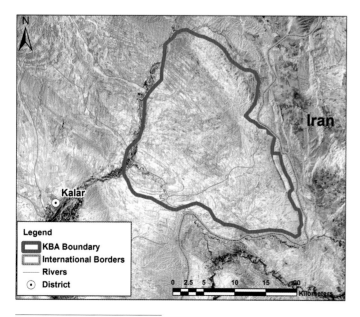

Legend
☐ KBA Boundary
☐ International Borders
— Rivers
⊙ District

0 2.5 5   10   15   20
Kilometers

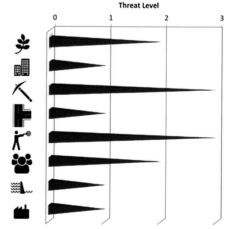

Threat Level
0   1   2   3

Jabal Makhool (IQ046)

Salah Ad-Din - 35.24066°N 43.332073°E

KBA Criteria: **V and Ia**
IBA Criteria: **A1**
IPA Criteria: **B1**

Area: **35257 ha** - Altitude: **113-468 m**
Ecoregion: **Mesopotamian Shrub Desert (PA1320)**
Status: **Unprotected**

©2012 A HALOOB/NI

Site Description: The site that borders the Tigris River north of the town of Baiji contains rocky hills with wooded slopes within the Irano-Turanian and Sahara-Sindian Desert biomes. It includes both the Makhool foothills that extend from the western borders of Al-Fatha (see below) toward Asherkat and the steppe situated on the eastern side of the main highway linking Baiji and Asherkat. The two habitat types were surveyed here were desert shrubs vegetation and steppe-herbaceous vegetation. The geology consists of marls, siltstone, sandstone, limestone, and gypsum; and the soil is clay.

The northeastern edge of the site borders the Tigris River and steppes of Kirkuk. A former presidential palace (Al-Abid) is situated at the top of Jabal Makhool. The site provided breeding habitat for the globally threatened Saker Falcon *Falco churrug* in the early 1990s (Allouse, 1960), and it was a traditional site for trapping species for Arabian falconry. It is also a suitable habitat for other birds of prey (Al-Sheikhly, 2011, 2012).

This sub-site includes an area called Al-Fattha "the Opening", which is a narrow gap between the Himreen foothills that extend from the southeastern edge of Jabal Makhool through which the Tigris River flows. Crossing the western ridge of Makhool the habitat changes frequently between the two biomes. Dry steppe and arid land is the dominant habitat of the site as viewed from the Himreen foothills on the highway road entering Kirkuk Governorate. The Baiji Oil and Gas Field Development Station, one of the largest oil and gas facilities in the northern part of Iraq is situated to the south of the site.

Zuwiyya Village is located near the Tigris River on the eastern edge of the site and this area features dry grasslands and arid country extending along the Makhool foothills on their northern and southern fronts. Several archeological sites, dating from the Assyrian to the Ottoman eras near Zuwiyya attract local visitors. Three sub-sites were visited in this area.

Key Biodiversity Area Criteria	Notes	
V. Vulnerability Criteria: *Presence of Critically Endangered and Endangered species – presence of a single individual or Vulnerable species– 30 individuals or 10 pairs.*		
Rafetus euphraticus	Reported by local people and fishermen from Tigris River	
Ia. Irreplaceability Sub-criterion: Restricted-range species based on global range		
Rafetus euphraticus	See above	
Important Bird Area Criteria	**Observations from Spring 2011**	
A1. Globally threatened species	**Breeding**	**Wintering/ Passage**
Egyptian Vulture *Neophron percnopterus* (Summer visitor)	2 pairs	
Important Plant Area Criteria		
B1. The site is a particularly species-rich example of a defined habitat type		
Foothills-desert shrubs vegetation habitat type and Steppe- herbaceous vegetation habitat type		

Additional Important Bird Observations: During the surveys 35 bird species were seen. Pallid Harrier *Circus macrourus* and European Roller *Coracias garrulus* (both Near Threatened) were recorded on passage, and three Irano-Turanian and eight Sahara-Sindian biome-restricted species were breeding, but these did not trigger inclusion under criterion A3.

Other Important Fauna: Local hunters and farmers have reported large packs of Grey Wolf *Canis lupus* near Makhool that attacked sheep at night, and they successfully hunted several during 2009 and 2010. Striped Hyena *Hyaena hyaena* have also been killed by locals and hunters have reported Goitered Gazelle *Gazella subgutturosa* (Vulnerable). A Wild Cat *Felis silvestris* was killed by a farmer near Baiji. The site is also rich in different reptile species, including many Lacertidae species such as, Arnold's Fringe-fingered Lizard *Acanthodactylus opheodurus*. The Levant Skink *Trachylepis aurata* was observed near the vegetated slopes of Al–Mussahag. The following important snakes were recorded:

Large Whip Snake *Dolichophis jugularis*, Collared Dwarf Snake *Eirenis collaris*, Persian Horned Viper *Pseudocerastes persicus*, and Desert Cobra *Walterinnesia morgani*.

Conservation Issues: The site was exposed to coalition bombing during 2003, and the remains of rockets and destroyed weapons are still present, including many unexploded mines and missiles. Future survey teams should proceed cautiously, especially at higher elevations. Despite the presence of munitions, the remains of Al-Abid Palace are a popular and crowded tourist site, especially during holidays. Scattered new settlements and development were observed near Zuwiyya Village and large sheep herds graze on the steppes. Pollution from the Baiji Oil and Gas Field is a very high threat to the site. Garbage left by local tourists was observed along the road toward Al-Abid Palace. Hunting and trapping of wildlife is practiced especially focused on birds of prey and carnivores.

Recommendations: The site is under the control of the Iraqi police so further research will require official permission to access. Iraqi literature contains little information about the ornithological and environmental importance of the site, as it was a restricted area during the previous regime. Further research and field surveys

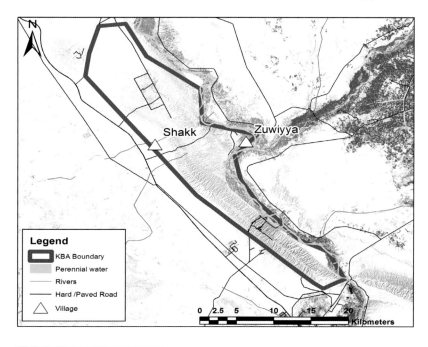

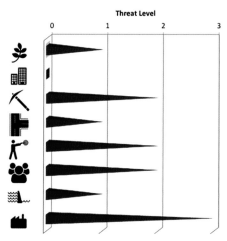

are recommended as only one visit to this area could be conducted. The development and implementation of an environmental management plan that outlines clear steps to reduce threats and pollution from the Baiji Oil and Gas Field is highly recommended. Clear enforcement of environmental regulations that protect the people and environment of the area is needed and regular monitoring is required to control pollution and related impacts. Managing hunting and trapping at a sustainable level will require cooperation with the local hunting associations and authorities. Al-Abid Palace will remain a tourist attraction but planning is needed to address waste from picnickers, and signage and interpretive programs could be developed to help educate visitors about both the environmental and cultural importance of the site.

Mahzam and Al-Alam Area (IQ047)

Salah Ad-Din - 34.715208°N 43.678417°E

KBA Criteria: **V and Ia**
IBA Criteria: **A2**
IPA Criteria: **B1**

Area: **2145 ha** - Altitude: **85-109 m**
Ecoregion: **Mesopotamian Shrub Desert (PA1320)**
Status: **Unprotected**

Site Description: Evans (1994) listed Mahzam and Tharthar as an important bird area (IBA007), but the recent surveys have split this into two independent sites due to their location and logistics. The site consists of homogeneous riparian habitats on both banks of the Tigris River as well as desert shrublands. There are also dense fruit and date palm orchards scattered amongst bush and thickets. The geology of the area is alluvial sediments and the land is arid and hemmed in along the western bank of the Tigris by elevated rocky cliffs that extend to Tikrit. The cliffs represent typical breeding habitat for many resident species of raptor, such as Common Kestrel *Falco tinnunculus*.

Many vegetable and fruit (largely grape) farms are distributed throughout the site. Bushes and shrubs thin out as the riverbanks turn to stone and gravel with thick poplar trees lining both sides. The eastern arm of the Tigris River is similar in habitat to Mahzam, although the Al-Alam region is characterized by date palms and orchards planted above wheat and barley fields. A few stony islands in the Tigris with scattered Tamarix plants provide nesting and roosting habitat for herons and waders. A number of gravel mines are also located on the Tigris riverbank near Al-Mahzam.

Key Biodiversity Area Criteria	Notes	
V. Vulnerability Criteria: *Presence of Critically Endangered and Endangered species – presence of a single individual or Vulnerable species– 30 individuals or 10 pairs.*		
Rafetus euphraticus	One adult was observed in spring 2011 in the Tigris River near Al-Alam.	
Ia. Irreplaceability Sub-criterion: Restricted-range species based on global range		
Rafetus euphraticus	See above	
Important Bird Area Criteria	**Observations made 2009-2011**	
A2. Restricted-range species	**Breeding**	**Wintering/ Passage**
Iraq Babbler *Turdoides altirostris* (Resident)	10 pairs (2009)	7 (count, 2011)
Important Plant Area Criteria		
B1. The site is a particularly species-rich example of a defined habitat type		
Inland running water-Riparian vegetation and Desert-shrub vegetation.		

Additional Important Bird Observations: In total, 120 bird species were seen. Pallid Harrier *Circus macrourus* and European Roller *Coracias garrulus* (both Near Threatened) were observed on passage. The site also held seven breeding Sahara-Sindian Desert biome-restricted species but did not trigger inclusion under criterion A3.

Other Important Fauna: Indian Gray Mongoose *Herpestes edwardsii* was observed and considerd the first record for Iraq (Al-Sheikhly and Mallon 2013). Also photographed at the site was the Indian Crested Porcupine *Hystrix indica*, which face heavy hunting pressure as they are a preferred food for locals in Salah Ad-Din. No fish survey was conducted.

©2010 A HALOOB/NI

Conservation Issues: Illegal hunting and fishing, primarily poisoning and occasionally electro-fishing, were considered the highest threat. Gravel mining is largely unregulated and has a high impact on the diversity of fish and invertebrates though disruption of spawning beds and rearing areas and may also impact overall water quality. Water pumping stations and electricity generators located on both sides of the river cause noise, oil and fuel spills and air pollution. In addition, oil spills upstream on the Tigris from the Baiji Oil and Gas Field have a very high impact on this site. The riverbank between Tikrit and Mahzam is a popular picnic spot in spring and summer causing pollution from garbage.

Recommendations: An environmental protection scheme, better regulation and enforcement of fishing and hunting, but also addressing gravel mining, water pumping, development, and waste management are recommended. Increased capacity for emergency oil spill response if also a critical need. Although the security situation in this region has dramatically improved, coordination with police or local councils should be pursued to reduce any danger faced by future survey teams.

Legend
- KBA Boundary
- Perennial water
- Hard /Paved Road
- Village

Khazmiya

0 0.375 0.75 1.5 2.25 3
Kilometers

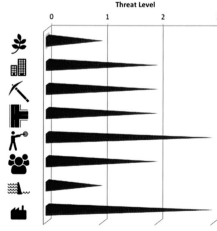

Ajeel Himreen Hills (IQ048)

Salah Ad-Din - 34.744864°N 44.034583°E

IBA Criteria: **A1**
IPA Criteria: **B1**

Area: **Area: 54261 ha**- Altitude: **130-310 m**
Ecoregion: **Mesopotamian Shrub Desert (PA1320)**
Status: **Unprotected**

©2010 A HALOOB/NI

Site Description: Arid hills (including a portion of the Himreen Hills) and steppe as well as some sandy areas are the main features of this site. Agricultural fields bordering the site to the southwest contain many hectares of irrigated fields of corn, wheat and barley. Two habitat types were observed, desert shrub and herbaceous vegetation-steppe.

The Al-Sder area, a steppe with some sand dune located in the southeastern part of the site has been a traditional spot for the hunting of Houbara (Macqueen's) Bustard *Chlamydotis macqueenii* and trapping of Saker Falcon *Falco cherrug* during winter migration.

Important Bird Area Criteria		Observations made 2009-2012	
A1. Globally threatened species		**Breeding**	**Wintering/ Passage**
Egyptian Vulture *Neophron percnopterus* (Summer visitor)		1 pair	2 (2010-2011)
Important Plant Area Criteria			
B1. The site is a particularly species-rich example of a defined habitat type			
Terrestrial Habitat- desert-shrub habitat type and Herbaceous vegetation-Steppe habitat type			

Additional Important Bird Observations: During the surveys 56 bird species were seen. In addition to Egyptian Vulture, Macqueen's Bustard *Chlamydotis macqueenii* (Vulnerable) was recorded but did not meet criterion A1. Pallid Harrier *Circus macrourus* and European Roller *Coracias garrulus* (both Near Threatened) were observed on passage. Six Sahara-Sindian Desert biome-restricted species were breeding but did not trigger inclusion under criterion A3. The endemic race of the Hooded Crow *Corvus cornix capellanus* also occurred.

Other Important Fauna: Important species seen during the surveys include Goitered Gazelle *Gazella subgutturosa* (Vulnerable), which were reported grazing on the grassy steppe near Al-Sder, but not in numbers that would meet

Vulnerability criteria. Also a Striped Hyena *Hyaena hyaena* (Near Threatened) killed by locals was examined by the team during the survey.

Conservation Issues: Very high threats include farming as well as grazing by large herds of fat–tailed sheep (Al-Awasi). Hunting pressures are also very high here. Several large local hunting companies are based in the southeast of the area and focus on taking people to hunt Houbara Bustard *Chlamydotis macqueenii* and trap Saker falcon *Falco cherrug* (Al-Skeikhly, 2011). The Albu Ajeel Oil Field, one of the region's main fields, is located to the northeast. This causes heavy pollution to the western part of the site but was considered a medium threat overall.

Recommendations: Intensive sureys in the eastern vicinity of Tikrit are recommended with additional efforts made to raise environmental awareness of hunting guide companies in coopertaion with the Iraqi Hunting Association

and Iraqi authorities. Stronger enforcement of hunting laws would help to reduce hunting and trapping pressure. Additional efforts here should focus on overgrazing in cooperation with the Ministry of Agriculture and pollution prevention from Albu Ajeel Oil Field with the assistance of the Ministry of Oil.

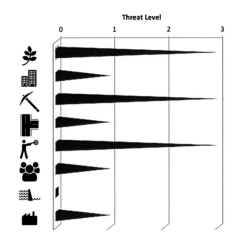

Qadissiya Lake (IQ049)

Anbar - 34.3575N 42.073333°E

KBA Criteria: **V and Ia**
IBA Criteria: **A1**
IPA Criteria **Not Assessed**

Area: 145230 ha - Altitude: **109-250 m**
Ecoregion: **Mesopotamian Shrub Desert (PA1320)**
Status: **Unprotected**

©2010 O AL-SHEIKHLY/NI

Site Description: Haditha Dam, which creates Qadissiya Lake, is an important strategic facility in northwestern Iraq located within Wadi Haditha in the Euphrates Valley. The towns of Anah and Rawa lie to the northwest. The dam has been functioning since the 1980s, providing electricity to Iraq's western regions. On the eastern side of the lake is an open steppe and barren semi-desert landscape called Abu Dalaia. The delineation for Qadissiya Lake includes an extension into this area where a Sociable Lapwing was satellite-tracked on the 15th and 19th of March of 2010. There was no botanical survey conducted here but wild vegetation consisted of *Acacia* sp, *Alhagi* sp, *Astragalus* sp, *Eucalptus* sp, *Halaxylon* sp, *Morus* sp, *Phragmites* sp, *Populus* sp, *Typha* sp, and *Zizphus* sp.

Qadissiya Lake is smaller than others lakes in western Iraq, such as Tharthar and Habbaniyah. The artificial lake features a muddy shore mixed with sand and gravel and some pockets of fresh grass and vegetation, while several gravel islands also appear near the eastern lakeshore. The river below the dam is moderately wide (30-40 m) with thick reeds covering the banks, extending to Haditha Wetlands and Baghdadi (IQ050), while a small number of date-palm trees and fruit farms are also distributed along both riverbanks. The wetland area is surrounded by arid steppes and dry sandy slopes. The majority of observations were made on the lake but the winter 2010 survey focused on an area downstream of the dam due to unfavorable weather conditions. The site remains closely protected by the army, who did not allow the team direct access to the dam area. Other parts of the lake are difficult to access due to poor security conditions, but the summer team was still able to survey much of the lake.

Key Biodiversity Area Criteria	Notes	
V. Vulnerability Criteria: *Presence of Critically Endangered and Endangered species – presence of a single individual or Vulnerable species– 30 individuals or 10 pairs.*		
Rafetus euphraticus	Euphrates Softshell Turtle was reported by fishermen in the reservoir.	
Ia. Irreplaceability Sub-criterion: Restricted-range species based on global range		
Rafetus euphraticus	See above	
Important Bird Area Criteria	**Observations from 2009-2010**	
A1. Globally threatened species	**Breeding**	**Wintering/ Passage**
Sociable Lapwing *Vanellus gregarius* (Rare Passage Migrant)		Satellite signal of tagged bird received on 15 and 19 March, indicating that the eastern extention of the site is a stop-over area for migration.
Egyptian Vulture *Neophron percnopterus* (Summer visitor)	1 pair (2009)	

Additional Important Bird Observations: During the surveys a total of 30 species were recorded. In addition, the site had breeding populations of two Sahara-Sindian Desert biome-restricted species but these did not trigger inclusion under A3 criterion.

Other Important Fauna: Many individuals of Golden Jackal *Canis aureus* were observed and photographed. No fish survey took place.

Conservation Issues: The site is a unique habitat in western Iraq with a wide spectrum of biodiversity but very little biological data has been collected so far. The highest threats observed were pollution and the Haditha Dam and management of water to generate electricity for the cities of Haditah, Anah, Rawa, Hit, Al-Qae'm and their vicinities. Huge turbines operate 20 hours a day. As a result industrial effluents (mainly machine oil and other chemicals) adversely affect water quality in the area near the dam. Garbage and solid waste, mainly non-degradable plastics, were observed at many localities. Also domestic wastewater and noise pollution from traffic as well as private electricity generators heavily affect the eastern parts of the lake. There are some small roads near the dam. Hunting was reported and a few boats fishing with nets were seen. The director of the dam indicated that fishing is prevented during the breeding season and the fishermen obey these rules. There is very limited recreational and tourism activity.

Recommendations: Further field surveys, with a specific focus on the biota and water quality are recommended. More cooperative work and awareness-raising activities with local fishermen to improve fish populations and sustainable fisheries is recommended. Logistically, the site is under the jurisdiction of several civilian and military agencies. Officially the site is managed by the Ministry of Electricity of Western Iraq, but security is controlled by the Iraqi National Army.

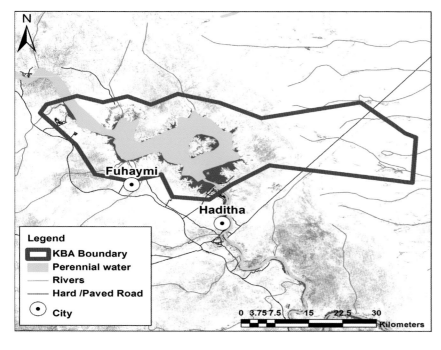

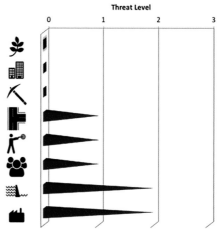

Haditha Wetlands & Baghdadi (IQ050)

Anbar - 33.905833N 42.532778E

KBA Criteria: **V and Ia**
IBA Criteria: **A1, A2 and A3**
IPA Criteria: **A4 and B1**

Area: **48274 ha** - Altitude: **65-187 m**
Ecoregion: **Mesopotamian Shrub Desert (PA1320)**
Status: **Unprotected**

©2009 MA SALIM/NI

Site Description: The site encompasses both banks of the Euphrates River in Baghdadi town and contains one of the most biologically important river valleys in Western Iraq, due to its unique habitat hosting important fish spawning areas during the spring months. The habitats surveyed in this area included desert shrublands and sparsely vegetated herbaceous vegetation in sparsely vegetated lands as well as riparian vegetation, reedbed and reedmace beds, and rooted submerged vegetation. The geology of the area is Euphrates limestone, shelly, dolomitized limestone.

The area is characterized by dense date palm orchards on either side of the river, in addition to citrus and other fruit trees. The shores of the Euphrates are grassy and muddy as the river flows towards Ramadi, with occasional rocks rising out of the river in the middle. These are considered an elevated extension of the river matrix, and one of the nesting sites for resident and migrant water birds. There is also occasionally submerged vegetation along the riverbank and some dense reed beds and marsh habitats.

Alongside the main highway there are desert and semi-arid habitats, sandy and open areas with rocky cliffs and scattered vegetation. Al-Haqlaniya is situated to the north of the site c. 10 km from the city of Haditha. Date farms and fruit orchards cover large areas along the east bank of the Euphrates but in dramatic contrast to the flat deserts beyond.

One of the oldest parts of Haditha is the shrine of Sheikh Hadid where a sinkhole is located. It is situated on the western bank of Euphrates and represents an important heritage landmark of the site. West of the site is Camp Al-Asad, one of the largest Iraqi military bases in Anbar Governorate, which regularly deploys troops and military vehicles for roadblocks and checkpoints in the surrounding area. An oasis is located on the base that is locally known as Abraham's Well. During 2010, a U.S. military official stationed at the base provided information on the biodiversity of the area but attempts by the KBA Team to visit the base were impossible.

Key Biodiversity Area Criteria	Notes	
V. Vulnerability Criteria: *Presence of Critically Endangered and Endangered species – presence of a single individual or Vulnerable species– 30 individuals or 10 pairs.*		
Rafetus euphraticus	Euphrates Softshell Turtle has been trapped by fisherman near Baghdadi.	
Garra widdowsoni	25 individuals found at the sinkhole at Sheikh Hadid Shrine in the spring of 2012	
Caecocypris basimi	Known historically at the Sheikh Hadid Shrine sinkhole but not observed in 2012	
Ia. Irreplaceability Sub-criterion: Restricted-range species based on global range		
Rafetus euphraticus	See above	
Important Bird Area Criteria	**Observations made 2010-2012**	
A1. Globally Threatened Species	**Breeding**	**Wintering/ Passage**
Sociable Lapwing *Vanellus gregarius* Passage Migrant		One bird reported to Nature Iraq at Al-Asad camp near Baghdadi.

A2. Range-restricted species	Breeding	Wintering/ Passage
Iraq Babbler *Turdoides altirostris* (Resident)	9 pairs (2011-2012)	7 (2010)
A3. Biome-restricted species		
Sahara-Sindian Desert biome	**Breeding**	**Wintering / Passage**
Cream-coloured Courser *Cursorius cursor* (Resident)	1 pair (2010)	
Spotted Sandgrouse *Pterocles senegallus* (Resident)	13 pairs (2010)	
Brown-necked Raven *Corvus ruficollis* (Resident)	2 pairs (2011)	
Hypocolius *Hypocolius ampelinus* (Summer vistor)	3 pairs (2010)	
Greater Hoopoe Lark *Alaemon alaudipes* (Resident)	1 pair (2012)	
Desert Lark *Ammomanes deserti* (Resident)	2 pairs (2012)	
Temminck's Horned Lark *Eremophila bilopha* (Resident)	2 pairs (2012)	
White-eared Bulbul *Pycnonotus leucotis* (Resident)	11 pairs	6 (2010)
Iraq Babbler *Turdoides altirostris* (Resident)	9 pairs (2011-2012)	7 (2010)
Dead Sea Sparrow *Passer moabiticus* (Resident)	10 pairs (2011)	38 (2010)
Important Plant Area Criteria		
A4. Site contains national endemic, near endemic, regional endemic and/or regional range-restricted species or infraspecific taxa		
Two near endemics *Allium vinicolor* (historically recorded) and *Onopordum canum* (seen on the KBA Survey) were found at this site.		
B1. The site is a particularly species-rich example of a defined habitat type		
Desert-desert shrubs habitat type; Herbaceous vegetation-sparsely vegetated land and Marsh vegetation- Helophytic vegetation- reedbeds and reedmace beds habitat type		

Additional Important Bird Observations: A total of 84 species was recorded. In addition to those listed in the table the following were observed on passage or in winter at levels that did not meet IBA criteria: in winter Marbled Duck *Marmaronetta angustirostris* (Vulnerable), Ferruginous Duck *Aythya nyroca* (Near Threatened), Eastern Imperial Eagle *Aquila heliaca* (Vulnerable) and Pallid Harrier *Circus macrourus* (Near Threatened).

Other Important Fauna: A local hunter reported on the rare hunting via electrocution of Eurasian Otter *Lutra lutra* near Haqlaniya. Golden Jackal, *Canis aureus* have been observed. The veterinary control center of the Coalition forces that were formerly stationed at Al-Asad military base have reported the trapping of adult male Striped Hyena *Hyaena hyaena* near the base perimeter as well as Rueppell's fox *Vulpes rueppelli* and Jungle cat *Felis chaus* in 2010. One amphibian species was observed at the site: Variable Toad *Bufotes variabilis.*

Fish: While the Euphrates River was not surveyed for fish, the site is important because the sinkhole at Sheikh Hadid Shrine contains two endemic cave-dwelling, blind fish species: *Garra widdowsoni* and *Caecocypris basimi* (both Critically Endangered). These two species were described in Iraq more than 50 and 32 years ago respectively by Trewavas (1955) & Banister and Bunni (1980). Also the site is reported to be an important spawning ground for important species such as *Luciobarbus esocinus* and *L. xanthopterus*, both Vulnerable species.

Conservation Issues: Baghdadi's habitat is uniquely representative of Western Iraq, and is a strong candidate for a nature reserve. Although the Ministry of Environment and Ministry of Agriculture have taken steps towards nominating the site for protected area status and including it in regular ministry visits and surveys, several fundamental issues must be resolved. Hunting and fishing is one of the highest impacts affecting the biodiversity. Waterfowl and game birds such as Black Francolin *Francolinus francolinus* and Macqueen's Bustard *Chlamydotis macqueenii* are the main bird species targeted by local hunters. The site was once an important area for Arabian Sand Gazelle that were previously reported to congregate in large groups drinking from the eastern side of Euphrates, but because of heavy hunting pressure are now rarely seen or reported (Al-Sheikhly, 2012). Electro-fishing was observed at many locations along the Euphrates River and local fishermen indicated that this technique had been imported from southern Iraq where it is widely practiced.

Agriculture is mainly represented by the extensive orchards along the river and by nomadic grazing and was considered a moderate threat; there were many new urbanization projects and tourism activities that are restricted to neighboring towns and villages of Anbar that influence the Euphrates River of Khan Al-Baghdadi; there are also a few new roads and only moderate activities of aircraft and helicopters heading toward Al-Asad Base; pollutants were mainly from urban wastewater, agricultural effluents, garbage, and noise.

Recommendations: This site should be declared a national protected area and requires additional, in-depth surveys (including botanical and fish surveys) to fully characterize its biological diversity. Protecting the annual fish spawning grounds especially the economically valuable species such as *Luciobarbus esocinus* and *L. xanthopterus* is a primary conservation issue. An examination of water quality is also needed. Priority actions include increasing awareness and enforcement of the current Iraqi hunting laws to reduce hunting pressure on threatened species. Pollution also deserves attention to improve sewage and solid waste handling and management to reduce agricultural pollutants. The strong military presence near the site warrants cooperation and communication with civilian and military authorities to ensure impacts on the remaining natural habitats, some of which are contained within the Al-Asad Military base, are minimized and access to important survey areas is allowed.

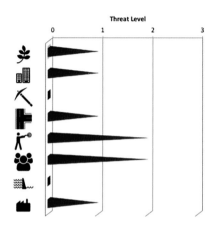

Tharthar Lake and Al-Dhebaeji Fields (IQ051)

Salah Ad-Din and Anbar -34.283889°N 43.183056°E

KBA Criteria: **V and Ia**

IBA Criteria: **A1, A2 and A3**

IPA Criteria: **B1**

Area: **340573 ha** - Altitude: **52-95 m**

Ecoregion: **Mesopotamian Shrub Desert (PA1320)**

Status: **Unprotected**

©2009 MA SALIM/NI

Site Description: This area, largely dominated by Tharthar Lake, is located on the border between the governorates of Salah Ad-Din (to the east) and Anbar (to the west) and between the Tigris and Euphrates Rivers. Regulators can control outflows from the lake to both these rivers. This area was list by Evans (1994) as an IBA site (IBA007). The habitats that were surveyed in the area included periodically flooded lands, reedbeds, rooted submerged vegetation as well as desert shrublands. The geology of the area is Lower Faris Series (marls, siltstones, gypsum/anhydrates, and limestone bands) and Euphrates limestone (shelly dolomitized limestone).

The landscape includes gravel hillsides covered in grass near the lake and a number of flat, sandy near-islands attached to the lake shore that extend out into the middle of the lake, which distinguish this area from the western part of Tharthar. The wide areas of open, arid steppe and cultivated areas of wheat and corn near the Al-Debe'e steppes harbor scattered halophytic vegetation and are considered one of the most important grazing areas in Iraq as a whole. While these are invaluable for cattle, they are also considered the main wintering grounds for many threatened species of birds such as Saker Falcon *Falco cherrug*, Houbara or McQueen's Bustard *Chlamydotis macqueenii*, and Sociable Lapwing *Vanellus gregarius*. Arabian Oryx *Oryx leucoryx* also occurred formerly. This habitat dominates the landscape of the north and northeast rib of the lake. Poor security conditions has been a problem for this site in the past but two sub-sites were surveyed.

Key Biodiversity Area Criteria	Notes	
V. Vulnerability Criteria: *Presence of Critically Endangered (CR) and Endangered (EN) species – presence of a single individual or Vulnerable species (VU) – 30 individuals or 10 pairs.*		
Rafetus euphraticus	There were two observations of Euphrates Softshell Turtle (EN): the first in fall 2010, a resting adult on the lake shore; the second in Spring 2011, an adult trapped and collected by fishermen.	
Ia. Irreplaceability Sub-criterion: Restricted-range species based on global range		
Rafetus euphraticus	See above	
Important Bird Area Criteria	**Observations made 2009-2012**	
A1. Globally threatened species	**Breeding**	**Wintering/ Passage**
Marbled Duck *Marmaronetta angustirostris* (Resident and winter visitor)	2-3 pairs (2009-2010)	43 (count 2010)
A2. Restricted-range species	**Breeding**	**Wintering/ Passage**
Iraq Babbler *Turdoides altirostris* (Resident)	1 pair (2012)	
A3. Biome-restricted species		
Sahara-Sindian Desert biome	**Breeding**	**Wintering/ Passage**
White-tailed Lapwing *Vanellus leucurus* (Resident)	1 pair (2009)	5 (count 2009)
Cream-coloured Courser *Cursorius cursor* (Resident)	1 pair (2012)	
Brown-necked Raven *Corvus ruficollis* (Resident)		4 (count 2009)
Greater Hoopoe-Lark *Alaemon alaudipes* (Resident)		2 (count 2010)
Desert Lark *Ammomanes deserti* (Resident)		2 (count 2010)
White-eared Bulbul *Pycnonotus leucotis* (Resident)	15 pairs (2010)	
Turdoides altirostris (Resident)	1 pair (2012)	
Dead Sea Sparrow *Passer moabiticus* (Resident)	8 pairs (2010)	
Important Plant Area Criteria		
B1. The site is a particularly species-rich example of a defined habitat type		
Inland standing water- Aquatic communities- rooted submerged vegetation habitat type; Flooded communities- periodically or occasionally flooded land habitat type; Marsh vegetation- Helophytic vegetation- reedbeds habitat type, and Terrestrial vegetation- desert-desert shrub habitat type		

Additional Important Bird Observations: During the surveys 54 bird species were seen. In addition to those listed in the table above Pallid Harrier *Circus macrourus*, European Roller *Coracias garrulus*, and Black-tailed *Godwit Limosa limosa* (all Near Threatened) were recorded on passage and the endemic race of Hooded Crow *Corvus cornix capellanus* was present.

Other Important Fauna: Golden Jackal *Canis aureus* was observed regularly. Striped Hyaena *Hyaena hyaena* was reported by locals and Caracal *Caracal caracal* was reported killed by local hunters near the lake edge. The Turkish Gecko *Hemidactylus turcicus* and Egyptian Spiny-tailed Lizard *Uromastyx aegyptia* (Vulnerable) were also observed.

Fish: Data were collected only in the winter of 2009 using a fisheries frame survey method, during which 10 species were observed. Fish caught in Tharthar tended to indicate good growth with some of the largest fish recorded for Iraq for that year. Fishing was by nets with mesh size of approximately 2 to 10 cm and a daily catch of about 30 kg/boat-day. About 100 boats were observed during this survey. The important species (listed with their catch ratios) were: *Carassius auratus* (3%), *Carasobarbus luteus* (12%), *Chondrostoma regium* (3%), *Cyprinus carpio* (15%, Vulnerable), *C. kais* (3%), *Leuciscus vorax* (15%), *Liza abu* (12%), *Luciobarbus xanthopterus* (31%, Vulnerable), *Mesopotamichthys sharpeyi* (3%, Vulnerable), and *Silurus triostegus* (3%).

Conservation Issues: There are many wheat and corn fields on the way to the survey area but mainly restricted to Al-Dhebaeji on the Salah Ad-Din side and Al-Geraeshi on the Anbar side of the lake. Center pivot irrigation was observed especially on the Salah Ad-Din side. The main fish landing sites for the local markets of Tikrit are located at Ein Al-Faras in Salah Ad-Din and near the former presidential complex at Al-Tharthar on the Anbar side. In both locations many fishermen were present, generally using legal fishing nets (neither small mesh-size nets or electro-fishing are allowed) but fishing was still considered a very high threat. The lake was deep, reaching nearly 80m at some areas, which renders illegal fishing procedures such as electro-fishing largely useless. But recent information regarding Iraq's long-range water resource planning indicate that after 2020 Tharthar may no longer receive waters from the Tigris except during flood conditions and from precipatation and with lower flow salinity will likely increase as the waters that do reach the lake evaporate. This is the result of reduced flows in Iraq overall and water resource allocation decisions. A high impact comes from pollution. Normally this was considered a low threat but there have been a number of oil spills on the Tigris River to the north and when these occur, oil ladden waters are diverted into Tharthar Lake to avoid these waters reaching population centers further south.

Recommendations: Given national water resource planning the future for Tharthar Lake itself indicates major ecological changes will be happening within the next decade, with the lake becoming smaller and more saline. This will represent significant socio-economic challenges in the region in addition to major shifts in the biodiversity of the area. For example, the fisheries of the lake will be greatly reduced and potentially eliminated. It is strongly recommended that a broad environmental action plan be developed to address these transitions particularly as they may affect agriculture, fishing and other uses of the area. Stronger fishing regulations and additonal fisheries surveys are needed to understand and plan for future changes to the lake. A national oil spill response program is also needed throughout the country and would benefit places such as Tharthar Lake. Finally, the surrounding steppe areas warrants further field surveys, which will require the cooperation with the national Iraqi police, army and local authorities to facilitate access.

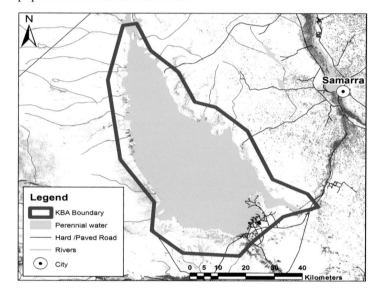

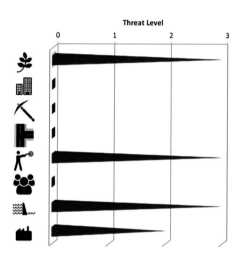

Samara Wetlands (IQ052)

Salah Ad-Din - 34.1925°N 43.852222°E

KBA Criteria: **V and Ia**
IBA Criteria: **A2**

Area: **4470 ha** - Altitude: **68-74 m**
Ecoregion: **Arabian Desert and East Sahero-Arabian Xeric Shrublands (PA1303)**
Status: **Unprotected**

Site Description: Samara Wetlands were listed originally as an Important Bird Area (IBA008) by Evans (1994). They are formed by the Samara Barrage, which was built in 1955 on the Tigris River, near the town of Al-Dure north of Samara City. The barrage regulates the water flow to Tharthar Lake (IQ051) through the Samara (Tharthar) Canal, which extends from Al-Dhloee'a town to the southeast edge of Tharthar Lake. The main purpose of the barrage was to divert

floodwaters to Tharthar but it also provides irrigation water and has a hydro-electric station. Sedimentation build up behind the barrage has helped to create the habitat of dense reed beds and submerged aquatic vegetation that are the main features of the site. This in turn attracts a significant number of migratory waterfowl and raptors.

No botanical survey was conducted here but there are beds of reeds, reedmace and carex with some terrestrial species such as Tamarix, Poplar and Eucalyptus bordering the marsh. Part of the site extends northwest along the main highway towards Tikrit, and is characterized by the same wetland habitat as the eastern side. Further to the west it transitions to arid steppe, covered with scattered xerophytic vegetation and a few fields of wheat, corn, and date palms. The side closer to Tikrit features dense reed beds that continue along the riverbank, in addition to scattered shrubs and thickets. Archeological ruins from the Abbasid

dynasty have been found close by, including Malewat, the ancient Samara mosque, and the Al-Ashiq palace to the west.

©2009 MA SALIM/NI

Key Biodiversity Area Criteria	Notes	
V. Vulnerability Criteria: *Presence of Critically Endangered and Endangered species – presence of a single individual or Vulnerable species– 30 individuals or 10 pairs.*		
Rafetus euphraticus	Euphrates Softshell Turtle was observed in Samarra Wetlands	
Ia. Irreplaceability Sub-criterion: Restricted-range species based on global range		
Rafetus euphraticus	See above	
Important Bird Area Criteria	Observations made 2009-2010	
A2. Restricted-range species	Breeding	Wintering/ Passage
Iraq Babbler *Turdoides altirostris* (Resident)	25 pairs (2009)	3 (count 2010)

Additional Important Bird Observations: In total, 70 bird species were seen. Marbled Duck *Marmaronetta angustirostris* (Vulnerable), Ferruginous Duck *Aythya nyroca* (Near Threatened), and European Roller *Coracias garrulus* (Near Threatened) were observed in the breeding season and Pallid Harrier *Circus macrourus* (Near Threatened) on passage. The site also held four breeding Sahara-Sindian Desert biome-restricted species but these did not trigger inclusion under A3 criterion.

Other Important Fauna: One Otter specimen was killed near the barrage by a farmer and deliverd to the Iraqi Natural History Museum in Baghdad. It was examined and identified as Eurasian Otter *Lutra lutra* (Near Threatened) by the author (Al-Sheikhly, 2012; Al-Sheikhly & Nader, 2013).

Fish: Data were collected in winter of 2009 using a fisheries frame survey method but at the time of the team visit fishing was officially not allowed for security reasons. However, that winter one person was seen using an electro-fishing device and he showed the survey team his catch of five species in the following catch ratios: *Carassius auratus* (25%), *Cyprinus carpio* (10%, Vulnerable), *Liza abu* (50%), *Mesopotamichthys sharpeyi* (5%, Vulnerable), and *Silurus triostegus* (10%). Though not in his catch, the fishermen indicated that two

others (*Luciobarbus esocinus* and *L. xanthopterus*, both Vulnerable species) were also present at the site.

Conservation Issues: This site is under strict military supervision due to the presence of the barrage. Political and military activities in this region may have led to its environmental preservation due to reduced civilian uses of the wetlands, particularly with regard to bird populations. However, significant environmental impacts were observed. Although hunting and fishing are prohibited around the dam, the presence of electro-fishing equipment in the winter of 2009 seems to indicate otherwise. The local people of Samara also hunt game birds and waterfowl in the eastern part of the wetlands after getting permisson letters from Iraqi authorities. Such activities are primarily practiced during the winter, targeting wintering and migrant birds. There is also increased construction especially the expansion of Samara city as well as the movement of military vehicles and troop training exercises. Solid wastes and garbage were observed in a few scattered locations, but a very high threat comes from the Baiji Oil and Gas Field upstream where oil spills often reach Samara and are redirected into Tharthar Lake (IQ051). Agriculture activities and cultivated areas were observed in places on the western edge of the wetlands. Agricultural practices in the Samara area in general, especially for seasonal vegetables and crops like wheat, corn and sunflower were considered a high threat.

Recommendations: A detailed survey and water quality study are highly recommended, as well as increased communication with officials and site authorities for greater access in future. The site needs active management and more research, in particular to estimate the capacity to

support sustainable fishing and other hunting activities. It is likely that the Ministry of Water Resources will dredge the area above the barrage and enlarge the barrage itself in the future. If so, it is important that the Ministry of Health and Environment be involved in this process to try to minimize impacts to the wetlands. Additionally, improved methods for swiftly handling oil spills are a critical need and enforcement of the hunting law is also essential. Further environmental management planning should guide agricultural expansion or human use of the site, especially as security and access to the site improve.

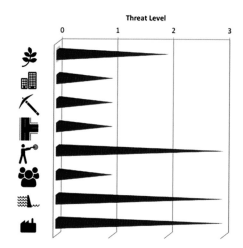

Himreen Lake (IQ053)

Diyala - 34.193056°N 45.003056°E

IBA Criteria: **A1**
IPA Criteria: **A4**

Area **80275 ha** - Altitude: **68-160 m**
Ecoregion: **Mesopotamian Shrub Desert (PA1320)**
Status: **Unprotected**

Site Description: This is an important lake in Iraq located near the Himreen Hills and known for its diverse bird life. The Himreen Dam (built in the late 1970s) formed the lake on the Diyala River that flows from the mountains of Iran and Iraqi Kurdistan. The lake is roughly rectangular in shape, with the dam on the southwestern edge. While no botanical

survey was conducted here the following plants are found: *Phragmites* sp, *Typha* sp, *Tamarix* sp, *Populus* sp, *Alhagi* sp, and *Astragalus* sp. Date palm and fruit orchards are also common in the area. The site is well protected by both the Iraqi Army and national police.

Important Bird Area Criteria	Observations from 2009-2011	
A1. Globally threatened species	**Breeding**	**Wintering/ Passage**
Lesser White-fronted Goose *Anser erythropus* (Winter visitor)		37 (2010)- See conservation issues section

Important Plant Area Criteria
A4. Site contains national endemic, near endemic, regional endemic and/or regional range-restricted species or infraspecific taxa
One endemic historically found in this site is *Ornithogalum iraqense*

©2010 O AL-SHEIKHLY/NI

Additional Important Bird Observations: During the surveys 48 bird species were seen. The following Vulnerable species were observed at the site in winter, though at levels that did not meet IBA criteria: Marbled Duck *Marmaronetta angustirostris*, Greater Spotted Eagle *Aquila clanga* and Eastern Imperial Eagle *Aquila heliaca*. The site also held one breeding Sahara-Sindian Desert biome-restricted species and the endemic race of Little Grebe *Tachybaptus ruficollis iraquensis*.

Other Important Fauna: Iraqi Mastigure *Saara loricata* were also seen at the site. Unfortunately, though this site has obvious importance for fish biodiversity, no fish surveys were conducted here.

Conservation Issues: This lake is an important water reservoir in Central Iraq and a favorable habitat for several threatened birds including Lesser White-fronted Goose (in 2010, 37 individuals of this species were found in the Baghdad market and the seller indicated that they had been trapped at Himreen Lake). Each year the site is visited by many hunters and trappers from nearby towns and cities and it is an important sites for waterfowl, birds of prey, and other wildlife used by the Iraqi hunting association based in Baghdad. Hunter convoys and camps are regularly found during winter on the nothern and northeastern edge of the lake. Hunting of Houbara (McQueen's) Bustard *Chlamydotis macqueenii* was observed in the open countryside close to the site. Fishing nets and boats were also observed and fishermen indicated that the lake provides many species of economic value and fishing is likely to increase when the security situation improves. Also Himreen Lake is an important recreation site for the people of Diyala. It becomes crowded especially near Al-Sudoor, in the southwestern part, near the water regulators.

Most visible pollution was solid waste such as plastic bags and other garbage resulting from improper waste management at nearby villages. Crop farms are largely restricted to the eastern and northeastern areas around the lake but appear to have a high impact. In addition, there is some housing and urban development but construction of new roads and service lines was limited.

Recommendations: Enforcing and strengthening hunting regulations is needed as well as efforts to raise awareness about species protection, through cooperation with Iraqi hunting associations at Himreen. Fishing and aquatic fauna surveys should be conducted and regulating fishing in coopereration with local fishermen and authorities in Diyala is important to sustain the local economy. Careful management, with support from local universities and Ministry of Agriculture should aim to control agricultural expansion around Himreen Lake in cooperation with local stakeholder groups to reduce negative impacts to biodiversity.

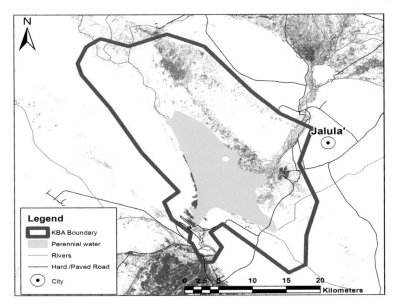

Legend
KBA Boundary
Perennial water
Rivers
Hard /Paved Road
City

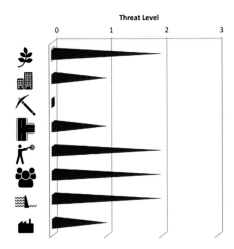

Threat Level

Mandli (IQ054)

Diyala - 33.766667°N 45.565278°E

KBA Criteria: **V**
IPA Criteria: **A4**

Area: **5386 ha** - Altitude: **135-240 m**
Ecoregion: **Mesopotamian Shrub Desert (PA1320)**
Status: **Unprotected**

©2010 O AL-SHEIKHLY/NI

Site Description: The area around the town of Mandli witnessed extensive fighting and was heavily bombed during the Iraq-Iran War, which destroyed much of its natural habitat and dramatically changed the biological characteristics of this unique site. This also includes, for example, the drying of some water bodies that originally flowed through this area. The site is located beneath a rocky extension of the Himreen hills. These form part of what is considered to be the national border of Iraq. Below these hills are open steppes with flat, sandy terrain covered in xerophytic and halophytic vegetation with occasional fresh and saline seasonal pools.

There was no botanical survey conducted but the site contains wetland habitat with *Typha* and *Phragmites* as well as terrestrial habitats featuring *Alhagi* and *Ziziphus* spp. and interesting tropical relict species such as *Acacia gerardii*. Dense fields of date palms are planted alongside fruit orchards and vegetable fields near the old city of Mandli to the south. A line of sandy cliffs passes through the site. These provide nesting sites for birds of prey, such as Barbary Falcon *Falco pelegrinoides* and Common Kestrel *Falco tinnunculus.*

Key Biodiversity Area Criteria	Notes
V. Vulnerability Criteria: *Presence of Critically Endangered and Endangered species – presence of a single individual or Vulnerable species– 30 individuals or 10 pairs.*	
Panthera pardus saxicolor	One Persian Leopard was killed by a villager from Mandli in 2009
Important Plant Area Criteria	
A4. Site contains national endemic, near endemic, regional endemic and/or regionalrange-restricted species or infraspecific taxa	

Two near endemics were historically found at this site including: *Cousinia stenocephala* and *Ergocarpon cryptanthum.*

Important Bird Observations: During the surveys 24 bird species were seen. The following species were also observed, but at levels that did not meet IBA criteria: European Roller *Coracias garrulus* (Near Threatened) and two Sahara-Sindian Desert biome-restricted breeding species.

Other Important Fauna: Few other observations were made at the site for non-avian fauna, but unconfirmed reports of the Vulnerable Persian Goitered Gazelle *Gazella subgutturosa* were obtained from local hunters.

Conservation Issues: Large numbers of date palms have been removed or died from the heavy bombing the site sustained during the Iran-Iraq war or as a result of military construction projects. Despite the high impacts the war had on the site, these issues are largely in the past and current human disturbance is caused by military activities and vehicle mobilization on the eastern borders of Mandli where an Iraqi military camp is based. Recently, Iran has constructed dams to block and redirect water that cross the border into Iraqi territory through Mandli, which has significantly impacted the area including harming local farming communities who now face water shortages. Because of the lack of water for irrigation, many farms and orchards have been abandoned and agriculture is on the decline.

The area has long been used for grazing, which may also contribute to its deterioration in certain localities. Mandli and its vicinity are also important areas for hunting, especially for the Vulnerable Persian Gazelle *Gazella subgutturosa subgutturosa* that are reported in the surroundings (Al-Sheikhly, 2012). Many bird species are trapped especially the White-eared Bulbul *Pycnonotus leucotis,* which is heavily targeted by locals. Many adults and juveniles were trapped and collected during the breeding season and brought to Baquba and Baghdad animal markets to be sold as pets. Other moderate threats are residential and commercial development and pollution from solid waste from nearby settlements.

Recommendations: Mandli's location near the Iranian border requires special security coordination to access. Cooperation with local officials and military authorities will continue to be a requirement for future surveys. Official dialogue between the Iraqi-Iranian authorities is recommended to solve the difficult water sharing and management issues. Implementation and enforcement of the Iraqi hunting law is essential and should be integrated with awareness raising efforts among local hunters and bird trappers about the importance of biodiversity conservation to Iraq's national heritage.

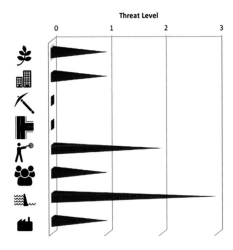

Ga'ara (IQ055)

Anbar - 33.501653°N, 40.436954°E

IBA Criteria: **A1**

IPA Criteria: **A4 and B1**

Area: **89558 ha** - Altitude: **465-655 m**
Ecoregion: **Mesopotamian Shrub Desert (PA1320)**
Status: **Unprotected**

Site Description: Ga'ara is a large desert site situated c. 60 km to the north of Rutba in northwestern Anbar province. The site is a natural depression in the western desert of Iraq and it surrounded on its southern and eastern parts by elevated rocky ground and cliffs (called Al-Afaeif), which form one of the main landmarks of the area. The two main habitats studied were desert shrublands and herbaceous vegetation-steppe lands. During the wet seasons the people of Rutba use part of the area for wheat and sunflower agriculture. Dam construction has been noted in several areas of Ga'ara but the team was not able to obtain any information on these projects.

©2010 A HALOOB/NI

Important Bird Area Criteria	Observations made 2012	
A1. Globally threatened species	**Breeding**	**Wintering/ Passage**
Egyptian Vulture *Neophron percnopterus* (Summer visitor)	<5 pairs	
Important Plant Area Criteria		
A4. Site contains national endemic, near endemic, regional endemic and/or regional range-restricted species or infraspecific taxa		
A near endemic species historically found here is *Allium vinicolor*		
B1. The site is a particularly species-rich example of a defined habitat type		
Desert-shrubland habitat type and Herbaceous vegetation-Steppe habitat type		

Additional Important Bird Observations: During the surveys a total of 24 species were recorded. In addition the site had breeding populations of six Sahara-Sindian Desert biome-restricted species but these did not trigger inclusion under criterion A3.

Other Important Fauna: Striped Hyaena *Hyaena hyaena* (Near Threatened) has been reported by locals. Gray Wolf *Canis lupus* attacks on local animal herds have also been reported which probably represents the Arabian subspecies. Reptiles observed included Egyptian Spiny-tailed Lizard *Uromastyx aegyptia* (Vulnerable); Blanford's Short-nosed Desert Lizard *Mesalina brevirostris,* and Arabian Horned Viper *Cerastes gasperettii.*

Conservation Issues: Very few human activities influence the site due to its remoteness. The village of Al-Ga'ara consists of only six houses. The people here are mainly Bedouin roaming the western desert of Iraq, cultivate wheat during the wet seasons (in spring), but only on a very small scale. The main impact is the hunting of birds, especially raptors, during the migration and the winter but this isolated site seems not to be affected by humans during the hot, dry seasons. More information is needed to understand and control the impacts of dam construction in this region.

Recommendations: Few true desert sites have been included in the KBA surveys and this remains a gap to be filled in future. We recommend more detailed surveys of the biodiversity of this site. Because it is remote and not fully secure, serious dialogue with Iraqi security authorities is needed to facilitate access.

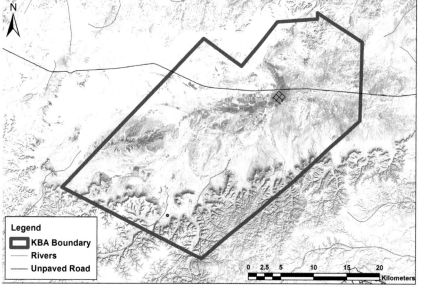

Legend
KBA Boundary
Rivers
Unpaved Road

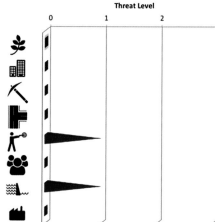

Habbaniyah Lake (IQ056)

Anbar - 33.196667°N 43.460556°E

KBA Criteria: **V and Ia**
IBA Criteria: **A1**
IPA Criteria: **Not assessed**

Area: **45390 ha** - Altitude: **43-65 m**
Ecoregion: **Arabian desert and East Sahero-Arabian xeric shrublands (PA1303)**
Status: **Unprotected**

©2009 MA SALIM/NI

Site Description: Habbaniyah Lake is located southeast of Ramadi, the capital of Anbar Governorate, and west of Baghdad. It is a large water reservoir constructed in 1982 and Evans (1994) included it in the original list of Important Bird Areas (IBA016). It receives excess floodwaters from the Euphrates in the summer through a small canal near Ramadi called Sin Al-Dhuban. The canal passes through Al-Saglawiya and the calcareous Al-Guss hills, which separate the canal from Habbaniyah. The excess floodwaters drain out on the southern edge of the lake through the narrow Al-Majarah Canal, which drains to Bahar Al-Milih and the northern part of Razzaza Lake (IQ058) in Karbala Governorate.

The main habitat is the lake itself and its wide, muddy shoreline. While no botanical survey was conducted at the site, species of *Phragmites* sp, *Typha* sp, *Achillea* sp, *Artemeisia* sp, *Acacia* sp and *Alhagi* sp were the most widely distributed plants.

There is a small elevation gain near the southern edge of the lake, and semi-desert forms the eastern and southwestern front of the lake with xeric and halophytic desert vegetation. The shore is widely exposed during the wintertime when the water levels are reduced to their minimum level. A few wetlands with submerged aquatic vegetation were observed near Al-Majarah water regulation canal, including a limited number of reed beds. The village of Al-Angoor is located on the southwestern edge of Habbaniyah and contains a few people who mainly practice fishing. The Habbaniyah tourism village is one of the most significant landmarks, located on the southeast edge of the lake. The habitat around this area features some dense shrubs and thickets.

Key Biodiversity Area Criteria	Notes	
V. Vulnerability Criteria: *Presence of Critically Endangered and Endangered species – presence of a single individual or Vulnerable species– 30 individuals or 10 pairs.*		
Rafetus euphraticus	*Euphrates Softshell Turtle was reported by local fishermen at the lake.*	
Ia. Irreplaceability Sub-criterion: Restricted-range species based on global range		
Rafetus euphraticus	See above	
Important Bird Area Criteria	**Observations from 2009-2011.**	
A1. Globally threatened species	**Breeding**	**Wintering/ Passage**
Marbled Duck *Marmaronetta angustirostris* (Resident, winter visitor)		60 (2009)

Additional Important Bird Observations: During the surveys a total of 49 species were recorded. The site also held breeding populations of four Sahara-Sindian Desert biome-restricted species but these did not trigger inclusion under criterion A3.

Other Important Fauna: One adult Rüppell's Fox *Vulpes rueppelli* was observed crossing the road that leads to the site near Al-Angoor Village. Egyptian Spiny-tailed Lizard *Uromastycs aegyptia* (Vulnerable) and Desert Cobra *Walterinnesia morgani* were observed.

Fish: Data were collected exclusively from interviews in the winter of 2009 as part of a fisheries frame survey because high winds did not allow fishermen to go out. There were about 100 boats operating on the lake with an estimated average daily catch of 20 kg/boat-day using floating gill nets with mesh sizes of 1–5cm. No electro-fishing was practiced at Habbaniyah.

The six important species, according to Coad (2010) that are found in the lake are provided here with their catch ratios: *Acanthobrama marmid* (20% of the catch), *Alburnus mossulensis* (10%), *Leuciscus vorax* (appear occasionally in the catch), *Carasobarbus luteus* (20%), *Liza abu* (50%), and *Luciobarbus xanthopterus* (the latter a Vulnerable species appears occasionally in the catch).

Conservation Issues: Water shortages are causing an increase in salinity, water stagnation and reducing water quality in some parts of the lake. The Habbaniyah tourism village highly impacts the local environment especially as the survey team witnessed new efforts at rehabilitation of the area by the Anbar authorities during 2011. Spring and summer sees the highest number of visitors and the most serious impact to the site, with large quantities of solid waste such as cans and plastic debris left behind, which spread rapidly throughout the site and ring the lake edges. Several small villages on the southern and eastern edges of the lake deposit sewage and other waste into the lake. Hunting and trapping of wildlife especially for large mammals, game birds, and raptors such as Saker Falcon and Peregrine Falcon during winter is another high threat (Al-Sheikhly, 2012).

Land near Habbaniyah is also used as an air force base. The frequent training flights result in widespread noise pollution and other environmental impacts sufficient to disturb and harm both resident and migrant species. Small farms growing annual crops are limited and restricted to the northern and northwestern edge of the site. A few villages and urban areas can be found on the western edge of the site near Al-Angoor. In 2015, a review done of Iraq's KBAs found that this site was affected by the conflict zone of the Islamic State of Iraq and Syria (ISIS).

Recommendations: The KBA fish surveys were limited here and more are needed as well as studies of other non-avian species. Water quality and flow conditions should also be more closely examined. Enforcing Iraq's current hunting laws and raising awareness among local hunters and fishermen in cooperation with local hunting groups or associations in Anbar can help reduce the potential for declines in wintering raptors and other fish and wildlife.

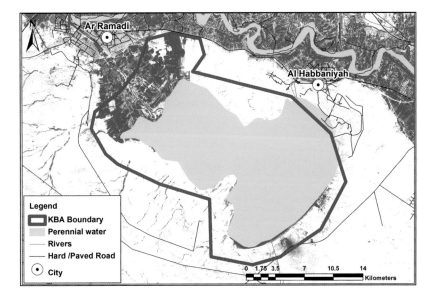

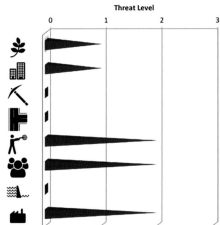

Jazman (Zurbatia) (IQ057)

Wasit - 33.116786°N 46.071816°E

KBA Criteria: **V**
IBA Criteria: **A1 and A3**
IPA Criteria: **A4 and B1**

Area: **155095 ha** - Altitude: **90-850 m**
Ecoregions: **Zagros Mountains Forest Steppe (PA0446); South Iran Nubo-Sindian Desert and Semi-Desert (PA1328), and Mesopotamian Shrub Desert (PA1320)**
Status: **Unprotected**

©2009 MA SALIM/NI

Site Description: Jazman (also known as Zurbatia) is a dry area with plant cover concentrated mainly in natural watercourses and agricultural canals. It is located northeast of the city of Kut, close to the Iranian border. The closest watercourse is Galala Badra that runs through the southern part of the site and feeds the dense orchards and farms of the area before reaching Badra town. There is an elevated area in the east and northeast parts of the site on the border with Iran where the site enters the Zagros Mountains Forest Steppe ecoregion. The area has many irrigation canals and small, natural watercourses. The geology is gravel, sands, silts, alluvium, sandstone, clay, and sandy gravel and the soil type was clay and sandy clay. The habitat is primarily desert shrub and sparsely vegetated herbaceous vegetation. Jazman provides good habitat for migratory raptors and passerines and is well known for gazelle herds that use the Iranian heights when they face pressure from hunting on the Iraqi side.

Key Biodiversity Area Criteria	Notes	
V. Vulnerability Criteria: *Presence of Critically Endangered and Endangered species – presence of a single individual or Vulnerable species– 30 individuals or 10 pairs.*		
Gazella spp.	Reported by locals and hunters frequently over the northern and northeastern areas. The species is probably *Gazella subgutturosa*.	
Important Bird Area Criteria	**Observation made 2005-2011.**	
A1. Globally threatened species	**Breeding**	**Wintering/ Passage**
Macqueen's Bustard *Chlamydotis macqueenii* (Winter visitor)		Reported frequently by locals and hunters. Large numbers hunted by falconers each winter.
A3. Biome-restricted species		
Sahara-SindianDesert biome	**Breeding**	**Wintering/ Passage**
Spotted Sandgrouse *Pterocles senegallus* (Resident)	300 pairs	3 (2009)
Pallid Scops Owl *Otus brucei* (Resident)	4 pairs	
Egyptian Nightjar *Caprimulgus aegyptius* (Summer visitor)	100 pairs	
Hypocolius *Hypocolius ampelinus* (Summer visitor)	200 pairs	
Pale Crag Martin *Ptyonoprogne obsoleta* (Summer visitor)	One pair (2009)	
Greater Hoopoe-Lark *Alaemon alaudipes* (Resident)	30 pairs	Present
Bar-tailed Lark *Ammomanes cinctura* (Resident)	20 pairs	
Desert Lark *Ammomanes deserti* (Resident)	100 pairs	Present
Temminck's Lark *Eremophila bilopha* (Resident)	10 pairs	
White-eared Bulbul *Pycnonotus leucoti* (Resident)	150 pairs	Present
Desert Finch *Rhodospiza obsoleta*	8 pairs (2009)	

Important Plant Area Criteria

A4. Site contains national endemic, near endemic, regional endemic and/or regional range-restricted species or infraspecific taxa

Historical near endemics at this site included: *Anthemis gillettii*, *Bupleurum leucocladum*, *Echinops armatus* var. *papillosus*, *E. tenuisectus* Rech.f., *Malabaila secacul* subsp. *aucheri* and *Plantago psammophila*

B1. The site is a particularly species-rich example of a defined habitat type

Herbaceous vegetation- Sparsely vegetated land habitat type

Additional Important Bird Observations: During the surveys, 22 bird species were recorded. Eastern Imperial Eagle *Aquila heliaca* and Greater Spotted Eagle *Aquila clanga* (both Vulnerable) were also seen and the area provides good habitat for wintering and migrant birds. Iraq Babbler *Turdoides altirostris*, a range-restricted species, was observed in 2010, but did not trigger inclusion under A2 criterion.

Other Important Fauna: According to field observations and frequent reports of locals and hunters, mammals include: Honey Badger *Mellivora capensis*, a wild cat *Felis silvestris/chaus* (species could not be distinguished), Red Fox/Ruppell's Fox *Vulpes* spp. (species not distinguished), Golden Jackal *Canis aureus*, and Grey Wolf *Canis lupus*. Gazelles are found in the eastern and northeastern hills and a few Gazelles were killed in 2008 close to the border. Some reptiles were seen but not identified.

Additional Plant & Habitat Information: This site is located in the junction of several ecoregions, so will have important habitat for plants and features of each of these ecoregions. It is under threat from aridification, grazing, oil development, pollution, human disturbance, and road development but it currently contains a good population of Date Palm *Phoenix dactylifera*, which is economically and culturally important. Multiple species of Wheat *Triticum* spp. and *Avena fatua* are present, which could be good genetic resources for wheat and oats respectively in Iraq.

Conservation Issues: Human disturbance from agriculture, hunting, and other activities, especially in the middle and lower parts of the site, is a very high threat. Agricultural intensification; road development; hunting (focused on the Vulnerable Houbara Bustard as well as Gazelle); the blockage of water resources from Iran and the potential development of oil resources are also threats to the site. Badra Oilfield also performs drilling and extraction activities in the southern parts of this site, and the expansion of this oilfield northward is expected.

Recommendations: Jazman is the only KBA site in Iraq located in an area where several ecoregions come together but it is still poorly known. More field research is recommended to gain information on its biodiversity and the threats facing it. It is also important to conduct meetings with the local government (including the border police) and stakeholders to coordinate conservation work. The expansion plans of the Badra Oilfield and related oil development should be examined and a dialogue initiated about biodiversity protection in case of an expansion northward.

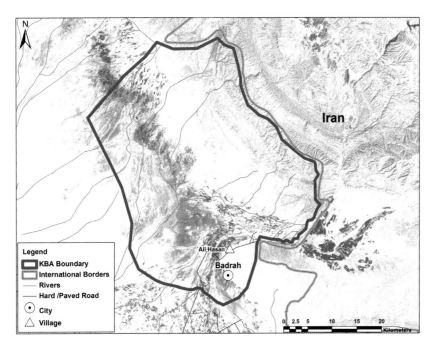

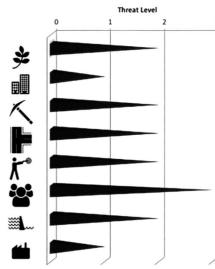

Razzaza Lake (IQ058)

Karbala & Anbar - **32.68333°N 43.66667°E**

IBA Criteria: **A1, A2, A3, A4i, and A4iii**
IPA Criteria: **A4 and B1**

Area: **156234 ha** - Altitude: **28-56m**
Ecoregion: **Arabian Desert and East Sahero-Arabian Xeric Shrublands (PA1303)**
Status: **Unprotected**

©2008 MA SALIM/NI

Site Description: In Evans (1994), this area is listed as "Bahr Al Milh" (IBA021) but it is more commonly referred to as Razzaza Lake. Evans states that the lake was formed in the 1970s as a storage reservoir to control floods on the Euphrates. It is connected to Habbaniyh Lake in the north by a narrow canal running through semi-desert, called Sin Al-Thibban Canal. Razzaza used to be a large, deep lake, but it has shrunk in size and is now characterized by very high salinity levels, which has increased during the past ten years due to the shortage of water and evaporation during Iraq's very dry, hot summers. Locals report that the lake is likely now only 5–10 m deep. The extensive shrinkage of the lake is graphically seen in the two satellite images taken in 2000 and 2013 (see below). These are not seasonal changes but are based on management and water allocation to Razazza via the Sin Al-Thibban Canal from Habbaniyah Lake. The delineated map above is based on the recent historical extent of the lake, which the lake would likely return to if more water is dedicated to the area.

Islands in the lake provide breeding areas for gulls and potential breeding ground for Greater Flamingo. By the last KBA Survey in 2010, the lake still seriously lacked water and most of the birds were found concentrated in the area of the sole source of water that brings sewage from Karbala. The Sin Al-Thibban Canal (that brings water from Habbaniyah) was still closed by the Anbar government. The geology of the area is marls, siltstone, gypsum/anhydrite, and limestone bands, mainly silts. Though there is little to no plant life on the lake itself and most plants are near the seasonal drainages and canals that bring sewage water to the lake, the main plants found include beds of *Phragmites australis*, *Juncus acutus*, *Aeluropus lagapoides*, *Salicornia herbacea*, and *Schoenoplectus littoralis* and there are desert shrublands with plants such as T*amarix aucherana*, *T. macrocarpa*, *Prosopis farcta*, *Zygophyllum fabago*, *Nitraria retusa*, and *Haloxylon salicornicum*.

Two main towns lie close to the lake: Ain Al-Tamr (to the southwest) and Al-Rahaliya (west). The area west of Ain Al-Tamr was part of the survey area. There are many date farms on the west edge of the site (all depending on well and rainwater) but the area to the south and east is shrublands. A paved road passes through the site. Al-Rahaliya, a small town situated near the western bank of the lake, is near a saline shallow-water area. The semi-desert areas around this features xerophytes and halophytes but also contains recently planted date palm trees and orchards near the town. Al-Taar area is a depression in the desert to the west of Karbala and south of the lake that collects rainwater from the relatively higher lands to the south and west. Four sub-sites were surveyed at Razzaza Lake.

March 2003   December 2013

Important Bird Area Criteria	Observations made 2008-2010.	
A1. Globally threatened species	**Breeding**	**Wintering/ Passage**
Greater Spotted Eagle *Aquila clanga* (Winter visitor)		35 (highest count)
Marbled Duck *Marmaronetta angustirostris* (Resident and winter visitor)	100 pairs	2,000-4,350 (counts)
A2. Restricted-range species	**Breeding**	**Wintering/ Passage**
Iraq Babbler *Turdoides altirostris* (Resident)	300 pairs (2008-2009)	Present
A3. Biome-restricted species		
Sahara-Sindian Desert biome	**Breeding**	**Wintering/ Passage**
Macqueen's Bustard *Chlamydotis macqueenii* (Winter visitor)	Few individuals seen in summer by locals and hunters	
White-tailed Lapwing *Vanellus leucurus* (Resident)	500 pairs	80-420 (count 2008-2010)
Cream-coloured Courser *Cursorius cursor* (Resident)	100 pairs	Present
Spotted Sandgrouse *Pterocles senegallus* (Resident)	200 pairs	
Egyptian Nightjar *Caprimulgus aegyptius* (Summer visitor)	100 pairs	
Brown-necked Raven *Corvus ruficollis* (Resident)	30 pairs	Seen in summer
Hypocolius *Hypocolius ampelinus* (Summer visitor)	300 pairs	
Bar-tailed Lark *Ammomanes cinctura* (Resident)	50 pairs	1 (count 2008-2010)
Desert Lark *Ammomanes deserti* (Resident)	Present, not counted	2 (2008-2010)
Temminck's Lark *Eremophila bilopha* (Resident)	50 pairs	2 (2008-2010)
Greater Hoopoe Lark *Alaemon alaudipes* (Resident)	200 pairs	6 (count 2008-2010)
White-eared Bulbul *Pycnonotus leucoti* (Resident)	200 pairs	8 (count 2008-2010)
Iraq Babbler *Turdoides altirostris* (Resident)	90 pairs	Present
Dead Sea Sparrow *Passer moabiticus* (Resident)	Present; suitable breeding habitat, but no nests were found.	200-600 (counts 2008-2010)
A4i. 1% or more of biogeographical population of a congregatory waterbird species		
	Breeding	**Wintering/ Passage**
Red-crested Pochard *Netta rufina* (Resident)	300 pairs	1,000-2,800 (counts)
Greater Flamingo *Phoenicopterus roseus* (Resident)	500 pairs, but breeding not confirmed.	1,500-3,500 (counts)
White-tailed Lapwing *Vanellus leucurus* (Resident)	500 pairs	80-420 (counts 2008-2010)
Kentish Plover *Charadrius alexandrinus* (Resident)	400 pairs	700-1,800 (counts)
Slender-billed Gull *Chroicocephalus genei* (Resident)	600-1,800 pairs (counts)	2,300-4,500 (counts)
Whiskered Tern *Chlidonias hybrida*		500-1,500 (counts)
A4iii. Holding congregations of 20,000 waterbirds or 10,000 pairs of seabirds of one or more species		
Congregatory Waterbirds		>28,000
Important Plant Area Criteria		
A4. Site contains national endemic, near endemic, regional endemic and/or regional range-restricted species or infraspecific taxa		
One nationally rare species was found: *Ephedra transitoria*		
B1. Site is a particularly species-rich example of defined habitat type		
Desert shrub habitat type		

Additional Important Bird Observations: During the surveys 42 bird species were observed. Razzaza Lake provides vast areas of mudflats that are suitable habitat for large numbers of migrant and wintering waterfowl and waders. There is a resident population of Greater Flamingo *Phoenicopterus roseus* that might use this wetland for breeding.

Important Other Fauna: The valleys and dense plant cover (including orchards) on the western side of the lake and the flat arid/semi-desert areas on the eastern and southern parts of the lake might harbor considerable wildlife diversity; however, these areas were not surveyed during the KBA surveys. According to local reports, mammals include: Rüppell's Fox *Vulpes rueppellii*, Golden Jackal *Canis aureus*, Indian Grey Mongoose *Herpestes edwardsii*, Jungle Cat *Felis chaus*, Wild Cat *Felis silvestris* and other common species.

Fish: A Fisheries Frame Survey was conducted at the site in 2009. Significant pressure previously existed with an estimated 300 boats catching approximately 50 kg/boat per day in the recent past. But at the time of the survey, only 20 to 25 boats were seen and no active fishing was taking place. At this time only one old fisherman could be found who reported that only one fish species (*Acanthopagrus cf. arabicus*) is found in the lake, which is a marine fish in origin and is stocked at the site by the government to support fishing. According to this fisherman, fishing was done using nets with mesh sizes of 0.5 to 3 cm and no unsustainable fishing methods were used but many fishermen have left the lake because of decreasing fish stocks and the declining water supply. After the last KBA survey, *Tilapia zillii* was later observed and reported by locals.

Additional Plant & Habitat Information: This area contains a threatened habitat (Inland standing water-pond or lake). It contains a good population of date palm *Phoenx dactylifera*, which is an economically and culturally important plant.

Conservation Issues: The main threat to Razzaza is the lack of water since the main source (from Habbaniyah Lake via the Sin Al-Thibban Canal) is completely blocked. Additional water in Razzaza would improve the circulation of water, reduce salinity of the lake and provide more habitat for birds, plants and fish. Unfortunately recent information on long-term planning indicates that Razzaza will continue to receive less water not more over the coming years. Pollution to the lake comes from the drainage canal that collects sewage and agricultural wastewater from the adjacent areas and is currently one of the only water input sources for Razzaza. Human intrusion (especially during the bird breeding season), movement of trucks causing dust and disturbance, gravel mining and the hunting of birds (particularly waterfowl) and fishing are significant impacts though the last is now in decline due to the declining water and increasingly salinity of the lake.

Recommendations: Razzaza Lake harbors considerable numbers of waterfowl (particularly the globally threatened Marbled Duck) and its mudflats attract large number of waders and shorebirds during their passage. It also has a number of inaccessible marshlands that are important for breeding birds, in addition to the islands that are perfect for breeding gull and tern species that occur in quite good numbers. For this lake to offer improved habitat for birds and fish, water must be released into it from Habbaniyah Lake, but regardless the site will always contain some waters that will provide support for biodiversity and continued surveys of Razzaza Lake and the surrounding area are recommended to study the threatened species that live in or regularly visit this wetland. This site was identified in 2013 as a proposed protected area by the National Protected Area Committee.

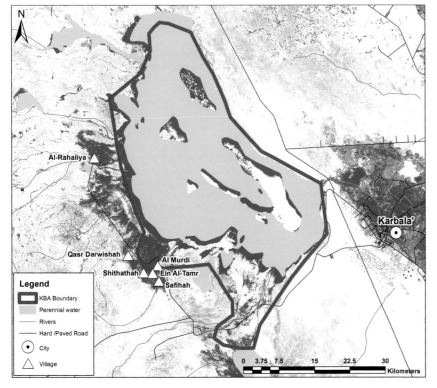

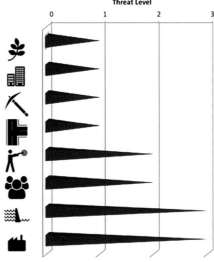

Musayab (IQ059)

Babil - 32.80806°N 44.27556°E

KBA Criteria: **V and Ia**
IPA Criteria: **Not assessed**

Area: **162 ha** - Altitude: **32-34 m**
Ecoregion: **Arabian desert and east Sahero-Arabian Xeric Shrublands (PA1303)**
Status: **Unprotected**

©2010 MA SALIM/NI

Site Description: Evans (1994) described this site (IBA018) as a wetland to the north of the city of Musayab based on field survey made by George and Mahdi (1969), but it seems that the marsh described by Evans does not exist now. During the 2009 surveys, the team surveyed the area to the north of Musayab, mainly along the eastern side of the Euphrates River. The original Musayab IBA site appears to be located on the eastern side of Euphrates close to a small island in the middle of the river. The area in general is composed of farms and orchards, and no undisturbed areas remain. The area includes some aquaculture pools, especially to the north but outside the delineated area.

Key Biodiversity Area Criteria	Notes
V. Vulnerability Criteria: *Presence of Critically Endangered and Endangered species – presence of a single individual or Vulnerable species– 30 individuals or 10 pairs.*	
Rafetus euphraticus	Euphrates Softshell Turtle was observed (and reported by locals) in considerable numbers over the entire length of the river.
Ia. Irreplaceability Sub-criterion: Restricted-range species based on global range	
Rafetus euphraticus	See above

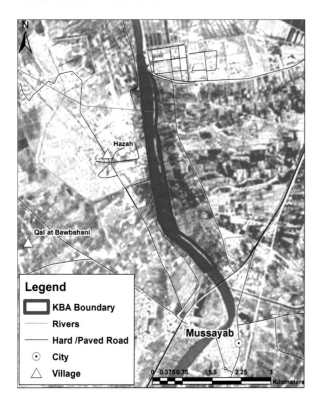

Legend
- KBA Boundary
- Rivers
- Hard /Paved Road
- ⊙ City
- △ Village

Additional Important Bird Observations: During the survey 27 species were observed. Pallid Harrier *Circus cyaneus*, and Black-tailed Godwit *Limosa limosa* (both Near Threatened) were recorded, and one restricted-range species, Iraq Babbler

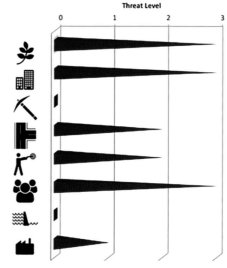

Threat Level

Turdoides altirostris, was found but this did not trigger inclusion under the A2 criterion. The site also held breeding populations of four Sahara-Sindian Desert biome-restricted species but these did not trigger inclusion under criterion A3. The Iraqi resident race of Little Grebe *Tachibaptus ruficollis iraquensis* breeds in the reedbeds along both sides of the river and the Iraqi race of Hooded Crow *Corvus cornix capellanus* (Mesopotamian Crow) occurs. Additional data on other (non-avian) fauna were not obtained.

Conservation Issues: The site was only visited in the winter of 2009 and could not be visited again for security reasons, so no clear image of its biodiversity has been developed as yet. Dense and extensive agricultural activities along both sides of the river, urban development, and human disturbance have major impacts within the area. There is active expansion of housing and urban areas into the natural areas of this site. Unsustainable hunting and fishing as well as road construction including dirt roads on the eastern side of the river are also important concerns.

Recommendations: This site includes a section of the Euphrates River and in general, surveys of both the Tigris and Euphrates should be done more extensively to identify areas of quality habitat and important biodiversity. Additional surveys are also needed at Musayab, particularly to cover areas to the northeast (Al-Qadiriya) as there may be marshes here that are the remains of the original wetlands described by Evans. Agricultural and urban development should be the focus of threat reduction activities in this area.

Hindiya Barrage (IQ060)

Babil - 32.73333°N 44.26667°E

KBA Criteria: **V and Ia**
IBA Criteria: **A1 and A2**
IPA Criteria: **B1 and B2b**

Area: **278 ha** - Altitude: **32-50 m**
Ecoregion: **Arabian desert and East Sahero-Arabian Xeric shrublands (PA1303) & Tigris-Euphrates alluvial salt marsh (PA0906)**
Status: Unprotected

©2008 MA SALIM/NI

Site Description: An original IBA site (IBA019) listed by Evans (1994), this site consists of a network of waterways on the Euphrates River created by the Hindiya Barrage, which was originally built in the early 1900s. Upstream of the barrage are several branching canals/rivers, chiefly the main stem of the Euphrates and the Shatt Al-Hilla rivers, which are bordered by marshes. Large numbers of waterfowl and various species of gulls were observed in this area. Hunting and fishing close to the barrage regulator are prohibited and entry to this area is restricted by police, which has protected the large assemblage of waterfowl.

The site is surrounded by dense palm orchards and a few open agricultural areas. The geology of the area is Mesopotamian alluvium, mainly silts and the habitats include marsh reedbeds of *Phragmites australis* and *Typha dominguensis*; rooted submerged vegetation and riparian vegetation, as well as shrub woodlands.

Key Biodiversity Area Criteria	Notes
V. Vulnerability Criteria: *Presence of Critically Endangered and Endangered species – presence of a single individual or Vulnerable species– 30 individuals or 10 pairs.*	
Rafetus euphraticus	Observed by the KBA team and reported frequently by locals.
Ia. Irreplaceability Sub-criterion: Restricted-range species based on global range	
Rafetus euphraticus	See above

Important Bird Area Criteria	All observations made 2009-2010.	
A1. Globally threatened species	**Breeding**	**Wintering/ Passage**
Basra Reed Warbler *Acrocephalus griseldis* (Summer visitor)	10 pairs	
A2. Restricted-range species	**Breeding**	**Wintering/ Passage**
Acrocephalus griseldis	10 pairs	
Iraq Babbler *Turdoides altirostris* (Resident)	30 pairs	7-11 (counts)
Important Plant Area Criteria		
B1. Site is aparticularly species-rich example of a defined habitat type		
Inland running water, river or canals- Riparian vegetation habitat type and Woodland- shrubs habitat type		
B2b.The site is a refuge for: biogeographically and bioclimatically restricted plants to 'retreat to' in the face of global climatic change.		
All the aquatic plants in the marshlands of Iraq are threatened with local extinction due to drying from climate change and the building of dams upstream. This site on the Euphrates River, which provides water to the southern marshlands, can function as a refuge for aquatic plants and assist in the future restoration of the marshlands.		

Additional Important Bird Observations: During the survey 28 bird species were observed. The site also supported five breeding Sahara-Sindian Desert biome-restricted species but these did not triggar inclusion under the A3 criterion. The endemic race of Little Grebe *Tachybaptus ruficollis iraquensis* and endemic race of Hooded Crow *Corvus cornix capellanus* (also known as Mesopotamian Crow) breed at the site. Hunters and locals reported that Marbled Duck *Marmaronetta angustirostris* breeds.

Other Important Fauna: No observations were made at this site but local reports suggested the presence of wild cats and foxes (exact species could not be distinguished). There

was no fish survey, but fishermen report the presence of the introduced species *Tilapia zillii*.

Additional Plant & Habitat Information: This site contains good populations of *Phragmites australis* and *Phoenix dactylifera*, which are important both economically and culturally.

Conservation Issues: Human disturbance at the site including farming activities have a very high impact on the site. Movement of the locals close to the barrage itself is prohibited, but further upstream these are a significant threat to the site. There is also high threats from expansion of farms and orchards; urban development in the city of Hindiya; roads development on both sides of the river; hunting and fishing, and pollution, including the continuous accumulation of floating trash and plastic carried by the flow of the river. This accumulates behind a boom fixed across the river to prevent garbage and other objects from entering the barrage area. An additional threat comes from the invasive fish *Tilapia zilli* and its potential effects on the biodiversity of the river.

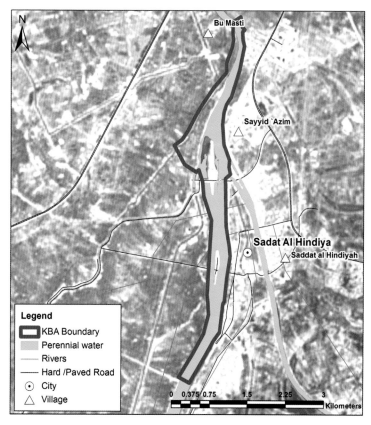

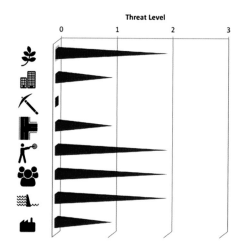

Recommendations: Activities should be coordinated with the local Environment Directorate in Babil Governorate and due to the high human use of the area awareness program that targets farmers hunters and local fishermen is needed to reduce the impacts on the biodiversity of the area. Fisheries surveys are also needed.

Shuweicha Marsh (IQ061)

Wasit - 32.70917°N 45.80889°E

IBA Criteria: **A1**
IPA Criteria: **C**

Area: **54938 ha** – Altitude: **7-18m**
Ecoregion: **Tigris-Euphrates alluvial salt marsh (PA0906) and Arabian desert and east Sahero-Arabian xeric shrublands (PA1303)**
Status: **Unprotected**

©2008 MA SALIM/NI

Site Description: Shuweicha was defined as an Important Bird Area (IBA020) by Evans (1994) and is a seasonal marsh with no plant cover other than a scattering of low-growing species, mainly over the margins of the seasonal water body. It receives water from the Iranian highlands to the east after the rainy season and discharges the water to the Tigris or lower depressions to the southeast of the Tigris (Hoshiya and Saroot (IQ066)). Shuweicha is directly affected by a recent lack of water with very few water patches found during the 2009-2010 survey due possibly to a regional lack of precipitation and/or blocked water sources in Iran.

Important Bird Area Criteria	Observation made 2009-2010.	
A1. Globally threatened species	**Breeding**	**Wintering/ Passage**
Lesser White-fronted Goose *Anser erythropus* (Winter visitor)		Frequent, historical and current, signals from tagged birds sent from this area.
Macqueen's Bustard *Chlamydotis macqueenii* (Winter visitor)		Reported frequently by locals and hunters.
Important Plant Area Criteria		
C. The site is identified as an outstanding example of a globally or regionally threatened habitat type.		

This site is located in a critically endangered ecoregion (Tigris-Euphrates alluvial salt marsh) and is a good example of the marshlands of southern Iraq, a globally significant wetland. These marshes are threatened by decreasing water levels and drought, grazing, and water pollution.

Additional Important Bird Observations: During the surveys, few birds were observed but this may be due to the timing of surveys. Although little water was seen in the last surveys, this wetland is very important for the Lesser White-fronted Goose (Vulnuerable) as satellite information shows frequent signals from tagged birds in this area and historical data as well as reports from locals and hunters indicate that this is an important wetland for wintering waders and waterfowl.

The open dry areas also provide good habitat for the Vulnerable Macqueen's Bustard *Chlamydotis macqueenii* and raptors such as Eastern Imperial Eagle *Aquila heliaca* (Vulnerable). The area might also be important for some passage passerines and other migrants and as well as some restricted-range species.

Other Important Fauna: According to the frequent reports of locals and hunters, the area contains suitable habitat for a variety of species. The reported animals are: Honey Badger *Mellivora capensis*, Striped Hyena *Hyaena hyaena* (Near Threatened), Red Fox/Ruppell's Fox *Vulpes* spp., Golden Jackal *Canis aureus*, and Grey Wolf *Canis lupus*.

Fish: Locals report that the area has supported a very economically important fishery but water management

decisions by the Iraqi government have not taking this into account. For example, after good rains during the 2011/2012 winter, the marsh supported abundant fish but was then drained early in the year resulting in a massive fish die off.

Conservation Issues: This site faces trans-boundary water management issues, as considerable amounts of the marsh's water comes from the Iranian highlands. The control of water by Iran and within Iraq has a very high impact on the site. Netting and hunting of large numbers of ducks in the winter is also a concern. There are oilfields and oil and gas pipelines in the area, and road development especially along the highway between the cities of Badra and Kut. Farms are in the southern parts of the area (though these appeared abandoned during the survey) and pollution is primarily from picnikers leaving plastic trash especially at the southern edges of Shuweicha.

Recommendations: As this is a seasonal marsh, Shuweich faces unique issues. It floods just before the arrival of migrant and wintering waterfowl and has the ability to support an important fishery thus conserving this area to support these species is crucial. The water resources feeding this area should be managed with the biodiversity of the site in mind and the seasonal wetlands should be allowed to follow their natural patterns. This will require coordination and cooperation with Iran as well as internally with Iraqi stakeholders, particularly with the Kut government and local water resource managers.

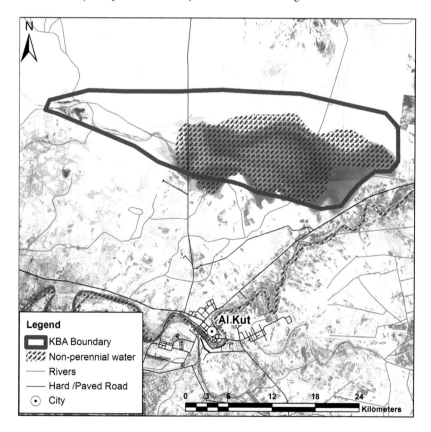

Legend
- KBA Boundary
- Non-perennial water
- Rivers
- Hard /Paved Road
- City

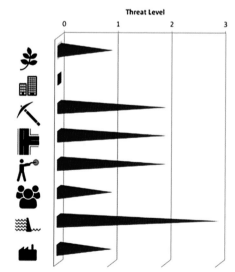

North Ibn Najm (IQ062)

Babil - 32.31528°N 44.40694°E

KBA Criteria: **V and Ia**
IBA Criteria: **A1, A2, and A4i**
IPA Criteria: **C**

Area: **1789 ha** - Altitude: **25-30 m**
Ecoregion: **Arabian Desert and East Sahero-Arabian Xeric Shrublands (PA1303)**
Status: **Unprotected**

PANORAMIC VIEW AFTER DRAINAGE - FACING SOUTH ©2010 MA SALIM/NI

Site Description: North Ibn Najm (locally known as Garrat Sayid Jafar) is a small, isolated marsh that receives water from the surrounding network of canals bringing drainage water from rice fields, farms and orchards that surround the site. The marsh (along with Ibn Najm IQ063)

FACING EAST ©2009 MA SALIM/NI

are the last remaining wetlands from the original Ibn Najm that covered a large area at the intersection of Najaf, Babil, and Qadissiya Governorates, which were described by Evans (1994) as a large freshwater marsh of possibly great importance for wintering waterbirds (IBA026). During the period of agricultural expansion between the 1970s-90s (particularly focused on rice production using local varieties) the original Ibn Najm was reduced to small patches among the rice fields and palm orchards. The geology of the area is Mesopotamian alluvium, mainly silts. Dense reed beds of *Phragmites* and *Typha* grow here, in addition to many species

of aquatic plants, but the site also contains shrub woodlands with *Tamrix* sp. and *Alhagi graecorum*. Large numbers of ducks were observed during the survey.

Major change happened in this marsh between the summer of 2009 and 2010. In 2010, the team found the area entirely dry except for some water patches in the canals and a few depressions. Most of the plant cover had died because of the lack of the water. The locals reported that because of the shortage of water in 2010, the Ministries of Agriculture and Water Resources did not allow the farmers to grow rice, which decreased the amount of water draining into the marsh.

Key Biodiversity Area Criteria	Notes	
V. Vulnerability Criteria: *Presence of Critically Endangered and Endangered species – presence of a single individual or Vulnerable species– 30 individuals or 10 pairs.*		
Rafetus euphraticus	Reported frequently throughout the area.	
Ia. Irreplaceability Sub-criterion: Restricted-range species based on global range		
Rafetus euphraticus	See above	
Important Bird Area Criteria	**All observations made 2009-2010.**	
A1. Globally threatened species	**Breeding**	**Wintering/ Passage**
Marbled Duck *Marmaronetta angustirostris* (Resident)	Not observed but likely present	400 (count 2009)
Basra Reed Warbler *Acrocephalus griseldis* (Summer visitor)	10 pairs	
A2. Restricted-range species	**Breeding**	**Wintering/ Passage**
Acrocephalus griseldis	10 pairs	
Iraq Babbler *Turdoides altirostris* (Resident)	25 pairs	Present but not found
A4i. 1% or more of biogeographical population of a congregatory waterbird species		
	Breeding	**Wintering/ Passage**
Marbled Duck *Marmronetta angustirostris*	Not observed but likely present	400 (count 2009)

Important Plant Area Criteria

C. The site is identified as an outstanding example of a globally or regionally threatened habitat type.

This site is located in a critically endangered ecoregion (Tigris-Euphrates alluvial salt marsh) and is a good example of the marshlands of southern Iraq, a globally significant wetland. These marshes are threatened by decreasing water levels and drought, grazing, and water pollution.

Additional Important Bird Observations: During the surveys 37 bird species were observed. The site also supported eight breeding Sahara-Sindian Desert biome-restricted species but these did not trigger inclusion under the A3 criterion. The endemic race of Little Grebe *Tachybaptus ruficollis iraquensis* and endemic race of Hooded Crow *Corvus cornix capellanu*s (Mesopotamian Crow) breed at the site.

Other Important Fauna: There were a few reports from locals of either Red Fox or Ruppell's *Fox Vulpes* sp. (species could not be distinguished) and Golden Jackal *Canis aureus*.

Fish: A Fisheries Frame survey was conducted here in 2009. A total of eight species were found. Ten fishing boats were observed using fixed nets (mesh sizes of 0.5-1 cm) and electro-fishing. The daily catch average was estimated at 3 kg/boat for net fishing and 7 kg/boat for electro-fishing. The following significant species were observed: *Alburnus mossulensis*, *Carasobarbus luteus*, *Carassius auratus*, *Cyprinus carpio* (Vulnerable), *Leuciscus vorax*, *Liza abu*, *Mesopotamichthys sharpeyi* (Vulnerable), and *Silurus triostegus*. The introduced invasive species *Tilapia zillii* was also reported by fishermen.

Additional Plant & Habitat Information: *Phragmites australis* is present, which is an important plant for local people for economic and cultural heritage reasons.

Conservation Issues: This site is highly variable due to often yearly changes in the water regime. But based on the results of these limited assessment visits, it appeared that the site can harbor threatened species such as *Marmaronetta angustirostris* and endemic species and subspecies such as *Tachybaptus ruficollis iraquensis*, *Turdoides altirostris* and *Corvus cornix capellanus* that breed regularly in or around this marsh. The lack of a stable supply to ensure adaquate water levels and quality is the highest threat here. Also over-exploitation due to hunting and over-fishing (primarily through electro-fishing); human intrusion, especially during the breeding season, and agricultural expansion have a very high impact.

Pollution threats here mainly stem from smoke from the asphalt factory adjacent to this marsh, but are also the result of plastic waste and other rubbish brought in either by the wind or by visitors. An additional concern is the presence of the invasive *Tilapia zillii*, which was reported by most of the fishermen in the area.

Recommendations: Restoration of this marsh will require a signficant effort to be made by the Babil Governorate. Coordination and cooperation with the Ministry of Water Resources and the local office of irrigation will be needed to fully address the lack and instability of water in order to restore and sustain at least a portion of North Ibn Najm and its plant cover.

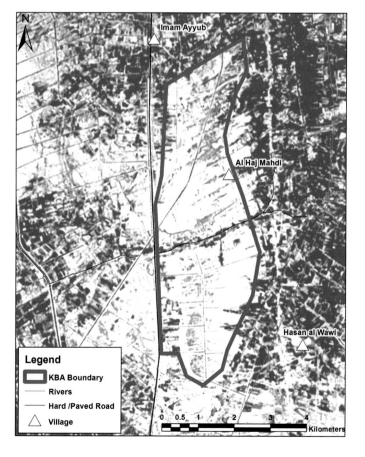

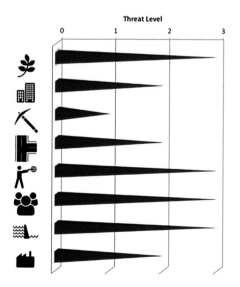

Ibn Najm (IQ063)

Babil and Najaf - 32.14917ºN 44.64167ºE

KBA Criteria: **V and Ia**
IBA Criteria: **A1 and A2**
IPA Criteria: **B1 and C**

Area: **4000 ha** - Altitude: **21-23 m**
Ecoregions: **Arabian Desert and East Sahero-Arabian Xeric Shrublands (PA1303)**
Status: **Unprotected**

Site Description: Evans (1994) provided little information on this site (IBA026). It was described as a seasonal freshwater lake lying east of the Euphrates River and c. 130 km south of Baghdad, in Babil Governorate, in the triangle between Babil, Qadissiya, and Najaf but the original body of water at Hor Ibn Najm had shrunk to scattered pools and marshes. After a period of intense agricultural expansion, these lands were used as rice farms and date-palm orchards but irrigation water drains to areas of well-developed reedbeds of Phragmites and Typha in addition to submerged vegetation such as *Hydrella verticillata*, *Myriophyllum verticillatum*, and *Ceratophyllum demersum* and free-floating plants such as *Lemna* sp. There are also shrub woodlands within the site. The geology of the area is *Mesopotamian alluvium*, mainly silts.

Some patches among the reedbeds form suitable habitat for birds to feed and shelter. During the summer 2008 survey,

FACING NORTH AT IBN NAJM MARSH ©2005 MA SALIM/NI

the marsh area was found to be almost dry (a network of canals and embankments has disfigured the original landscape) and the reed and Typha areas were generally dry, except close to the drainage canals. The site has been under severe threat of drought since a large drainage canal was dug in the middle of the area. No water has returned to the marsh since; it remains dry and the reed beds also started to dry up. It seems that a large regulator was under construction at the newly dug draining canal at the time of the summer 2010 visit, but no further information was gained after that survey.

Key Biodiversity Area Criteria	Notes	
V. Vulnerability Criteria: *Presence of Critically Endangered and Endangered species – presence of a single individual or Vulnerable species– 30 individuals or 10 pairs.*		
Rafetus euphraticus	Reported frequently throughout the area, and individuals were seen.	
Ia. Irreplaceability Sub-criterion: Restricted-range species based on global range		
Rafetus euphraticus	See above	
Important Bird Area Criteria	**Observations made 2005-2010.**	
A1. Globally threatened species	**Breeding**	**Wintering/ Passage**
Marbled Duck *Marmaronetta angustirostris* (Resident)	Present, not counted	180-220 (counts 2005-2010)
Basra Reed Warbler *Acrocephalus griseldis* (Summer visitor)	30 pairs	
A2. Range-restricted species	**Breeding**	**Wintering/ Passage**
Iraq Babbler *Turdoides altriostris* (Resident)	22 (counts)	
Basra Reed Warbler	30 pairs	
Important Plant Area Criteria		
B1. Site is a particularly species-rich example of defined habitat type		
Desert shrub habitat type and Marsh vegetation- Helophytic vegetation- Reedbed, Reedmace bed, or Schoenoplectus bed habitat type		

C. The site is identified as an outstanding example of a globally or regionally threatened habitat type.

This site is located in a critically endangered ecoregion (Tigris-Euphrates alluvial salt marsh) and is a good example of the marshlands of southern Iraq, a globally significant wetland. These marshes are threatened by decreasing water levels and drought, grazing, and water pollution.

Additional Important Bird Observations: A total of 54 bird species were observed. Additionally, the site supported seven breeding Sahara-Sindian Desert biome-restricted species but this did not triggar inclusion under the A3 criterion. The endemic race of Little Grebe *Tachybaptus ruficollis iraquensis* and endemic race of Hooded Crow *Corvus cornix capellanus* (also known as Mesopotamian Crow) breed regularly at the site. Although not an IBA species, it is noteworthy that Black-shouldered Kite *Elanus caeruleus*, which appears to be slowly colonising Iraq, has recently started to bred in this area (Salim 2002).

Other Important Fauna: Little information was collected at this site on other fauna, but the team observed one Small Asian Mongoose *Herpestes javanicus*.

Fish: Information was collected for 2008 & 2009, during which 14 species were observed. According to Coad (2010) significant species were: *Acanthopagrus cf. arabicus, Acanthobrama marmid, Alburnus mossulensis, Carasobarbus luteus, Carassius auratus, Ctenopharyngodon idella, Cyprinus carpio* (Vulnerable), *Leuciscus vorax, Liza abu, Heteropneustes fossilis, Luciobarbus xanthopterus* (Vulnerable), *Mesopotamichthys sharpeyi* (Vulnerable), and *Siluru striostegu.* In 2009, a Fisheries Frame Survey was conducted and found that fishing was conducted using fixed gill nets (mesh sizes of 0.5-1 cm) and electro-fishing. About 10 boats were recorded using both methods with an estimated catch of approximately 7 kg/boat per day. In later visits to the site after 2010, *Tilapia zillii* was reported in increasing numbers.

Additional Plant & Habitat Information: *Polypogon maritimus*, which is very rare in Iraq and has been previously reported only once (Townsend, 1968, p. 314) is present at Ibn Najm.

Conservation Issues: This area was once a true marsh habitat that harbored a considerable diversity of birds including threatened and endemic species as well as quite large numbers of wintering waterfowl, waders and raptors. But this historic marsh has sadly been slowly eradicated as more and more of the area is converted for agriculture and human habitation. Continued efforts to drain the wetland for agricultural uses and the large drainage canal that cuts the area into two parts has severely affected the natural status of this wetland and may eventually turn it into a dry area. Unsustainable hunting and overfishing, human disturbances, especially during the breeding season, and agricultural expansion have a significant impact. The introduced invasive fish (*Tilapia zillii*) was found in the area as well as frequently reported by locals.

Recommendations: Restoration of this marsh would require large-scale coordination and more control of the agricultural expansion in the area between the environment departments of the three relevant districts (Babil, Qadissiya and Najaf), in addition to the related Water Resources and Agricultural Departments. Local residents who are dependent on the marsh and the area for their livelihood should also be involved. Agriculture and other developments and human activities should be restricted to areas away well away from the remaining marsh areas.

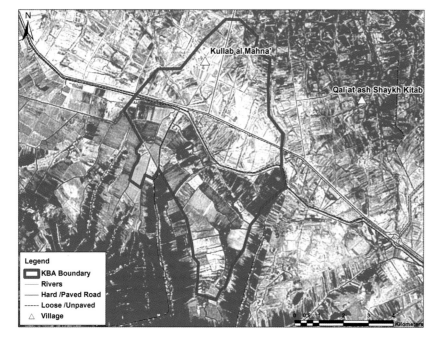

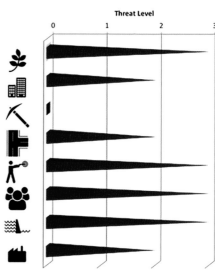

Dalmaj (IQ064)

Al-Qadissiya and Wasit - 32.199998°N 45.46667°E

KBA Criteria: **V and Ia**
IBA Criteria: **A1, A2, A3, A4i,A4ii, A4iii**
IPA Criteria: **B1 and C**

Area: **92076 ha** - Altitude: **14-23 m**
Ecoregions: **Tigris-Euphrates alluvial salt marsh (PA0906) and Arabian desert and East Sahero-Arabian xeric shrublands (PA1303)**
Status: **Unprotected, but proposed as a protected area**

©2006 MA SALIM/NI

©2006 MA SALIM/NI

Site Description: Originally listed as IBA032 by Evans (1994), Dalmaj wetland contains both terrestrial habitats ranging from arid areas to true desert with sand dunes, and a large body of water that can be divided into an open-water lake reaching depths exceeding 2 m and true marshes with dense reedbeds and shallower water (less than 1 m). The geology of the area is Mesopotamian alluvium, mainly silts.

The aquatic areas are fed by and drain into the Main Outfall Drain (MOD), a large agricultural drainage canal, and the water level in Dalmaj is unstable because of the control of the Ministry of Water Resources over the flow of the MOD. Embankments surround the marsh to contain the body of water. The Ministry of Agriculture promotes fish farming in the marsh, giving management rights to local

investors. Some have developed large enclosed pools near the edge of the lake, while others release fingerlings (particularly *Bunni Mesopotamichthys sharpeyi*) from a local hatchery, which is the dominant farmed species. Dalmaj has one of the few hatcheries known to produce these species and there appears to be a relatively healthy population in the marsh due to protection and the absence of unsustainable fishing methods such as electro-fishing and poison.

The southern section of Dalmaj is mainly mudflats, featuring Phragmites and Typha reedbeds in addition to submerged plants with occasional dry ground scattered with bushes and terrestrial species. Many waders and waterfowl were observed at the site in large numbers in addition to passerines, most of which were observed breeding. The eastern part of the site includes much of the open and deeper parts of Dalmaj Lake, which is a favorable habitat for gulls, terns, and fish. To the east of the embankment there are shallow, salty marshes with a strip of dense reedbeds and *Tamarix* bushes. The freshwater marshes in the northern part of the site are defined by rich plant cover, such as *Phragmites* and *Typha* reedbeds and *Tamarix* in drier areas. These marshes have clear, transparent waters and submerged plants, which provide excellent protection for juvenile fish and offer high oxygen production.

Dalmaj is rich in biodiversity, being a wintering ground for numerous waterfowl and a main breeding area for Marbled Duck *Marmaronetta angustirostris*, Ferruginous Duck *Aythya nyroca*, and Red-crested Pochard *Netta rufina*, three of the four known breeding ducks in Iraq, and a major breeding site for the endemic Basra Reed Warbler. In addition to its biological importance, Dalmaj is also home to ancient (potentially Sumerian) historical sites along the Nahr An-Neel, the ancient route of the Euphrates. The Iraqi State Board of Antiquities and Heritage has recently begun studying these sites and initiated excavations.

Six sub-sites were surveyed regularly from 2005 to 2010 during KBA surveys in Dalmaj area. Data from previous surveys since 1990 are also integrated with this account.

Key Biodiversity Area Criteria	Notes	
V. Vulnerability Criteria: *Presence of Critically Endangered and Endangered species – presence of a single individual or Vulnerable species– 30 individuals or 10 pairs.*		
Rafetus euphraticus	Large numbers of this endangered turtle were seen during the surveys in the area especially the southern part of the area.	
Ia. Irreplaceability Sub-criterion: Restricted-range species based on global range		
Rafetus euphraticus	See above	
Important Bird Area Criteria	**Observations made 2005-2011 and 2013**	
A1. Globally threatened species	**Breeding**	**Wintering/ Passage**
Lesser White-fronted Goose *Anser erythropus* (Winter visitor)		14 (count 2011)
Red-breasted Goose *Branta ruficollis* (Rare Winter visior, rare)		9 (count 2013)
Marbled Duck *Marmaronetta angustirostris* (Resident)	700 pairs	135-2,780 (counts 2007, 2009, 2010)
White-headed Duck *Oxyura leucocephala* (Winter visitor)		4 (count 2011)
Basra Reed Warbler *Acrocephalus griseldis* (Summer visitor)	500 pairs	
Macqueen's Bustard *Chlamydotis macqueenii* (Winter visitor; may breed)	A few individuals recorded in summer, but breeding not confirmed.	44 (count 2011)
A2. Restricted-range species	**Breeding**	**Wintering/ Passage**
Iraq Babbler *Turdoides altirostris* (Resident)	300 pairs	
Basra Reed Warbler *Acrocephalus griseldis*	500 pairs	
A3. Biome-restricted species		
Sahara-Sindian Desert biome	**Breeding**	**Wintering/ Passage**
White-tailed Lapwing *Vanellus leucurus* (Resident)	400 pairs	180-960 (counts 2009-2010)
Chlamydotis macqueenii (Winter visitor; may breed)	A few individuals recorded in summer, but breeding not confirmed.	44 (count 2011)
Cream-coloured Courser *Cursorius cursor* (Resident)	40 pairs	18-88 (counts 2009-2010)
Pallid Scops Owl *Otus brucei*	One pair 2005	
Egyptian Nightjar *Caprimulgus aegyptius* (Summer visitor)	100 pairs	
Hypocolius *Hypocolius ampelinus* (Summer visitor)	200 pairs	
Bar-tailed Lark *Ammomanes cinctura*	One pair 2005	
Greater Hoopoe-Lark *Alaemon alaudipes* (Resident)	50 pairs	Present
White-eared Bulbul *Pycnonotus leucotis* (Resident)	50 pairs	12 (count 2010)
Basra Reed Warbler *Acrocephalus griseldis* (Summer visitor)	500 pairs	
Iraq Babbler *Turdoides altirostris* (Resident)	300 pairs	Present
Dead Sea Sparrow *Passer moabiticus* (Resident)	1,000 pairs	20 (count 2010)

A4i. 1% or more of biogeographical population of a congregatory waterbird species		
	Breeding	**Wintering/ Passage**
Marbled Duck *Marmaronetta angustirostris* (Resident)	700 pairs	135-2,780 (2007 & 2009-2010)
Eurasian Coot *Fulica atra* (Resident & winter visitor)		250,000
Black-tailed Godwit *Limosa limosa* (Passage migrant & winter visitor)		>20,000
A4ii. 1% or more of global population of a congregatory seabird or terrestrial species		
Passer moabiticus (Resident)	1,000 pairs	20 (2010)
A4iii. Holding congregations of 20,000 waterbirds		
Congregatory waterbirds (waders and wildfowl)		>500,000
Important Plant Area Criteria		
B1. The site is a particularly species-rich example of a defined habitat type		
Marsh vegetation- Helophytic vegetation- Reedbed, Reedmace bed, or Schoenoplectus bed habitat type		
C. The site is identified as an outstanding example of a globally or regionally threatened habitat type		
This site is located in a critically endangered ecoregion (Tigris-Euphrates alluvial salt marsh) and is a good example of the marshlands of southern Iraq, a globally significant wetland. These marshes are threatened by decreasing water levels and drought, grazing, and water pollution.		

Additional Important Bird Observations: A total of 140 species have been observed. In addition to those mentioned in the table the Vulnerable Eastern Imperial Eagle *Aquila heliaca* and the Near Threatened Eurasian Curlew *Numenius arquata* are regularly seen on migration and in winter but in sub-IBA threshold numbers. The Irano-Turanian biome-restricted Menetries's Warbler *Sylvia mystacea* is resident. The endemic race of Little Grebe *Tachybaptus ruficollis iraquensis* and the Iraqi race of Hooded Crow *Corvus cornix capellanus* (also known as Mesopotamian Crow) both occur.

Other Important Fauna: This site boasts significant fauna diversity in comparison with other sites of similar area. Though mammals were not consistently the subject of the survey effort, 19 species were found or reported in Dalmaj including Honey Badger *Mellivora capensis*, Caracal *Caracal caracal*, Striped Hyena *Hyaena hyaena* (Near Threatened), Ruppell's Fox *Vulpes rueppellii*, Gray Wolf *Canis lupus*, Otter *Lutra lutra* (Near Threatened), Wild Boar *Sus scrofa* (in quite large herds) and wild cat (either F*elis silvestris* or *F. chaus*). According to reports by locals and hunters, the last group of gazelles in Dalmaj was seen in the late 1990s. These were likely Goitered Gazelle *Gazella subgutturosa* (Vulnerable). Due to the lack of observations, and increasing disturbance, the likelihood of gazelle presence in Dalmaj is now thought to be low. A notable reptile observed at Dalmaj was the Desert Monitor *Varanus griseus*.

Fish: Dalmaj is one of the most important wetlands in Iraq for fish. The Bunni in Dalmaj appears to be the last healthy stock found in southern Iraq and might be an important source for the re-introduction of this species into the southern marshes of Iraq. Fish data were collected from 2006 to 2009, when nine species were reported. According to Coad's (2010), the following significant species were: *Carassius auratus*, *Cyprinus carpio* (Vulnerable), *Heteropneustes fossilis*, *Carasobarbus*

luteus, *Leuciscus vorax*, *Liza abu*, *Mesopotamichthys sharpeyi* (Vulnerable), and *Silurus triostegus*. Additionally, in recent years introduced *Tilapia zillii* have become increasing reported by fishermen.

In 2009, fishing practices in the marshes were examined through a Fisheries Frame survey, which found that fishing was conducted using fixed gill nets with mesh sizes of 2-3 cm; cast nets with a mesh size of 1cm; floating gill nets with mesh sizes of 2-3 cm and seine nets with a fine mesh size of 0.5 cm. The approximate number of fishermen active in the area at the time of this survey was 150 boats with an estimated daily catch of 10 kg/boat, although in spring the daily catch is reported to increase to approximately 70 kg/boat.

Additional Plant & Habitat Information: This site contains a good population of *Phragmites australis*, which is important economically and for local heritage.

Conservation Issues: The Dalmaj area faces several very high threats. A key issue is hunting and poaching (large numbers of ducks, other waterfowl, and fish are poached each year through the use of clap-nets, shotguns, and fishing nets). Agriculture is also a very high threat as most of the dry land inside the site is used for wheat farming and very little remains untouched. Significant numbers of domestic animals are found in the area, including Water Buffalo *Bubalus bubalus* raised by the local Ma'dan (marsh dwellers), and dromedaries, sheep and goats raised by local Bedouin. However, the lack of fresh water for irrigation has forced many farmers to abandon their farms. Additionally human intrusion is a very high threat because Dalmaj is a popular weekend hunting and picnic area for residents of Diwaniya city, which does not have many open areas. As Dalmaj is mainly dependent on drainage water from the MOD, it is highly threatened by fluctuations and lack of water caused by upstream dams and diversions. While the MOD should be a permanent source of water, no alternative is available if it dries up, and water levels are currently not managed with the biodiversity of Dalmaj in mind. Additional high threats come from road construction, which is somewhat mitigated because of the relatively low scale of construction. There is also a high threat from commercial development. In addition, potentially problematic invasive species such as *Telapia zillii* and Namaqua Dove are present.

Recommendations: Due to the biological diversity of the marsh and the many threats it faces, Dalmaj would particularly benefit from more formal protection and in 2013,

this site was identified as a proposed protected area by the National Protected Area Committee (NPAC). A local group (Friends of Dalmaj) is interested in protecting this area and they should be encouraged to involve younger members and develop their communication and networking abilities. Dalmaj is still poorly studied and further scientific work is required to understand the biodiversity and the relationships among the biotic and abiotic factors in the area. In particular, research is needed on the distribution and numbers of flora and fauna to refine the delineation of the area. More studies should be conducted on invasive and introduced species such as *Tilapia* and others. A long-term goal is to re-introduce species, particularly gazelles and Arabian Oryx *Oryx leucoryx* to the area.

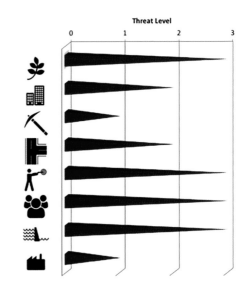

Gharraf River (IQ065)

Wasit & Thi Qar - 31.94389°N 45.97306°E

KBA Criteria: **V and Ia**
IBA Criteria: **A1 and A2**
IPA Criteria: **B1, B2b, and C**

Area: 50461 ha - Altitude: **8-23 m**
Ecoregion: **Tigris-Euphrates alluvial salt marsh (PA0906)**
Status: **Unprotected**

Site Description: Evans (1994) described this site (IBA029) as "a series of small ponds and seasonal wetlands along the 125+ km of the Shatt Al-Gharraf waterway between Kut and Shatra." Now, the majority of these ponds have been turned into agricultural farms and orchards, and very few of them remain on either side of the main motorway between Kut and Shatra. The trunk of the Gharraf River that branched from the Kut regulator consists of some 'unused' strips that still include good shelter (mainly reedbeds) for birds and mammals, while the river itself

contains considerable diversity of fish and other water-related flora and fauna.

Habitats at the site include the submerged vegetation of the river itself; floating vegetation and beds of *Phragmites australis* and *Typha domingensis* along the margins; also riparian vegetation and lastly shrub woodlands with species such as *Capparis spinosa*, *Tamrix* sp., *Suaeda* sp, and *Prosopis farcta*. The geology of the area is Mesopotamian alluvium, mainly silts. Four sub-sites were visited at various times during the survey effort.

©2006 MA SALIM/N

Key Biodiversity Area Criteria	Notes	
V. Vulnerability Criteria: *Presence of Critically Endangered and Endangered species – presence of a single individual or Vulnerable species– 30 individuals or 10 pairs.*		
Rafetus euphraticus	Euphrates Softshell Turtle was observed (and reported by locals) in considerable numbers over the entire length of the river.	
Lutrogale perspicillata	According to local reports, Otters are found frequently in the river, which could be the Smooth-coated Otter *L. perspicillata maxwelli*	
Ia. Irreplaceability Sub-criterion: Restricted-range species based on global range		
Rafetus euphraticus	See above	
Important Bird Area Criteria	**Observation made 2005-2006**	
A1. Globally threatened species	**Breeding**	**Wintering/ Passage**
Basra Reed Warbler *Acrocephalus griseldis* (Summer visitor)	30 pairs	
A2. Restricted-range species	**Breeding**	**Wintering/ Passage**
Basra Reed Warbler *Acrocephalus griseldis* (Summer visitor)	30 pairs	
Iraq Babbler *Turdoides altirostris* (Resident)	40 pairs	22 (count 2006)
Important Plant Area Criteria		
B1. Site is a particularly species-rich example of a defined habitat type		
Inland running water, rivers or canals- submerged river or canal vegetation habitat type and Inland running water, river or canals- Riparian vegetation habitat type		
B2b.The site is a refuge for: biogeographically and bioclimatically restricted plants to 'retreat to' in the face of global climate change.		
All the aquatic plants in southern Iraq are threatened with extinction if the marshlands dry out due to climate change and/or the building of dams upstream. Therefore, because this site is a freshwater river and is one of the sources for water to the southern marshlands after 2003, it can serve as a refuge for the future restoration of the marshlands and aquatic plants.		
C. The site is identified as an outstanding example of a globally or regionally threatened habitat type.		
This site is located in a critically endangered ecoregion (Tigris-Euphrates alluvial salt marsh) and is a good example of the marsh-lands of southern Iraq, a globally significant wetland. These marshes are threatened by decreasing water levels and drought, grazing, and water pollution.		

Additional Important Bird Observations: During the surveys 34 bird species were observed. Eastern Imperial Eagle *Aquila heliaca* and Marbled Duck *Marmaronetta angustirostris* (both Vulnerable) were also found wintering, as were two Near Threatened species, Pallid Harrier *Circus macrourus* and Black-tailed Godwit *Limosa limosa* (passage and winter), all in sub-IBA threshold numbers. The site also held breeding populations of five Sahara-Sindian Desert biome-restricted species but these did not trigger inclusion under criterion A3. The Iraqi race of Little Grebe *Tachibaptus ruficollis iraquensis* and Iraqi race of Hooded Crow *Corvus cornix capellanus* (Mesopotamian Crow) were found breeding.

Other Important Fauna: The main stem of the river and its branches provide good habitat for the Endangered *Rafetus euphraticus* where it resides and breeds. Being a permanent waterbody, this area might have served as a refuge for the

Vulnerable Smooth-coated Otter *Lutrogale perspicillata* during the droughts in other areas of the southern marshes. Some mammal species were found or reported frequently by locals and hunters during the KBA surveys, such as Golden Jackal *Canis aureus* and Jungle/Wild Cat *Felis chaus/F. silvestris*.

Fish: Sixteen species were recorded. Significant species, according to Coad (2010), were: *Arabibarbus grypus* (Vulnerable), *Carasobarbus luteus*, *Carassius auratus*, *Ctenopharyngodon idella*, *Cyprinus carpio* (Vulnerable), *Heteropneustes fossilis*, *Leuciscus vorax*, *Liza abu*, *Luciobarbus xanthopterus* (Vulnerable), *Mesopotamichthys sharpeyi* (Vulnerable), and *Silurus triostegus*. *Ablennes hians* was one marine fish species observed.

Additional Plant & Habitat Information: This site contains good populations of *Phragmites australis* and *Phoenix dactylifera*, which are important both economically and for local heritage.

Conservation Issues: Dense, continuous agricultural activities found along both sides of the river have a very high impact on the Gharraf and disturbance caused by the various human activities is also very high. High threats were the continuous housing and urban expansion into natural areas; road construction; hunting and fishing; and water management in the area, mainly for agricultural activities.

Recommendations: It is not likely that the agricultural activities and housing expansion in the area can be stopped, but biodiversity protection and environment-friendly techniques should be considered to decrease their negative impacts. These include better agricultural practices, integrated pest management and more efficient water use, zoning, etc. Such activities, as well as other awareness raising initiatives should target farmers and locals in and around the main towns (Nassr, Rifaee, and Shatra).

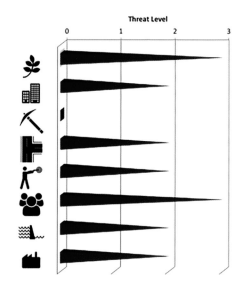

Legend

KBA Boundary
Rivers
Hard /Paved Road
City

Hoshiya & Saaroot (IQ066)

Missan - 32.33333°N 46.83333°E

IBA Criteria: **A1 and A4iii**
IPA Criteria: **C**

Area: **33560 ha** - Altitude: **8-17 m**
Ecoregions: **Tigris-Euphrates Alluvial Salt Marsh (PA0906) and South Iran Nubo-Sindian Desert and Semi-desert (PA1328)**
Status: **Unprotected**

©2006 MA SALIM/NI

Site Description: Part of this site, Hor Saaroot (IBA024 in Evans, 1994) is a seasonal marsh on a flat plain running in a narrow band along the eastern part of the Tigris between the towns of Ali Al-Gharbi and Amara. The approximate length is about 60 km and the width approximately 8 km. The area consists of lowland that acts as a catchment for rainwater flowing from the Iranian highlands to the east. Most of the area dries out during the dry seasons and water becomes confined to scattered patches in the depressions. A high soil embankment was constructed parallel to the Amara-Kut motorway to prevent flooding, which contains openings with regulators that discharge water to the eastern bank of the Tigris River. Most of the water in the lower part of this area discharges south towards Sinnaaf Seasonal Marshes (IQ069).

Hoshiya is a flat plain south of Saaroot, which floods during the rainy season. It was listed by Evans (1994) as IBA028 but is really a part of Saaroot (IBA024). This area also receives some waters from the Tigris when the water level exceeds a certain point, via a canal with a barrage. Some embankments and unpaved roads exist around the area, which are used during summer only. No agricultural or institutional features were observed at the site. The site has remained largely unchanged since Scott visited it in the 1970s and like Saaroot it attracts considerable numbers of waterfowl.

The geology of the entire area is Mesopotamian alluvium, mainly silts. There are no reedbeds or dense plant cover except in the depressions and canals. The dominant plant cover can be described as scattered annual herbs and bushes. There were three sub-sites surveyed in this area.

Important Bird Area Criteria	Observation made 2005-2007.	
A1. Globally threatened species	**Breeding**	**Wintering/ Passage**
Macqueen's Bustard *Chlamydotis macqueenii* (Winter visitor)		Reported frequently by locals and hunters, and considerable numbers are hunted each season.

A4iii. Holding congregations of 20,000 waterbirds or 10,000 pairs of seabirds of one or more species		
	Breeding	Wintering/ Passage
Congregatory Waterbirds (mainly Waders and Waterfowl)		> 20,000 wintering Waders and Waterfowl

Important Plant Area Criteria

C. The site is identified as an outstanding example of a globally or regionally threatened habitat type.

This site is located in a critically endangered ecoregion (Tigris-Euphrates alluvial salt marsh) and is a good example of the marshlands of southern Iraq, a globally significant wetland. These marshes are threatened by decreasing water levels and drought, grazing, and water pollution.

Additional Important Bird Observations: During the surveys, 39 bird species were observed in this seasonal marsh area. The area also provides good habitat for large numbers of wintering waterfowl during the rainy season. The open plain areas also provide good habitat for the Vulnerable Macqueen's Bustard *Chlamydotis macqueenii* and raptors, e.g Eastern Imperial Eagle *Aquila heliaca* (Vulnerable). Even though few Marbled Duck *Marmaronetta angustirostris* (Vulnerable) were seen in winter, the area might harbor many more of this species. The restricted-range Iraq Babbler *Turdoides altirostris* was found breeding in this area, but did not trigger the A2 criterion. The site also held breeding populations of six Sahara-Sindian Desert biome-restricted species but this did not trigger inclusion under A3 criterion.

Important Other Fauna: According to frequent reports by locals and hunters, the reported animals are: Honey Badger *Mellivora capensis*, Striped Hyena *Hyaena hyaena* (Near Threatened), Red Fox/Ruppell's Fox *Vulpes* sp., Golden Jackal *Canis aureus*, and Grey Wolf *Canis lupus*. Some reptiles were observed, but not identified. No fish data were collected.

Conservation Issues: Control and manipulation of the water coming from Iran was very high threat. The netting of large numbers of ducks in the winter and human disturbance from the movement of locals picnicking as well as the movement of heavy vehicles have high impacts on the site. Farms in the central part of the area may be a threat but appeared abandoned during the survey; there may be threats caused by the roads between stone qarries to the east of the area and the main highway; and pollution such as plastic trash and litter left by visitors accumulates in the site.

Recommendations: As Hoshiya and Saaroot is a seasonal marsh, this site faces unique issues. It floods just before the arrival of migrant and wintering waterfowl, so conserving this area to support these migrants is crucial. But the site faces transboundary issues as part of its waters come from the Iranian highlands. The water resources feeding this area should not be controlled and the seasonal wetlands should be allowed to follow their natural

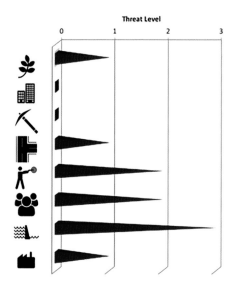

patterns. This will require coordination and cooperation with Iran and should address areas to the east of Saaroot and Hoshiya marshes as well. Bird netting is a serious issue in this area and it and many other threats to the site need to be addressed through a combination of both awareness-raising and enforcement of environmental laws. The government of Amara and local NGOs will need to play an important role in this.

Teeb Oasis & Zubaidaat (IQ067)

Missan - 32.38333°N 47.36667°E

IBA Criteria: **A1 and A3**
IPA Criteria: **B1**

Area: **28578 ha** - Altitude: **34-180 m**
Ecoregion: **South Iran Nubo-Sindian desert and semi-desert (PA1328)**
Status: **Unprotected, but proposed as a protected area**

TEEB OASIS PANORAMA ©2009 MA SALIM/NI

Site Description: This site is a combination of two survey areas, Teeb Oasis and Zubaidaat, both in Missan Governorate. Teeb Oasis is situated in a semi-desert area and was surveyed in the winter of 2009. It is located in the foothills that rise to the mountainous border with Iran to the east. The oasis contains a freshwater spring used by shepherds and locals to fill their water tanks. In comparison with the surrounding area, plant cover is considered rich. There are some trees within the site in addition to thorny shrubs and grasses with the dominant habitats being desert shrubs and desert herbaceous vegetation. Approximately 60% of the area is unvegetated. The geology of the area is gravel, sand, silt and alluvium. Another oasis, Ein Al-Kibreet, lies several kilometers to the southwest and contains water rich in sulfur. These are some of several oases in the Al-Teeb Desert northeast of Amara city and near the Iranian border. The permanent water and plant cover create an attractive habitat for desert birds as well as migratory species such as passerines.

Zubaidaat is characterized by dry and hilly terrain, with many valleys (wadis). It is located close to the Iraq-Iran border, with several oil fields, pipeline networks and a small number of paved and dirt roads. It is uninhabited except for nomadic grazers who pass through the area and has been generally devoid of human activity, rendering much of its wildlife undisturbed. This area is also an important migration corridor for passerines and other birds. Many parts of the area contain the remains from the Iraq-Iran War of the 1980s (bombs, shells, land-mines, etc.). During the summer 2010 visit, signs of new oil well development were seen in the area and some excavators were observed digging.

Important Bird Area Criteria	All observations made 2009-2010.	
A1. Globally threatened species	**Breeding**	**Wintering/ Passage**
Macqueen's Bustard *Chlamydotis macqueenii* (Winter visitor)		Reported frequently by locals and hunters, but no birds found during the 2009-2010 surveys.
A3. Biome-restricted species		
Sahara-Sindian Desert biome	**Breeding**	**Wintering/ Passage**
Cream-coloured Courser *Cursorius cursor* (Resident)	2 pairs (count 2009)	
Spotted Sandgrouse *Pterocles senegallus* (Resident)	30 pairs (count 2009)	2 (2009)
Hypocolius *Hypocolius ampelinus* (Summer visitor)	100 pairs	
Greater Hoopoe-Lark *Alaemon alaudipes* (Resident)	1 pair	1(2010)
Bar-tailed Lark *Ammomanes cinctura* (Resident)	1 pair	

Desert Lark *Ammomane deserti* (Resident)	50 pairs	
White-eared Bulbul *Pycnonotus leucotis* (Resident)	50 pairs	Present
Dead Sea Sparrow *Passer moabiticus* (Winter visitor)	Present	30 (2009-2010)
Desert Finch *Rhodopechys obsoletus*	Present	

Important Plant Area Criteria

B1. The site is a particularly species-rich example of a defined habitat type

Desert- Herbaceous vegetation

ZUBAIDAAT PANORAMA ©2010 MA SALIM/NI

Additional Important Bird Observations: Locals report large numbers of raptors passing through the area during migration and in winter. The site also held three Irano-Turanian species but these did not trigger inclusion under A3 criterion.

Additional Plant & Habitat Information: Two poorly known endemics (*Arnebia linearifolia subsp. Desertorum* and *Eremurus rechingeri*) are listed in the Flora of Iraq with a known distribution at Jabal Al-Muwilih not far from Wadi Teeb (Tib). More work is needed to determine their distribution.

Other Important Fauna: According to reports by locals and hunters, mammals include: Honey Badger *Mellivora capensis*, Striped Hyena *Hyaena hyaena* (Near Threatened), Grey Wolf *Canis lupus*, plus wild cats *Felis* sp. and foxes *Vulpes* sp. Gazelles used to occur in the hilly parts of this area where a few were reported killed close to the border. Two significant reptile species observed at this site were Desert Monitor

Varanus griseus and Egyptian Spiny-tailed *Lizard Uromastyx aegyptia* (Vulnerable). No fish data were collected.

Conservation Issues: Heavy hunting pressure on game birds (See-see Partridge and doves) and gazelles in the higher areas is the biggest concern here. Overgrazing is also an important issue as is the potential for oil development to affect this area, bringing with it road development, construction of pipelines, and additional development along the main road (particularly between Amara city and the picnic area in Teeb Oasis). Human disturbance (from movements of people collecting spring water, picnic activities occuring mainly in Teeb Oasis, and military exercises by the border police) is also a threat here, as is pollution from abandoned munitions and garbage left by picnickers.

Recommendations: Further surveys are recommended to confirm whether some bird species of conservation concern breed at the site and identify areas important for endemic plants. A comprehensive management plan should be devised and implemented with local stakeholder involvement as this site (along with Teeb Seasonal Wetlands, IQ068) was identified in 2013 by the National Protected Area Committee as a proposed protected area.

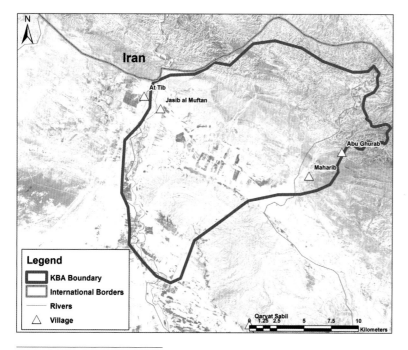

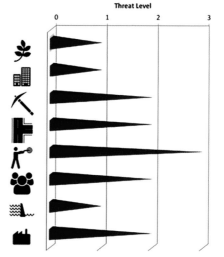

Teeb Seasonal Wetlands (IQ068)

Missan - 32.16667°N 47.38333°E

IBA Criteria: **A1 and A4iii**
IPA Criteria: **B1**

Area: **14827 ha-** Altitude: **9-25m**
Ecoregion: **Tigris-Euphrates Alluvial Salt Marsh (PA0906)**
Status: **Unprotected, but proposed as a protected area**

©2008 MA SALIM/NI

Site Description: This site is a seasonal wetland, with patches of water that remain throughout the year in scattered depressions. The area is sedimentary with clay soil types and the habitats are desert shrub and helophytic marsh vegetation with about 60% of the area unvegetated. Historically, the area received water from the Iranian uplands mainly through the Dwairij River, and from other small drainages from highlands in Iraq and Iran, and deep pools stretched over many kilometers. However, the general lack of rain recently and the closure of the site's primary water source due to an embankment dam built on the Dwairij River by Iran have resulted in drought conditions at this site.

The wetlands that do remain provide very good habitat for large numbers of geese and ducks, especially in the western, more remote areas of the site. A main road crosses the site and traces of water were noticeable on both sides of the road during the surveys. A few new, small oil wells and oil exploration developments were observed during the summer 2010 visit.

Important Bird Area Criteria	Observations made in 2009-2010.	
A1. Globally threatened species	**Breeding**	**Wintering/ Passage**
Lesser White-fronted Goose *Anser erythropus* (winter visitor)		70 (counts 2010)
Macqueen's Bustard *Chlamydotis macqueenii* (Winter visitor)		Reported frequently by locals and hunters. Considerable numbers hunted by falconers
A4iii. Holding congregations of 20,000 waterbirds or 10,000 pairs of seabirds of one or more species		
	Breeding	**Wintering/ Passage**
Congregatory Waterbirds		> 30,000 wildfowl species.
Important Plant Area Criteria		
A4. Sites contains national endemic, near endemic, regional endemic and/orregional range-restricted species or infraspecific taxa		

Eremurus rechingeri, an endemic listed in the Flora of Iraq (Townsend & Guest, Vol 8, P. 65) had a known distribution at Jabal Al Muwilih near Wadi Teeb (Tib), which has a similar habitat.

B1. The site is a particularly species-rich example of a defined habitat type
Desert- Desert shrublands habitat type

Additional Important Bird Observations: During the surveys a total of 26 bird species were recorded. Eastern Imperial Eagle *Aquila heliaca* and Greater Spotted Eagle *A. clanga* (both Vulnerable) were also seen frequently. Hunters report "large numbers" of Spotted Sandgrouse *Pterocles senegallus*. Nine Sahara-Sindian Desert biome-restricted species were observed but these did not trigger inclusion under criterion A3.

Other Important Fauna: According to the frequent reports of locals and hunters, mammals include Honey Badger *Mellivora capensis* and Grey Wolf *Canis lupus*, as well as wild cats and foxes (species unknown). Goitered Gazelle *Gazella subgutturosa* (sub-species unknown) used to occur in the eastern hilly parts of this area and a few were reported killed close to the border. Some reptiles were seen but not identified.

Conservation Issues: The Bazirgan, Abu Gharab and Fakka oilfields are nearby and oil development seems to be extending into the site. This and changes in the water management by Iran are some of the highest concerns at this site. Other issues include agricultural and livestock grazing. There is also extensive movement of water trucks by locals collecting water from springs and there is some new road development particularly for the the oilfields and oil export pipelines as well as along the main road between Amara city and the picnic area in Teeb Oasis (part of IQ067) to the north. Hunting of birds occurs fall through spring and the movement of bedouin, vehicle traffic (especially for the oil industry), and people picnicking causes significant disturbance within the site.

Recommendations: Surveys were limited here and more work is recommended to understand breeding, migrating and wintering bird populations. A conservation or restoration plan is needed especially for the two Vulnerable species at the site: Lesser White-fronted Goose and Macqueen's Bustard as well as for Gazelle. Water management issues here will require negotiations with Iran to resolve but in 2013, this site (along with Teeb Oasis and Zubaidaat (IQ067)) was proposed by the National Protected Area Committee for official protection and water resources, along with oil development, will be two of the key issues for the committee to address.

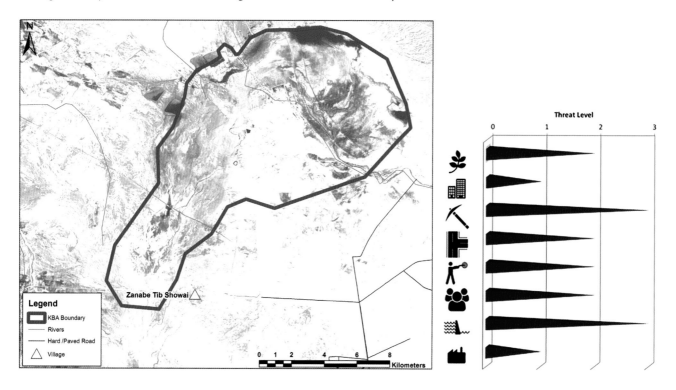

Sinnaf Seasonal Wetlands (IQ069)

Missan - 31.86667°N 47.31667°E

IBA Criteria: **A1, A4i, and A4iii**
IPA Criteria: **B1**

Area: **26049 ha** - Altitude: **3-12 m**
Ecoregions: **Tigris-Euphrates alluvial salt marsh (PA0906)**
Status: **Unprotected**

©2006 MA SALIM/NI

Site Description: This site consists of a large complex of marshes to the east of the River Tigris and north of Umm An-Ni'aaj in Hawizeh Marsh (IQ073), a large part of which was described by Evans (1994) as an Important Bird Area (IBA030). Sinnaf is a seasonally flooded wetland that receives water during the rainy season in winter from uplands in Iran located to the north and east, which has made the area important for large numbers of passage waders using the shallow mudflats and large numbers of ducks. Many globally, regionally and restricted range species have been recorded here. The geology of the area is gravel, sand, silt and alluvium, mainly sandy clay and it is generally flat in all directions. The habitats are salt pioneer swards vegetation and desert herbaceous vegetation with approximately 50% of the area unvegetated. Most of the area contains saline soils dominated by halophytic plants.

Approximately 2-3 years ago, Iran began blocking water from the Iranian highlands, much of which reached the area via the Dweireej and Teeb Rivers. This, as well as a reported decline in rainfall, has reduced the water supply to Sinnaf. The area still receives a good amount of water during the rainy season but now mainly only from Iraqi lands. During the KBA surveys, three sub-sites were visited to cover the eastern, middle (what Evans called Haur Chubaisah (IBA030)) and western parts of Sinnaf. Only a portion of the western part has permanent water due to the presence of sewage pipes from the city of Amara, which attracts large numbers of waders and gulls. Parts of the area have been converted into farms, and others between the lower embankment and the main Amara-Msharah road, were converted into fish farms that receive water from the Msharah River that runs along part of the southern side of the site.

Important Bird Area Criteria	All observations made 2005-2010.	
A1. Globally threatened species	**Breeding**	**Wintering/ Passage**
Marbled Duck *Marmaronetta angustirostris* (Resident)	Not found	200 (counts 2006)
A4i. 1% or more of biogeographical population of a congregatory waterbird species		
	Breeding	**Wintering/ Passage**
Slender-billed Gull *Chroicocephalus genei* (Resident)		> 3000
A4iii. Holding congregations of 20,000 waterbirds or 10,000 pairs of seabirds of one or more species		
Congregatory Waterbirds		More than 35,000 individuals (waterfowl and waders)
Important Plant Area Criteria		
B1. The site is a particularly species-rich example of a defined habitat type		
Salt pioneer swards vegetation		

Additional Important Bird Observations: During the survey 46 species were observed. Eastern Imperial Eagle *Aquila heliaca* (Vulnerable) occurred in winter and Black-tailed Godwit *Limosa limosa* (Near Threatened) on passage and in winter, but both in numbers below the IBA threshold. The site also held five Sahara-Sindian Desert biome-restricted species breeding but this did not trigger inclusion under criterion A3. The endemic race of Little Grebe *Tachybaptus ruficollis iraquensis* is resident. The Sinnaf area is also important for Slender-billed Gull *Chroicocephalus geneias* and large numbers were found after the breeding season — see the table above. This population might breed elsewhere in the marshlands of southern Iraq, but appears to use this site frequently.

Other Important Fauna: Several mammal species were found or reported in Sinnaf, including Honey Badger *Mellivora capensis*, Striped Hyena *Hyaena hyaena* (Near Threatened), Gray Wolf *Canis lupus* and Wild cat (either *Felis silvestris* or *F. chaus*). Gazelles were reported to the north of Sheeb (east of Sinnaf — outside the delineated area) but none were found during the surveys.

Conservation Issues: The most serious threat Sinnaf Seasonal Wetlands face today is the lack of water caused by diversion and/or blocking of water from Iran. Human disturbance such as intrusion by border police or those visiting the area for recreation and oil development in the area also have a very high impact. Hunting of birds poses a serious threat to waterfowl populations thought this is mainly practiced in the wetlands of western Sinnaf. The southwestern part of the site where the water is permanent is protected by bird hunters, who mainly hunt flamingos and ducks each winter. Farm fields and fish farms are present in the southern part of the site. Although the sewage water has helped to attract many birds and probably other species to the area, dumping of this untreated water poses a long-term health risk to these species and to people who consume them.

Recommendations: In order to fully restore the much larger area of Sinnaf, negotiated agreements with Iran will be required to allow waters to flow once again from the eastern uplands across the border. Monitoring of this site on a regular basis is recommended and any educational campaigns and other conservation actions should focus on protection of the key mudflats that harbor large numbers of migrant and wintering birds and areas in the southwestern part of Sinnaf that have high threats due to various human impacts.

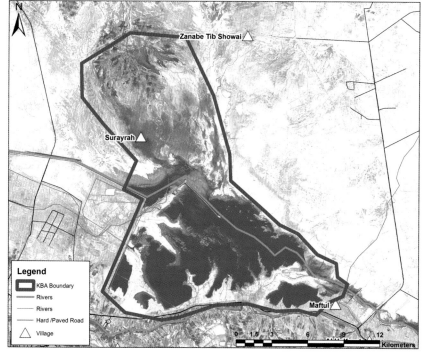

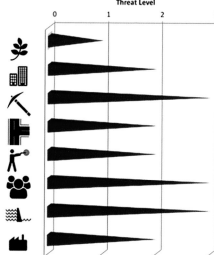

Wadi Al-W'eir & Sh'eeb Abu-Talha (IQ070)

Najaf - 31.06667°N 44.01667°E

IBA Criteria: **A1 and A3**
IPA Criteria: **B1 and B2b**

Area: **142755 ha** - Altitude: **60-380 m**
Ecoregion: **Arabian desert and east
Sahero-Arabian Xeric shrublands (PA1303)**
Status: **Unprotected**

©2010 MA SALIM/NI

Site Description: This site in the southwestern desert consists of a large wadi (seasonal watercourse) system, which includes Sh'eeb Abu-Talha to the south/southwest that travels northeast and is joined by Wadi Al-W'eir. Wadi Al-W'eir (also known as Sh'eeb Al-W'eir) is a relatively large wadi close to the strategic oil pipeline. Sh'eeb Abu-Talha likely received its name from a unique tree species called 'Talh' (Acacia gerrardi), which was only found at this site. Abu-Talha literally means "father or land of the Talh." The area has a rocky desert habitat with relatively good plant cover that vary from desert shrubs and desert herbaceous vegetation to salt pioneer sward vegetation to marsh reedbeds. The non-vegetated area covers approximately 70% of the area. The geology is limestone and marls with rocky ground and sandy soil.

In the upper part of the wadi, there is a fenced area for gazelles created by the Ministry of Agriculture after 2003. There are many archaeological features, including an ancient road (known as 'Darb Zubaida') and periodic khans (destroyed forts and rest areas from the Abbasid era, when this area was used for pilgrims traveling to Mecca) extending from Wadi Al-W'eir to the Iraqi-Saudi Arabian border. An unpaved road is located beside the ancient road. This wadi system brings seasonal rainwater from the southern and western areas of the desert to a wetland called Bahr An-Najaf. Two sub-sites were visited in the survey.

Important Bird Area Criteria	Observations made in 2010.	
A1. Globally threatened species	**Breeding**	**Wintering/ Passage**
Egyptian Vulture *Neophron percnopterus* (Passage migrant)		Seen in the area but the locals reported that it is not uncommon. Satellite tracked bird was retrieved (dead) from this area.
Macqueen's Bustard *Chlamydotis macqueenii* (Winter visitor)		Reported frequently by locals and hunters. Feathers of killed birds were found in 2010. A satellite-tagged adult (from Kazakhstan) was killed close to this site.
A3. Biome-restricted species		
Sahara-Sindian Desert biome	**Breeding**	**Wintering/ Passage**
Macqueen's Bustard *Chlamydotis macqueenii* (Winter visitor)		See above
Cream-coloured Courser *Cursorius cursor* (Resident)	50 pairs	17 (count)
Spotted Sandgrouse *Pterocles senegallus* (Resident)	100 pairs	11 (count)

Egyptian Nightjar *Caprimulgus aegyptius* (Summer visitor)	50 pairs	
Brown-necked Raven *Corvus ruficollis* (Resident)	30 pairs	
Hypocolius *Hypocolius ampelinus* (Summer visitor)	50 pairs	
Greater Hoopoe-Lark *Alaemon alaudipes* (Resident)	20 pairs	8 (count)
Desert Lark *Ammomanes deserti* (Resident)	100 pairs	9 (count)

Important Plant Area Criteria
B1. The site is a particularly species-rich example of a defined habitat type
Desert-Desert shrub habitat type and Desert- Herbaceous vegetation habitat type
B2b.The site is a refuge for: biogeographically and bioclimatically restricted plants to 'retreat to' in the face of global climate change.
This is a refugia site - an isolated relict of tropical vegetation that refers to previous climate optimum.

Additional Important Bird Observations: Despite the fact that a high number of bird species were not observed during the 2010 surveys, the site seems to provide good habitat because of the watercourses and relatively rich plant cover. In addition to those mentioned in the table, and according to the reports of locals, large numbers of raptors are present, indicating the general health of this area. The flat, non-rocky parts (with scattered desert vegetation) provide good habitat for the Vulnerable Macqueen's Bustard *Chlamydotis macqueenii*.

Other Important Fauna: According to the reports by locals and hunters, Sh'eeb Abu-Talha (including part of Sheeb Al-Mhari) provides suitable habitat for a variety of mammal and reptile species. Mammals include Honey Badger *Mellivora capensis*, Striped Hyena *Hyaena hyaena* (Near Threatened), and Grey Wolf *Canis lupus*. Locals state that Indian Crested Porcupine *Hystrix indica* occurs in considerable numbers, especially in Wadi Al-W'eir. Gazelles used to be found in the lower parts of this area until the last couple of decades but it is not clear whether this population still exists. Reptiles observed were Desert Monitor *Varanus griseus* and Egyptian Spiny-tailed Lizard *Uromastyx aegyptia*. Both are large lizards that are often targeted by hunters and *U. aegyptia* is a Vulnerable species and listed in the CITES Appendix II.

Additional Plant & Habitat Information: According to the Flora of Iraq, 30 trees of the very rare *Acacia gerrardi* subspecies *negevensis* (synonym *iraqensis*) were recorded in 1947 in Wadi Abu Talha (called Wadi Al-Mhari in the Flora of Iraq), but only one tree was found there in the 1960s-1970s (Townsend and Guest 1974). This tree was not observed during the KBA survey, possibly because the survey team did not visit the exact locality. *Rhazya stricta* is also mentioned in the Flora of Iraq surveys (and in older botanical reports), and is a plant found in Iraq only in the southern desert and mostly in Najaf province. It is recorded at different locations along the ancient "Darb Zubaida" road between Najaf and Shbicha, which is part of this site.

Conservation Issues: This site is considered relatively botanically rich for its biogeographic

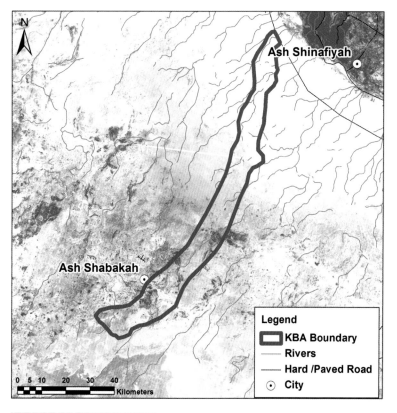

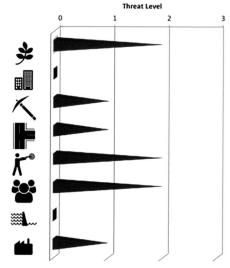

zone, with many plant species recorded. The primary threats were hunting of birds and heavy grazing by livestock and camels. Human intrusion from both of these activities as well as the movement of traffic along the unpaved road are also a concern. The impacts of sand mining in a small area in the northern part of this site, and pollution due to war remains and garbage were considered medium threats. The local government also plans to develop a road between Najaf and Shbicha cities, which will pass through the site and greatly increase traffic in the area.

Recommendations: Surveys should continue to investigate the possible presence of species of conservation concern including *Acacia gerrardi*. Historically, Shbicha (on

the southwestern edge of the site) was a hotspot for desert vegetation and many plant specimens were collected in that area during the Flora of Iraq project in the 1960s and 1980s, as well as during the late 19th and early 20th centuries. Therefore, it is recommended to conduct more studies at Shbicha to try and locate the rare or endemic plants that were previously recorded there.

Bird hunting and overgrazing are pressing issues to address as well. It is also recommended to liaise with the local government of Najaf Governorate to make them aware of the importance of the biodiversity here and the impact of any development plans, especially related to development of the road between Najaf and Shbicha.

Sawa Lake and Area (IQ071)

Muthanna - 31.349362°N 44.948980°E

KBA Criteria: **V and Ia**
IBA Criteria: **A1**
IPA Criteria: **B1**

Area: **20058 ha** - Altitude: **20–30 m**
Ecoregion: **Arabian Desert and East Sahero-Arabian Xeric Shrublands (PA1303)**
Status: **Unprotected**

©2010 MA SALIM/NI

Site Description: Sawa Lake is fed by groundwater that originates from the higher western desert areas and has no outlet. It is located on the eastern edge of Iraq's southern desert close to the Euphrates River, which runs through and borders the site. Most of the delineated area is desert and semi-desert with scattered shrubs, while the lake itself forms a small portion of the site. The geology is sedimentary with

sandy-clay soil. Locals reported that the size and depth of the lake has greatly reduced over time and the current conditions are likely due to declining underground water resources within the region. There was a small area of construction for a recreational center on the southeastern corner of the lake but this complex appeared to be abandoned by the time of the survey.

Key Biodiversity Area Criteria	Notes
V. Vulnerability Criteria: *Presence of Critically Endangered and Endangered species – presence of a single individual or Vulnerable species– 30 individuals or 10 pairs.*	
Rafetus euphraticus	Locals report that large numbers are found in the Euphrates River
Ia. Irreplaceability Sub-criterion: Restricted-range species based on global range	
Rafetus euphraticus	See above

Important Bird Area Criteria	Observations from 2010.	
A1. Globally threatened species	**Breeding**	**Wintering/ Passage**
Marbled Duck *Marmaronetta angustirostris* (Resident)		140 (counts 2010)
Macqueen's Bustard *Chlamydotis macqueenii* (Winter visitor)		Reported frequently by locals and hunters in the western edge
Important Plant Area Criteria		
B1. Site is a particularly species-rich example of defined habitat type		
Desert- Desert shrub habitat type		

Additional Important Bird Observations: During the surveys 25 species were observed in Sawa Lake and the surrounding areas. The site also supported seven Sahara-Sindian Desert biome-restricted species but this did not trigger inclusion under the A3 criterion. The site held large numbers of waterfowl, mainly ducks and Coot *Fulica atra*, but this also did not trigger inclusion under IBA criteria. The endemic race of Little Grebe *Tachybaptus ruficollis iraquensis* and the Iraqi race of Hooded Crow *Corvus cornix capellanus* (Mesopotamian Crow) occur, as well as the near-endemic Hypocolius *Hypocolius ampelinus*. Locals and hunters reported the frequent occurrence of "different kinds of raptors" especially in spring and autumn, so the site may be important as a staging area.

Other Important Fauna: The desert area on the western side of the lake, the wetland strip along the western branch of the Euphrates River (including adjacent orchards), and the flat arid/semi-desert areas over the southern parts of the lake might harbor considerable wildlife diversity, but were not well surveyed by the KBA team. According to local reports, mammals present include Rüppell's Fox *Vulpes rueppellii*, Striped Hyena *Hyaena hyaena* (Near Threatened), Honey Badger *Mellivora capensis*, Indian Grey Mongoose *Herpestes edwardsii*, as well as other common desert species. No fish survey was conducted, but local fishermen reported a variety of species in the lake.

Conservation Issues: Because the lake is the only waterbody available for the city of Samawa and its surroundings, picnic activities in the area cause significant human disturbance and accumulation of garbage. There appears to be no effort by visitors or government to manage or remove garbage from the site. Likely unsustainable fishing and bird hunting is also a very high threat. Because there is no plant cover on the lake (reedbeds, etc), water birds have few places to shelter from visitors or hunters. Also a proposed investment project for the area may impact the site due to construction around the lake.

Recommendations: As Sawa Lake represents a unique, closed water body in Iraq it is an important site for its scientific, educational and biodiversity value. Further detailed ecological studies of the lake and surrounding desert are recommended. Awareness-raising initiatives should coordinate with Muthanna University and the environment directorate to target locals from Samawa who use the area. This site was identified in 2014 as a proposed protected area by the National Protected Area Committee (NPAC) and in September of 2015, the lake was listed as a Ramsar Site.

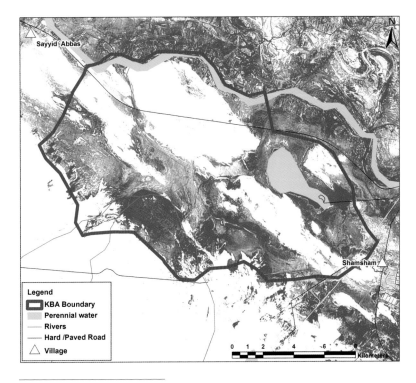

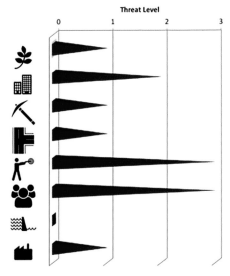

Auda Marsh (IQ072)

Missan - 31.64139°N 46.85139°E

KBA Criteria: **V and Ia**
IBA Criteria: **A1 and A2**
IPA Criteria: **B1 and C**

Area: **19241 ha** - Altitude: **less than 8 m**
Ecoregion: **Tigris-Euphrates alluvial salt marsh (PA0906)**
Status: **Unprotected**

©2005 MA SALIM/NI

Site Description: This original IBA site (IBA034) as described by Evans (1994) historically flowed to the Central Marshes (not Hammar as stated by Evans). It was drained along with other marshland areas in the 1990s but was reflooded after 2003. Currently, Auda is a brackish marsh that has significant problems with water quality due to lack of flow-through. The geology of the area is Mesopotamian alluvium, mainly silts.

Previously, and during the drought period, the predominant plants were *Tamarix* and *Suaeda* and the site still includes shrub woodlands with *Capparis spinosa, Tamrix* sp., and *Suaeda* sp. but with reflooding marshland species appeared such as *Phragmites australis, Typha domingensis,* and *Schoenoplectus littoralis* along with rooted submerged vegetation. During the summer surveys, the marsh was semi-dry and the water concentrated in the old drainages canals but some flooded patches close by provide good foraging habitat for waders and herons.

Key Biodiversity Area Criteria	Notes	
V. Vulnerability Criteria: *Presence of Critically Endangered and Endangered species – presence of a single individual or Vulnerable species– 30 individuals or 10 pairs.*		
Rafetus euphraticus	Reported by locals frequently over the permenant water areas.	
Lutrogale perspicillata	The site is one of the few areas where an isolated population of this species was found before the area was affected by drainage	
Ia. Irreplaceability Sub-criterion: Restricted-range species based on global range		
Rafetus euphraticus	See above	
Important Bird Area Criteria	**Observations made 2005-2009**	
A1. Globally threatened species	**Breeding**	**Wintering/ Passage**
Marbled Duck *Marmaronetta angustirostris* (Resident)		18 (count 2006)
Basra Reed Warbler *Acrocephalus griseldis* (Summer visitor)	20 pairs	
A2. Restricted-range species	**Breeding**	**Wintering/ Passage**
Basra Reed Warbler	20 pairs	
Iraq Babbler *Turdoides altirostris* (Resident)	35 pairs	

Important Plant Area Criteria
B1. Site is a particularly species-rich example of defined habitat type
Marsh vegetation- helophytic vegetation- reedbeds; reedmace beds; and Schoenoplectus beds habitat type; Aquatic communities- Rooted submerged vegetation habitat type, and Woodland- shrubs habitat type
C. The site is identified as an outstanding example of a globally or regionally threatened habitat type.
This site is located in a critically endangered ecoregion (Tigris-Euphrates alluvial salt marsh) and is a good example of the marshlands of southern Iraq, a globally significant wetland. These marshes are threatened by decreasing water levels and drought, grazing, and water pollution.

Additional Important Bird Observations: During the surveys 32 bird species were observed. In addition to those listed above, Greater Spotted Eagle *Aquila clanga* (Vulnerable) was found wintering as well as Ferruginous Duck *Aythya nyroca*, Pallid Harrier *Circus macrourus*, and Black-tailed Godwit *Limosa limosa* (all Near Threatened) on passage and in winter, but in sub-IBA threshold numbers. The site also held breeding populations of four Sahara-Sindian Desert biome-restricted species but these did not trigger inclusion under criterion A3. The endemic race of Little Grebe *Tachybaptus ruficollis iraquensis* breeds in the area.

Other Important Fauna: The site might harbor the Smooth-coated Otter *Lutrogale perspicillata* (Vulnerable). The isolated population in the Lower Mesopotamian marshlands is regarded as a separtate subspecies (*L. p.maxwelli*) but its status and distribution have been unclear due to confusion with the Eurasian Otter *Lutra lutra* (Near Threatened), which also occurs in the region. Recent surveys (Omer et al., 2012, Al-Sheikhly and Nader, 2013) have confirmed the presence of smooth-coated otter in the southern marshes for the first time since the 1950s-1960s and it is likely that this species occurs in Auda Marsh, as well as other parts of the Tigris wetlands and marshes on both sides of the Iran-Iraq border.

Fish: Data were collected for the years 2006 and 2007 only, during which 14 species were reported. According to Coad

(2010), the following significant species were: *Carasobarbus luteus, Carassius auratus, Ctenopharyngodon idella, Cyprinus carpio* (Vulnerable), *Heteropneustes fossilis, Leuciscus vorax, Liza abu, Luciobarbus xanthopterus* (Vulnerable), *Mesopotamichthys sharpeyi* (Vulnerable), and *Silurus triostegus.* Species for which there is a lack of information are *Eleutheronema tetradactylum* and *Nemipterus bleekeri* (Coad, 2010).

Additional Plant & Habitat Information: This site contains a good population of *Phragmites australis*, which is important both economically and culturally.

Conservation Issues: The situation of this marsh in summer is quite poor and the water level has declined continuously because of reduced inflow. Due to the lack of outlets, its waters tend to stagnate. Agricultural expansion, over-exploitation of species through hunting (mainly birds) and fishing, (including electro-fishing), and human intrusion at the site were all considered high threats at the site. Two medium threats were road construction and pollution. Despite its somewhat poor condition, the site still attracts a number of birds that are of conservation concern, as the reedbeds provide suitable shelter.

Recommendations: To improve this marsh, it will be necessary to construct outlets to re-establish water circulation and improve water quality and to control water levels. There were plans to do this by the MoWR and these should be implemented as soon as possible.

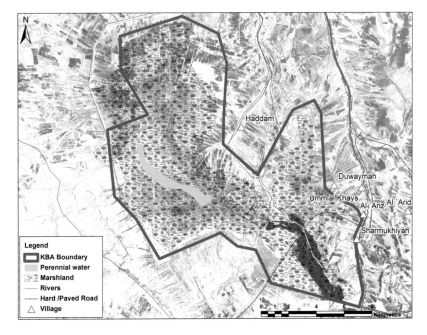

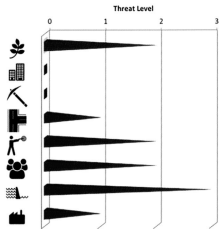

Hawizeh (IQ073)

Missan & Basrah 31.577177°N 7.684901°E

KBA Criteria: **V and Ia**
IBA Criteria: **A1, A2, A4i and A4iii**
IPA Criteria: **B1, B2b, and C**

Area: **164023 ha** - Altitude: **4-11 m**
Ecoregion: **Tigris-Euphrates alluvial salt marsh (PA0906)**
Status: **Ramsar Site**

Site Description: Hawizeh Marsh lies to the east of the Tigris River, straddling the Iran-Iraq border. As a transboundary wetland approximately 75-80% is located in Iraq and the remainder in Iran where it is known as Hor Al-Azim and is fed primarily by the Kharkeh River. In Iraq, this marsh is fed by two main distributaries from the Tigris River near Amarah, known as Al-Musharah and Al-Kahla'a. Evans (1994) listed Hawizeh as two separate IBA sites (IBA032 & IBA036) but the KBA surveys consider them to be a single site. The marsh extends from Birkat Al-Udheim and Umm An-Ni'aaj Lakes in the north and is bordered by the Iran-Iraq frontier on the northeast

©2009 MA SALIM/NI

and the Majnoon Oilfield to the southeast. Villages and agricultural lands site on the western edge of Hawizeh and include many roads, dykes and causeways such as Al-Sheeb, Al-Musharah, and Al-Kahla'a to the northwest, and Al-Uzayr and Al-Khana (Dasim) in the southwestern extremity. Small linear sections follow the wetland waterways of the Al-Kassarah Canal in the west and along the Iranian border area in the east. Numerous dykes topped by paved roads also bisect the site including the Lisan E'jayrda Causeway running west to east. The geology of the area is Mesopotamian alluvium; mainly silts and the habitats range from submerged and emergent marshland vegetation to riparian vegetation of *Salix* spp. to shrub woodlands of *Tamarix* sp. During the KBA

surveys, eight sub-sites were surveyed in winter and summer between 2005 and 2010.

Hawizeh Marsh became the first Ramsar site in Iraq, designated when Iraq signed the Convention for Protecting Internationally Important Wetlands in 2008 (Ramsar 2011). The total area of the designated Ramsar site is 137,700 ha. The estimated 1973 pre-drainage area of Hawizeh Marsh was 243,500 ha, and about 56% of the original area was included in the Ramsar site designation. The Iraqi Ramsar site and the KBA site delineations exclude the marsh areas that extend into Iran. The Ramsar site only includes the water bodies with narrow margins around the marshlands, but the KBA delineation for Hawizeh is somewhat larger because a larger

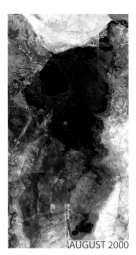
AUGUST 2000

MAY 2005

FEBRUARY 2008

SEPTEMBER 2010

DECEMBER 2013

buffer zone is proposed including some terrestrial areas. This is to ensure the biological requirements of the key species of this site. The area is dominated by freshwater to brackish marshes, both permanent and seasonal, with reedbeds and open areas of shallow water. Hawizeh was one of the only Mesopotamian marshlands that never fully dried out in the 1990s because it continued to receive water from Iran through the Kharkeh River. Nevertheless the re-flooding that took place throughout the marshes after the war in 2003 greatly benefitted Hawizeh Marsh as well (see the 2000 and 2008 Satellite images of Hawizeh below).

Unfortunately, the wetlands continues to face problems and is becoming increasingly restricted northwards as dry land extends from the southern edges of the Majnoon oilfield to the north of Lisan E'jayrda. This is largely due to the completion of a 90-km long embankment in 2011 on the Iranian side of the border that now impedes water entering the marshes from Iran (this is shown in the southern portion of the 2010 and 2013 satellite images below). In addition, water resources from the Tigris are declining due to upstream dam and diversion projects.

Key Biodiversity Area Criteria	Notes	
V. Vulnerability Criteria: *Presence of Critically Endangered and Endangered species – presence of a single individual or Vulnerable species– 30 individuals or 10 pairs.*		
Rafetus euphraticus	Euphrates Softshell Turtle was observed in considerable numbers over the entire area with a concentration in northern areas of the site.	
Lutrogale perspicillata	The site is one of the few areas where an isolated population of this species was found.	
Ia. Irreplaceability Sub-criterion: Restricted-range species based on global range		
Rafetus euphraticus	See above	
Important Bird Area Criteria	**Observation made 2005-2010.**	
A1. Globally threatened species	**Breeding**	**Wintering/ Passage**
Marbled Duck *Marmaronetta ngustirostris* (Resident)	150 pairs	180-1,300 (counts 2005-2009)
Basra Reed Warbler *Acrocephalus griseldis* (Summer visitor)	500 pairs (based on sample counts and 2,250 ha of reed-beds suitable for breeding)	
White-headed Duck *Oxyura leucocephala* (Winter visitor)		38 (count 2005)
A2. Restricted-range species	**Breeding**	**Wintering/ Passage**
Iraq Babbler *Turdoides altirostris* (Resident)	800 pairs (based on sample counts and 3,000 ha of reed-beds suitable for breeding)	Present
Acrocephalus griseldis	500 pairs	
A4i. 1% or more of biogeographical population of a congregatory waterbird species		
	Breeding	**Wintering/ Passage**
Marmaronetta angustirostris (Resident)	150 pairs (2005-2009)	180-1,300 (counts 2005-2009)
A4iii. Holding congregations of 20,000 waterbirds		
Congregatory waterbirds		> 35,000
Important Plant Area Criteria		
B1. The site is a particularly species-rich example of a defined habitat type		
Inland Standing Water-Aquatic communities-Rooted Submerged Vegetation, Rooted floating vegetation, and Free Floating Vegetation habitat type; Marsh vegetation- Helophytic vegetation-Reed beds, Reed mace beds, and Schoneplectus beds habitat type; Riparian vegetation habitat type, and Woodland- Shrubs habitat type		
B2b.The site is a refuge for: biogeographically and bioclimatically restricted plants to 'retreat to' in the face of global climate change.		
This is the only marsh area that was not completely drained in the 1990s. It is therefore likely to have higher plant bidoversity than other areas. As such it has the potential to act as a source of genetic material for reflooded areas.		
C. The site is identified as an outstanding example of a globally or regionally threatened habitat type.		
This site is located in a critically endangered ecoregion (Tigris-Euphrates alluvial salt marsh) and is a good example of the marsh-lands of southern Iraq, a globally significant wetland. These marshes are threatened by decreasing water levels and drought, grazing, and water pollution.		

Additional Important Bird Observations: During the 2005-2010 surveys 94 bird species were observed in Hawizeh. Two Vulnerable species, Eastern Imperial Eagle *Aquila heliaca* and Greater Spotted Eagle *A. clanga*, were also found wintering at this site, as were three Near Threatened species, Ferruginous Duck *Aythya nyroca* (summer and winter), Pallid Harrier *Circus cyaneus*, and Black-tailed Godwit *Limosa limosa* (passage and winter), all in sub-IBA threshold numbers. The Iraqi race of Little Grebe *Tachibaptus ruficollis iraquensis* and the Iraqi race of Hooded Crow *Corvus cornix capellanus* (also known as Mesopotamian Crow) breed here. Additionally, the site supported eight breeding Sahara-Sindian Desert biome-restricted species but these did not trigger inclusion under the A3 criterion. Hawizeh is the only wetland in Iraq that holds a breeding population of African Darter *Anhinga rufa* (of the Middle East race chantrei) and African Sacred Ibis *Theskiornis aethiopicus*. According to frequent reports of locals and hunters, the Goliath Heron *Ardea goliath* occurs in the northern part of the marshes, but in small numbers.

Other Important Fauna: Data were collected in 2005-2010 at various sites in Hawizeh. The southen marshes lie at the centre of the distribution of an isolated subspecies of Smooth-coated Otter *Lutrogale perspicillata maxwelli*. Its status and distribution have been unclear due to confusion with the Eurasian Otter *Lutra lutra* (Near Threatened), which also occurs in the region. Recent surveys (Omer et al. 2012, Al-Sheikhly and Nader 2013) have confirmed the presence of Smooth-coated Otter in parts of the southern marshes for the first time since the 1950s-1960s and it is likely that this species occurs in Hawizeh, as this is one of the few areas in southern Iraq that was not completely drained in the 1990s. Some key carnivore species found or reported during the KBA surveys include Jungle Cat *Felis chaus* and Wild Cat *Felis silvestris*. In 2012, Grey Wolf *Canus lupus*, Golden Jackal *Canis aureus*, and Wild Cat *Felis silvestris* were camera trapped in Majnoon.

Fish: Data were collected from 2005 through 2007, and in 2009, during which 15 species were found. Significant species according to Coad (2010) were: *Acanthobrama marmid*, *Acanthopagrus arabicus*, *Alburnus mossulensis*, *Carasobarbus luteus*, *Carassius auratus*, *Cyprinus carpio* (Vulnerable), *Heteropneustus fossilis*, *Leuciscus vorax*, *Liza abu*, *L. carinata*, *Luciobarbus xanthopterus* (Vulnerable), *Mesopotamichthys sharpeyi* (Vulnerable) and *Silurus triostegus*.

In 2009, a Fisheries Frame survey was conducted at three of the sub-sites within Hawizeh to the north, center and south (Umm An-Ni'aaj (HZ1), E'jayrda (HZ4), & Majnoon (HZ8) respectively). Fishing was by floating and fixed gill nets ranging in size from 0.5, 1, 2, 3 and 4 cm. The daily catch in HZ1 was estimated to be approximately 8 kg/boat each day for gill nets from about 20 boats observed but extensive electro-fishing was practiced at the site with about 150 boats using this method in an estimated daily catch of about 26 kg/boat-day. In HZ4, approximately 100 fishing boats were working in the area with a daily catch of about 13kg/boat

each day. Here electro-fishing is practiced in addition to net fishing. HZ8 had about 100 fishing boats working with a daily catch of about 8 kg/boat per day. But at HZ8 no electro-fishing was reported during the survey.

Additional Plant & Habitat Information: This site contains a good population of *Phragmites australis*, which is important both economically and culturally.

Conservation Issues: Hawizeh is one of the most important water bodies in Iraq as it harbors a number of very key species that breed in the region's dense reed beds. However, many of the Hawizeh sub-sites are now severely affected by the Iranian embankment that runs 90 km along the Iraq/Iran border, bisecting the entire length of Hawizeh Marsh and directly impacting the water supply to the Iraqi wetlands from Al-Karkha River. The most urgent priority is to remove this embankment to renew water supplies. During recent years, the habitat has changed dramatically and most of this site is suffering from serious drought that began in 2009, negatively impacting the biodiversity of the area. Unfortunately, many parts that were once extremely rich in waterfowl (based both on historical information and the KBA 2005-2009 surveys) have dried up in less than two years.

Oil development is primarily focused in the northern part of the site near Umm An-Ni'aaj and the southern part in Majnoon area, and was considered a very high threat due to previous pollution from well sites as well as plans for further development. Human intrusion, both from past damaging military campaigns and current movement of people and equipment across the southern part for oil development was also considered a very high threat. Additional high threats result from urban and commercial development and expansion occurring largely close to Umm An-Ni'aaj; unsustainable fishing and bird hunting that is concentrated at the northern sub-sites (as well as E'jayrda and Majnoon before 2008); and pollution from sewage coming from upstream cities and towns as well as petrochemical pollution from oil facilities in southern Hawizeh. In addition, in 2010/2011, Iran used mobile pumps to pump highly saline water (more saline than waters of the Gulf) directly into Hawizeh with damaging effects upon these wetlands and local agriculture. The Iraqi Ministry of Water Resources began construction on a new embankment in 2013 in the southern part of Hawizeh (in Basrah Governorate) in an attempt to protect Iraqi lands from such activities in the future.

Recommendations: With the listing of Hawizeh as Iraq's first Ramsar Site, a management plan was developed but remains largely unimplemented (Nature Iraq, 2008a & b). In addition, in 2016 the Hawizeh Marsh was incorporated, along with other wetland areas, into the Ahwar of Southern Iraq, a UNESCO World Heritage Site. It is critical to update the management plan and implement the program of work and the key recommendations it made. Here we address only recommendations for the critical threats now facing Hawizeh since the completion of the management plan. Due to the serious decrease in water levels in Hawizeh because of the embankment built by Iran, liaison with the relevant Iranian authorities is necessary and the solution to this problem will be atleast partly political. This site must also be protected from human activities that affect the fauna, such as over-hunting of birds and electro-fishing. These can be addressed by the local governments in Basrah and Missan Governorates prohibiting or curtailing fishing and hunting in the breeding/spawning seasons as well as the bird migration seasons.

Additional research should be conducted of the local fish population to determine the sustainable catch for the area, and a community program devised to regulate the fisheries. Frequent field education activities should be conducted to raise local awareness about threatened and key species and how to protect them. It is noteworthy that Shell Oil Company (contracted by the Iraqi government) is developing the Majnoon Oifield, which overlaps Hawizeh Marshes. Shell consulted Wetlands International, Flora and Fauna International, Nature Iraq and Mott MacDonald as well as other conservation organizations and consulting companies to develop a Biodiversity Action Plan for Hawizeh Marsh. The draft action plan was completed in the spring of 2013 and aims to minimize the negative impacts of oil development on the biodiversity of the area.

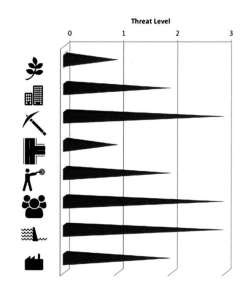

Suwaibaat (or Sleibaat) (IQ074)

Muthanna & Thi-Qar - 30.80611°N 45.96639°E

IBA Criteria: A1 and A4iii

Area: **84753 ha - Altitude: 4-20 m**
Ecoregion: **Arabian desert and East Sahero-Arabian xeric shrublands (PA1303) & Tigris-Euphrates alluvial salt marsh (PA0906)**
Status: **Unprotected**

Site Description: Suweibaat (also known as Sleibaat or sometimes Sleibeekhaat) is a large depression and seasonal wetland approximately 54km long and 15km wide. It collects rainwater from the southwestern desert (Al-Shamiya) via branched wadis that carry water seasonally, of which the main ones are: Rijlat Al-Shalhoobiya, Sh'eeb Al-'Ash'ali, Sh'eeb Al-Qaseer, and Rijlat Al-Safawi. These wadis are not deep but contain relatively good vegetation. Suwaibaat also receives a small amount of water from the western bank of the Euphrates River through Al-Qadissiya Canal, which was originally built in the 1990s to divert water away from the marshlands. In addition to the seasonal waterbody, Suwaibaat includes vast desert and arid areas with desert shrub habitat. Approximately 90% of the land is unvegetated and the geology is marls, siltstone, gypsum, and limestone bands.

The waterbody itself provides good habitat and shelter for quite large numbers of migrant and wintering waterfowl and waders. Eridu, an ancient Sumerian city, lies close to but outside the delineated site and surrounded by a large soil embankment. The Ziggurat of Eridu emerges as a large hill and a paved road leads to the ruins. The KBA team covered only a small portion to the southeast of the area near Eridu and was not able to survey the rest of this site. Most of the information was collected from reports from locals and hunters that regularly visit the area.

©2014 SALWAN ALI/IOCN

Important Bird Area Criteria	Observations from 2010.	
A1. Globally threatened species	**Breeding**	**Wintering/ Passage**
Macqueen's Bustard *Chlamydotis macqueenii* (Winter visitor)	May breed	Reported in 'large numbers'
Marbled Duck *Marmaronetta angustirostris* (Resident)		Frequently reported by hunters in 'large numbers'
A4iii. Holding congregations of 20,000 waterbirds or 10,000 pairs of seabirds of one or more species		
	Breeding	**Wintering/ Passage**
Congregatory Waterbirds (mainly Waders and Waterfowl)		> 20,000 waterbirds

Additional Important Bird Observations: In addition to the species listed in the table above, the site also held seven Sahara-Sindian Desert biome-restricted species that might breed at the site but these did not trigger inclusion under criterion A3.

Other Important Fauna: According to local reports, mammals include: Honey Badger Mellivora capensis, Striped Hyena *Hyaena hyaena* (Near Threatened), and Grey Wolf *Canis lupus* among other common species. Several reptiles were observed and reported but species were not identified and fish were not studied.

Conservation Issues: Oil development has a high impact on the area as does birds hunting, overgrazing, and human disturbance due to Bedouin activities and other users of the site (for example, hunters driving 4x4 vehicles in search of game birds). In addition there were several reports that bird hunters used poisoned bait to collect as many birds (especially ducks) as possible. The western and southern parts of Suwaibaat appear to have the most severe hunting pressure on Houbara. Agricultural intensification is occurring in the northern parts of the site and wastewater containing agricultural chemicals drains into Suweibaat from neighbouring farms.

Recommendations: Additional surveys are recommended to investigate the presence of other species of conservation concern. Bird

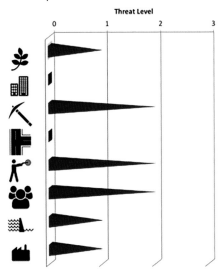

Legend
- KBA Boundary
- Rivers
- Hard /Paved Road
- City
- Village
- Archaeological Site

hunting is a particularly important issue to address in this area with a focus on local stakeholders in Muthanna and Thi-Qar who use Suweibaat. The presence of the Eridu Archeological Site offers opportunities for integrated cultural

and environmental activities. The National Protected Area Committee identified this site in 2014 as a proposed protected area.

Central Marshes (IQ075)

Thi-Qar - 30.958880°N 46.988837°E

KBA Criteria: **V and Ia**
IBA Criteria: **A1, A2, A3, A4i, A4ii, and A4iii**
IPA Criteria: **B1 and C**

Area: **131780 ha** - Altitude: **less than 6 m**
Ecoregion: **Tigris-Euphrates alluvial salt marsh (PA0906)**
Status: **A portion of this area has protected status as a national park.**

Site Description: Historically, the Central Marshes comprised a vast complex of mostly permanent freshwater marshes, with scattered areas of open water, located to the west of the River Tigris and north of the River Euphrates (within Basrah, Thi-Qar, and Missan provinces) and represented many diverse habitats from seasonal to permanent marshes. Evans (1994) listed it as an Important Bird Area (IBA038) fed by both rivers but today, even after

©2009 MA SALIM/NI

re-flooding of some areas many of the connections from the Tigris remain severed. The Central Marshes are now restricted to two areas: Chibaish Marshes, north of the city of Chibaish, and Abu Zirig Marsh to the northwest; the two areas are still connected and the geology of both areas is Mesopotamian alluvium, mainly silts.

Chibaish Marshes were formerly freshwater marshes with many open areas and deep water (more than 3m) but nowadays are brackish with very dense submerged and emergent vegetation and shallow water. Eutrophication has led to oxygen depletion and fish mortality in the last few years. The habitat consists of dense reedbeds and areas of *Typha domingensis* and *Schoenoplectus litoralis*. Submerged plants cover much of the sediments in open water areas. Seven inlets from the Euphrates feed Chibaish Marshes and they serve as an outlet when the water level in the Euphrates falls below that in the marshes. Due to the sharp reduction in water levels in the Euphrates during 2009 and 2010, the Ministry of Water Resources, in collaboration with the local governments, constructed an embankment across the river in early summer 2010 to divert the water flow completely towards the Central Marshes as well as West Hammar (IQ076).

Abu Zirig, on the western side of the Central Marshes is a freshwater marsh with relatively good water flow and discharge. Al-Gharraf River, a branch of the Tigris controlled by the Kut regulator provides water to Abu Zirig. Water depth ranges from 0.5 to 1.5 m or more in the main waterways. It is close to Al-Islah city, southeast of Nasiria city. This area suffers from significant human disturbance such as fishing, bird hunting and reed harvesting though it has dense vegetation in some areas, mainly *Phragmites australis* and *Typha domingensis*. Fieldwork concerning the bird and fish populations at this site has been conducted since 2004 as a part of the New Eden Project.

The Central Marshes, with their mix of landscapes (including open water, dense reedbeds and mudflats) are very important for bird and animal life. They are a key area for wintering waterbirds, especially the globally Vulnerable Marbled Duck *Marmaronetta angustirostris* and an important breeding area for the restricted-range endemic Basra Reed Warbler *Acrocephalus griseldis*, Iraq Babbler *Turdoides altirostris*, and the Iraqi subspecies of Little Grebe *Tachybaptus ruficollis iraquensis*.

The 13 sub-sites in the Central Marshes were surveyed in different seasons from 2005–2010. Some sites were visited in 2005–2010, while others were not visited due to the difficulties in reaching them when they were turned into either dry or muddy areas.

Key Biodiversity Area Criteria	Notes	
V. Vulnerability Criteria: *Presence of Critically Endangered and Endangered species – presence of a single individual or Vulnerable species– 30 individuals or 10 pairs.*		
Rafetus euphraticus	Observed by the KBA team and reported by locals	
Lutrogale perspicillata	An isolated population of this species was found before the area was affected by drainage.	
Ia. Irreplaceability Sub-criterion: Restricted-range species based on global range		
Rafetus euphraticus	See above	
Important Bird Area Criteria	**Observations made 2005-2010**	
A1. Globally threatened species	**Breeding**	**Wintering/ Passage**
Marbled Duck *Marmaronetta angustirostris* (Resident)	100-150 pairs (counts)	1,800-12,000 (counts)
Basra Reed Warbler *Acrocephalus griseldis* (Summer visitor)	2,000 pairs (based on sample counts and c.7,000 ha of reedbeds being suitable for breeding).	
A2. Restricted-range species	**Breeding**	**Wintering/ Passage**
Acrocephalus griseldis (Summer visitor)	2,000 pairs	
Iraq Babbler *Turdoides altirostris* (Resident)	1,000 pairs	Present
A3. Biome-restriced species		
Sahara-Sindian Desert biome	**Breeding**	**Wintering/ Passage**
White-tailed Lapwing *Vanellus leucurus* (Resident)	50 pairs	100-180 (counts)
Spotted Sandgrouse *Pterocles senegallus* (Resident)	10 pairs	Present but not counted
Egyptian Nightjar *Caprimulgus aegyptius* (Summer visitor)	10 pairs, not seen 2006-2008	
Hypocolius *Hypocolius ampelinus* (Summer visitor)	20 pairs (counts 2006-2007)	
White-eared Bulbul *Pycnonotus leucotis* (Resident)	10 pairs (counts 2005-2010)	6-12 (counts 2006-2009)
Acrocephalus griseldis (Summer visitor)	2,000 pairs	
Turdoides altirostris (Resident)	1,000 pairs	Present
Dead Sea Sparrow *Passer moabiticus* (Resident)	2,000 pairs	7000 (counts); not seen in 2007
A4i. 1% or more biogeographical population of a congregatory waterbird species		
	Breeding	**Wintering/ Passage**
Marmaronetta angustirostris (Resident)	100-150 pairs (counts)	1,800-12,000 (counts)
A4ii. 1% or more of global population of a congregatory seabird or terrestrial species.		
Passer moabiticus (Resident)	2,000 pairs	7,000 (counts); not seen in 2007
A4iii. Holding congregations of 20,000 waterbirds or 10,000 pairs of seabirds of one or more species.		
Congregatory Waterbirds		Up to 74,000 (counts)
Important Plant Area Criteria		
B1. Site is a particularly species-rich example of defined habitat type		
Inland Standing Water- Aquatic communities- Rooted Submerged Vegetation habitat type; Inland Standing Water- Aquatic communities- Free Floating Vegetation habitat type, and Marsh vegetation- Helophytic vegetation- Reedbed, Reedmace bed, or Schoenoplectus bed habitat type		
C. The site is identified as an outstanding example of a globally or regionally threatened habitat type		
This site is located in a critically endangered ecoregion (Tigris-Euphrates alluvial salt marsh) and is a good example of the marshlands of southern Iraq, a globally significant wetland. These marshes are threatened by decreasing water levels and drought, grazing, and water pollution.		

Additional Important Bird Observations: During the surveys 94 bird species were observed but the species count since 2010 has reach 120 species (Fazaa & Whittingham, 2014). In addition to those listed in the table, two Vulnerable species, Greater Spotted Eagle *Aquila clanga* and Eastern Imperial Eagle *A. heliaca* were found wintering, as well as five Near Threatened species: Ferruginous Duck *Aythya nyroca* (summer and winter), Pallid Harrier *Circus cyaneus,* Eurasian

Curlew *Numenius arquata*, Black-tailed Godwit *Limosa limosa* (passage and winter) and Cinereous Vulture *Aegypius monachus* (seen in winter 2013), but in sub-IBA threshold numbers. The endemic race of Little Grebe *Tachybaptus ruficollis iraquensis* breeds widely.

Other Important Fauna: Data were collected 2005-2010, but with more focused attention in 2007. The southen marshes lie at the centre of the distribution of an isolated subspecies of smooth-coated Otter *Lutrogale perspicillata maxwelli*. Its status and distribution have been unclear due to confusion with the Eurasian Otter *Lutra lutra* (Near Threatened), which also occurs in the region. Recent surveys (Omer et al., 2012, Al-Sheikhly and Nader, 2013) have confirmed the presence of smooth-coated otter in parts of the southern marshes for the first time since the 1950s-1960s and it is likely that this species occurs in the Central Marshes too, as well as other parts of the Tigris wetlands and marshes on both sides of the Iran-Iraq border.

Some notable mammals found during the surveys (and according to local reports) were: Rüppell's Fox *Vulpes rueppellii*, Jungle Cat *Felis chaus* and Wild Cat *Felis silvestris*. Due to the availablility of food and shelter, the Central Marshes provide good habitat for the Endangered *Rafetus euphraticus*.

Fish: Data were collected from 2005 through 2007 and again in 2009, during which 18 species were found. Significant species according to Coad (2010) were: *Acanthopagrus* cf. *arabicus*, *Acanthobrama marmid*, *Alburnus mossulensis*, *Carasobarbus luteus*, *Carassius auratus*, *Cyprinus carpio* (Vulnerable), *Cyprinion kais*, *Heteropneustes fossilis*, *Leuciscus vorax*, *Liza abu*, *L. klunzingeri*, *Luciobarbus esocinus* (Vulnerable), *L. xanthopterus* (Vulnerable), *Mesopotamichthys sharpeyi* (Vulnerable), *Tilapia zillii* and *Silurus triostegus*. The latter appears to be increasing.

In 2009, a Fisheries Frame survey was conducted at 3 sub-sites within the Central Marshes (CM1, CM10 & CM16). Fish catches in South Baghdadiya (CM1) were 5 kg/boat per day using gill nets (mesh sizes of 0.5-4 cm) and 7 kg/boat per day using electro-fishing. In North Fuhood (CM10), fixed gill nets of 1-4 cm were used but electro-fishing was also used and the total average catch was 15 kg/boat per day. In Abu Zirig (CM16), where electro-fishing is prohibited and fishing is done by fixed and floating gill nets (mesh size 0.5-4 cm), the average catch was 12 kg/boat per day.

Additional Plant & Habitat Information: This site contains a good population of *Phragmites australis*, important both economically and for local heritage.

Conservation Issues: Natural systems modification, human disturbance, and over-exploitation were ranked as the very highest threats. Construction of dams upstream on the Tigris and Euphrates rivers have a serious effect on this site, which suffers from a severe lack of water, especially in 2009-2010 when the water dropped to critical levels, destroying habitats by causing wide areas of wetland and reedbeds to dry out. Hunting pressure is severe in some areas, and despite the extensive area of dense reed beds where ducks can hide hunting still causes high levels of disturbance. In addition, overfishing is common throughout the entire area along with the use of unsustainable fishing practices such as electro-fishing and fishing during spawning season. Other pressure of over-exploitation and consumption of biological resources is represented by unsustainable reed harvesting.

Three threats were considered high: urban intensification, energy production, and pollution. Agricultural intensification and road construction were scored as medium threats. Though the KBA team did not assess the threats caused by invasive species, during surveys between 2005 and 2010, it was obvious from examining the regular fish catch that introduced and invasive *Tilapia* spp. (Family: *Cichlidae*) have been released or somehow reached the area. By the end of the survey period, *Tilapia* spp. formed the majority of the catch for most fishermen in the Central Marshes, displacing many of the native species.

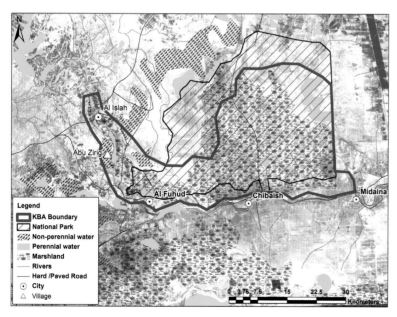

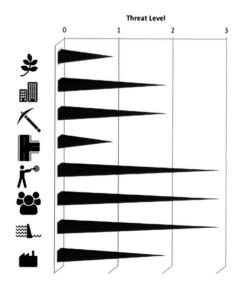

Recommendations: This site, especially the proposed National Park area (New Eden Group, 2010), is a key part of the southern marshes complex and needs a stable and adequate water supply and effective protection from hunting and disturbance to maintain its biodiversity value. In July of 2013, the Iraq Council of Ministers approved the designation of a portion of the Central Marshes as Iraq's first National Park. In addition, in 2015 the Central Marshes were designated a Ramsar Wetland Site of International Importance and then in 2016, along with other wetland areas, this site was listed as a UNESCO World Heritage Site. Key to the protection of these wetlands will be the full implementation of the management plan for the area. Thorough surveys of the flora and fauna should also continue. It is also highly recommended to strengthen the relationship with the locals of the area and pay more attention to those who are keen in protecting their marsh by encouraging the formation of Local Conservation Group(s) (LCG) for better implementation of any conservation activities.

West Hammar (IQ076)

Thi-Qar - 30.83333N° 46.71667E°

KBA Criteria: **V and Ia**
IBA Criteria: **A1, A2, A3, A4i, A4ii, and A4iii**
IPA Criteria: **B1 and C**

Area: **136326 ha** - Altitude: **less than 8 m**
Ecoregions: **Tigris-Euphrates alluvial salt marsh (PA0906)**
Status: **Unprotected**

©2009 MA SALIM/NI

Site Description: Evans (1994) described the Hammar Marshes (IBA039) as an area of wetlands about 3,500 km² in size with the largest open water bodies in the lower Euphrates including a shallow, eutrophic lake approximately 60 km long by up to 20 km wide. The Euphrates River forms its border to the north and the Shatt Al-Arab on the east, to the south and west are desert regions. However, the historical Hammar marsh has now been bisected by the West Qurna and North Rumaila oilfields and is split into two main parts: East Hammar (IQ077 dealt with now in a separate account) and West Hammar. The area is now very different due to intensive building of canals, ditches, embankments, and roads, especially around the edges of the site, as well as considerable agricultural work inside it. Yet it remains a sizable wetland complex extending from Suq Ash-Shuyukh in the west to an embankment running north-south from the Euphrates just east of Chibaish, with the Euphrates River to the north and the Main Outfall Drain (MOD) Canal to the south. Two sources of water feed the West Hammar marshes: the Euphrates River (three openings in the river embankment allow water to enter from the north) and the MOD Canal (via a connecting canal that brings the water northward to the site). Today water depth averages 0.5 to 2m at the deepest point.

As stated above, the Hammar marshes as a whole and particular West Hammar are divided by embankments, canals, and roads into several sub-sites. There are open water areas such as Rashid lake and Buhaira Al-Hilwa; marsh vegetation like Teena, Umm Nakhla, Kermashia, Abu Hedeeda, and Abu Ajaj; and terrestrial areas such as Ghabishiya. The water is brackish and there is fishing using nets and electro-fishing, and hunting of birds. There are rice farms and date palm plantations especially at Umm Nakhla. Many reed houses exist on the embankments. During 2008-2010, drought in the Hammar marshes caused most of the people (M'adan) living on the embankments to leave the area and a large canal was dug near Al-Khamissiya to bring water from the MOD northwards to flood the northwestern parts of the site.

The geology of the area is Mesopotamian alluvium, mainly silts. The habitats at the site include: rooted-submerged vegetation; free floating vegetation; marsh beds of *Phragmites australis*, *Typha dominguensis* and *Schoenoplectus littoralis*; seasonally or occasionally flooded lands, and desert shrub vegetation. Twenty-one sub-sites were surveyed within West Hammar.

Key Biodiversity Area Criteria	Notes	
V. Vulnerability Criteria: *Presence of Critically Endangered and Endangered species – presence of a single individual or Vulnerable species– 30 individuals or 10 pairs.*		
Rafetus euphraticus	Observed by the KBA team and frequently reported by locals	
Ia. Irreplaceability Sub-criterion: Restricted-range species based on global range		
Rafetus euphraticus	See above	
Important Bird Area Criteria	**All observations made 2005-2010.**	
A1. Globally threatened species	**Breeding**	**Wintering/ Passage**
Marbled Duck *Marmaronetta angustirostris* (Resident)	40 pairs (2005); 700 pairs (2007-2010)	1,800-4,400 (counts)
Basra Reed Warbler *Acrocephalus griseldis* (Summer visitor)	1,000 pairs (based on sample counts and c.2,000 ha of reed-beds suitable for breeding)	
A2. Restricted-range species	**Breeding**	**Wintering/ Passage**
Iraq Babbler *Turdoides altirostris* (Resident)	2,000 pairs	2-29 (counts 2007-2010)
Acrocephalus griseldis (Summer visitor)	1,000 pairs	
A3. Biome-restricted species		
Sahara-Sindian Desert biome	**Breeding**	**Wintering/ Passage**
White-tailed Lapwing *Vanellus leucurus* (Resident)	1,000 pairs	240 (highest count)
Spotted Sandgrouse *Pterocles senegallus* (Resident)	40 pairs	Present
Pallid Scops Owl *Otus brucei* (Resident)	20 pairs	
Egyptian Nightjar *Caprimulgus aegyptius* (Summer visitor)	300 pairs	
Hypocolius *Hypocolius ampelinus* (Summer visitor)	500 pairs	
Greater Hoopoe-Lark *Alaemon alaudipes* (Resident)	10 pairs	Present - see summer
White-eared Bulbul *Pycnonotus leucotis* (Resident)	200 pairs	41 (counts 2006-2010)
Acrocephalus griseldis (Summer visitor)	1,000 pairs	
Iraq Babbler *Turdoides altirostris* (Resident)	2,000 pairs	Present
Dead Sea Sparrow *Passer moabiticus* (Resident)	1,300 pairs	800-3,000 (counts 2007-2010)
A4i. 1% or more of biogeographical population of a congregatory waterbird species		
	Breeding	**Wintering/ Passage**
Marmaronetta angustirostris	700 pairs	1,800-4,400 (counts)
Kentish Plover *Charadrius alexandrinus* (Resident)	1,500 pairs	3,200 (highes count)
Slender-billed Gull *Chroicocephalus genei* (Resident)	3,600 pairs	880 (highest count)
Whiskered Tern *Chlidonias hybrida* (Resdient)	1,800 pairs	266 (highest count)
A4ii 1% or more of global population of a congregatory seabird or terrestrial species		
Passer moabiticus	1,300 pairs	800-3,000 (counts 2007-2010)
A4iii. Holding congregations of 20,000 waterbirds		
Congregatory Waterbirds	>50,000	
Important Plant Area Criteria		
B1. The site is a particularly species-rich example of a defined habitat type		

Inland Standing Water- Aquatic communities- Rooted Submerged Vegetation habitat type; Inland Standing Water- Aquatic communities- Free Floating Vegetation habitat type; Marsh vegetation- Helophytic vegetation- Reedbed, Reedmace bed, or Schoenoplectus bed habitat type and Inland standing water- Flooded communities- Periodically or occasionally flooded land habitat type

C. The site is identified as an outstanding example of a globally or regionally threatened habitat type.

This site is located in a critically endangered ecoregion (Tigris-Euphrates alluvial salt marsh) and is a good example of the marshlands of southern Iraq, a globally significant wetland. These marshes are threatened by decreasing water levels and drought, grazing, and water pollution.

Additional Important Bird Observations: A total of 110 species were recorded. Eastern Imperial Eagle *Aquila heliaca* (Vulnerable), Eurasian Curlew *Numenius arquata* and Ferruginous Duck *Aythya nyroca* (both Near Threatened) were also seen on migration and in winter, but in sub-IBA threshold numbers; the site also held a widespread breeding population of Ferruginous Duck. The endemic race of Little Grebe *Tachybaptus ruficollis iraquensis* and Hooded Crow *Corvus cornix capellanus* (also known as Mesopotamian Crow) were present.

Other Important Fauna: Data were collected in 2007 only and the only mammals found were Rüppell's fox *Vulpes rueppellii* and Golden Jackal *Canis aureus*. No significant reptiles were found.

Fish: Data were collected in 2005-2007 and in 2009, when 17 species were reported. Significant fish according to Coad (2010) were: *Acanthobrama marmid, Acanthopagrus arabicus, Alburnus mossulensis, Arabibarbus grypus* (Vulnerable — observed only in 2006), *Carasobarbus luteus, Carassius auratus, Ctenopharyngodon idella, Cyprinus carpio* (Vulnerable), *Heteropneustes fossilis, Leuciscus vorax, Liza abu, L. klunzingeri, Luciobarbus xanthopterus* (Vulnerable), *Mesopotamichthys sharpeyi* (Vulnerable), and *Silurus triostegus*. One marine species *Bathygobius fuscus* was observed.

Additional Plant & Habitat Information: This site contains a good population of *Phragmites australis* and *Typha domingensis*, which are economically and culturally important.

Conservation Issues: This marsh harbours considerable numbers of waterfowl and shorebirds during migration. Large parts of the site are relatively inaccessible, making them important breeding sites for birds. The main threat to the Hammar Marshes in general is the lack of water because of dam building upstream and changes in water management. Also channelization of waterways (for road building and agriculture) has resulted in some areas, particularly in the southern and eastern edges of West Hammar, becoming completely dry. Also some of the soil embankments that were built to dry out the Hammar Marshes are still blocking waterways. As West Hammar Marsh is adjacent to two major oilfields, West Qurna and Rumaila, the embankments surrounding these oilfields are barriers to full restoration of the original marshes.

Hunting and electro-fishing are practiced heavily throughout much of the area. Residential & commercial development; human disturbance (especially during the bird breeding season), and water pollution from sewage and agricultural wastewater are also of concern as well as the build up of salinity in the marshes overall. Also during two surveys between 2009 and 2010 the introduced and invasive *Tilapia zillii* were recorded and by the end of the survey period they formed the majority of the catch for most fishermen in the Hammar marshes, displacing many of the native species.

Recommendations: In 2015, West and East Hammar were jointly designated a Ramsar Wetland Site of International Importance and then in 2016, along with other wetland areas, West Hammar was listed as a UNESCO World Heritage Site. Developing and implementing a comprehensive management plan will be an important next step.

Urgent actions are needed to protect the area. The most important issue to address is the water supply. Enough water needs to be allocated for the southern marshes in general and Hammar in particular. The New Eden Master Plan for Integrated Water Resource Management in the Marshland Areas (New Eden Group 2006) outlined ways to achieve this goal. The plan has been at least partially adopted by the Ministry of Water Resources, but a more concerted effort by the MoWR and the Iraqi government is needed to confirm a share of water for the marshes.

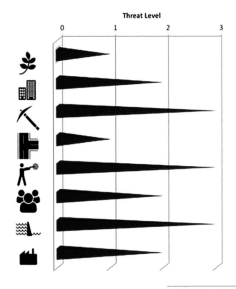

Hammar Marshes is an important area for flora and fauna and the Iraqi government should highlight the importance of this site and develop and implement the plan to protect it. For example, a campaign is needed in the oilfields to adapt technologies that have less harmful effects on the biodiversity of the area. Also more indepth studies of the fisheries are needed and should examine the impact of the invasive *Tilapia zillii*.

East Hammar (IQ077)

Basrah- 30.71667°N 47.46667°E

KBA Criteria: **V and Ia**
IBA Criteria: **A1, A2, A4i, and A4iii**
IPA Criteria: **B1 and C**

Area: **82968 ha** - Altitude: **less than 7 m**
Ecoregion: **Tigris-Euphrates alluvial salt marsh (PA0906) and Arabian Desert and East Sahero-Arabian Xeric Shrublands (PA1303)**
Status: **Unprotected**

FACING NORTH, EAST HAMMAR IN SHILEICHIYA ©2009 MA SALIM/NI

Site Description: Hor Al-Hammar (or Hammar Marshlands) was listed by Evans (1994) as an Important Bird Area (IBA039) meeting most of the IBA criteria and harbored very high numbers of wintering and passage waterfowl and waders as well as other birds. Historically, Hammar used to extend as a continuous waterbody for more than 55 km west of the Shatt Al-Arab River and south of the Euphrates. But due to the West Qurna Oilfield that was established in the middle of Hammar by draining the area and surrounding it with soil embankments, Hammar has become two separate areas: East and West Hammar. The current account deals with East Hammar (West Hammar (IQ076) will be dealt with in a separate account).

East Hammar is an extensive area of wetlands. It is located at the upper corner formed by the meeting of the Euphrates and Shatt Al-Arab Rivers and extends west to the oilfields of West Qurna. The Shatt Al-Arab flows southwards along the eastern edge of the site. Hammar Marshes were originally drained in a systematic campaign over the 1980s to early 2000s for various reasons. They were only reflooded in 2004 when the embankments blocking the water from entering the marshes were removed or destroyed by local communities. Gradually the plant cover (mainly reedbeds and reedmace) returned and birds and other wildlife came back to the site. The geology is Mesopotamian alluvium, mainly silts. Habitats vary from desert shrublands to aquatic communities of submerged and free floating vegetation, reedbeds, salt or brackish mudflats with *Salicornia herbacea*, and periodically flooded lands.

The different parts of the site that were visited in the KBA surveys (there were six sub-sites surveyed in different seasons from 2005 to 2010) are described separately. Al-Nagarah is located west of Al-Deer town (sub-district on the west side of Shatt Al-Arab, 30 km northwest of Basrah). This is a large marsh fed by the Euphrates River; it is tidally affected and is characterized by the presence of dense vegetation with distinct water flow. It is bounded by reed, cattail, & club-rush on two sides, and about 500 m to the north, the river expands to form a wide shallow area (locally called "Barga"). The area is affected by tides, which creates good habitat for waders and a high diversity of birds. There is a high level of hunting and fishing throughout this area.

Al-Shileichia area is located southeast of the Rumaila oil fields. The Main Outfall Drainage (MOD) runs through it and the main road parallels the MOD along a high embankment. There is very little water south of the MOD save for scattered ponds with reed cover. The plant cover mainly consists of Phragmites and stretches of semi-arid adapted plants. There is some grazing and hunting in this area. The southern area below the road and MOD is subject to tidal action. Al-Shileichia is characterized by the presence of dense floating algae and the existence of diverse birds and fish species. Mudskippers were seen in sub-tidal habitats. The quality of water, according to the locals, has improved through 2010 and this may be due to the decreasing flow of the MOD to the Gulf via the Khor Al-Zubayr canal caused by the closing of the Khor Al-Zubayr regulator. This is the main regulator on the Shatt Al-Basrah Canal and it regulates the water level between Khor Al-Zubayr (where the tidal effect is very active) and the lower parts of Al-Hammar. In 2009, the Ministry of Water Resources decided to close this regulator to improve the water level in this area.

FACING NORTH, EAST HAMMAR IN NAGGARA ©2008 MA SALIM/NI

Slein (or Ghatra) area is located about 30 km west of Shileichiya. The area is cut by the MOD that runs eastward to join the Al-Basrah Canal. It is a new area that was added to the KBA site in 2008 as it regularly harbors considerable numbers of wintering waterfowl including threatened species such as Marbled Duck. A paved road and embankment run parallel to the MOD linking Shileichiya with the Rumaila Oilfields. Large numbers of waterfowl were observed in the Slein area but most of them were unidentifiable because of the distance and inaccessibility of the southern margins of this area due to vast mudflats. The water level is shallow with scattered reed beds.

Three other areas within East Hammar include Al-Mas'hab, Al-Sallal and Shaafi. Al-Mas'hab area is located in the middle of the site and is tidally influenced by Mas'hab River. Phragmites australis is the prevailing plant species, which is utilized for grazing and cutting. Al-Sallal is a large marsh fed by the Shatt Al-Arab River, close to Mas'hab. The area contains good vegetation cover and is tidally affected with terrestrial plants along the fringes. Shaafi is a wetland located northwest of Basrah. The reedbeds here are dense in most of this area but with some open-water patches. High hunting pressure was observed here.

Key Biodiversity Area Criteria	Notes	
V. Vulnerability Criteria: *Presence of Critically Endangered and Endangered species – presence of a single individual or Vulnerable species– 30 individuals or 10 pairs.*		
Rafetus euphraticus	Reported frequently throughout the area	
Lutrogale perspicillata	The site represents one of the few areas where an isolated population of this species was once found before the area was affected by drainage.	
Ia. Irreplaceability Sub-criterion: Restricted-range species based on global range		
Rafetus euphraticus	See above	
Important Bird Area Criteria	**Observations made 2005-2012, additional visit in 2012**	
A1. Globally threatened species	**Breeding**	**Wintering/ Passage**
Basra Reed Warbler *Acrocephalus griseldis* (Summer visitor)	300 pairs (2007-2010)	
Marbled Duck *Marmaronetta angustirostris* (Resident)	Present, not counted	400-800 (counts 2009-2010)
A2. Restricted-range species	**Breeding**	**Wintering/ Passage**
Iraq Babbler *Turdoides altirostris* (Resident)	400 pairs (2008-2010)	Present
Acrocephalus griseldis	300 pairs (2007-2010)	
A4i. 1% or more of biogeographical population of a congregatory waterbird species		
Marbled Duck *Marmaronetta angustirostris*		400-800 (counts 2009-2010)
Slender-billed Gull *Chroicocephalus genei* (Resident)	3,000 pairs	1,070 (count 2009)
A4iii. Holding congregation of >20,000 waterbirds		
Congregatory Waterbirds		>30,000
Important Plant Area Criteria		
B1. The site is a particularly species-rich example of a defined habitat type		
Inland Standing Water- Aquatic communities- Rooted Submerged habitat type; Inland Standing Water- Aquatic communities-Free Floating habitat type; Marsh vegetation- Helophytic vegetation- Reedbed, Reedmace bed, or *Schoenoplectus* habitat type; Marsh vegetation- pioneer communities growing on salt or brackish habitat type and Inland standing water- Ponds or lakes-Unvegetated standing water habitat type		
C. The site is identified as an outstanding example of a globally or regionally threatened habitat type.		
This site is located in a critically endangered ecoregion (Tigris-Euphrates alluvial salt marsh) and is a good example of the marshlands of southern Iraq, a globally significant wetland. These marshes are threatened by decreasing water levels and drought, grazing, and water pollution.		

Additional Important Bird Observations: A total of 81 species were observed during the 2005-2010 surveys. Greater Spotted Eagle *Aquila clanga* and Eastern Imperial Eagle *A. heliaca* (both Vulnerable) were also recorded in winter but in numbers below the IBA threshold. The endemic race of Little Grebe *Tachybaptus ruficollis iraquensis* and endemic race of Hooded Crow *Corvus cornix capellanus* (Mesopotamian Crow) breed here. The site also held breeding populations of seven Sahara-Sindian Desert biome-restricted species but these did not trigger inclusion under the A3 criterion.

Other Important Fauna: Mammal data were collected in 2005-2010 and six species have been observed and/or reported in the area, including Golden Jackal *Canis aureus* and Grey Wolf *Canis lupus*. The southen marshes lie at the centre of the distribution of an isolated subspecies of Smooth-coated Otter *Lutrogale perspicillata maxwelli*. Its status and distribution have been unclear due to confusion with the Eurasian Otter *Lutra lutra* (Near Threatened), but recent surveys (Omer et al. 2012, Al-Sheikhly and Nader 2013) have confirmed the presence of smooth-coated otter in parts of the southern marshes for the first time since the 1950s-1960s and it is likely that this species occurs in Hammar Marshes too.

Fish: Data were collected at this site for the years 2005-2007 and 2009, during which 25 species were reported. According to Coad (2010), significant species were: *Acanthobrama marmid*, *Acanthopagrus arabicus*, *Alburnus mossulensis*, *Arabibarbus gyrpus* (Vulnerable), *Carassius auratus*, *Carasobarbus luteus*, *Ctenopharyngodon idella*, *Cyprinus carpio* (Vulnerable), *Cyprinion macrostomum*, *Gambusia holbrooki*, *Heteropneustes fossilis*, *Leuciscus vorax*, *Liza abu*, *L. subviridis*, *Luciobarbus xanthopterus* (Vulnerable), *Mesopotamichthys sharpeyi* (Vulnerable), *Silurus triostegus*, and *Tenualosa ilisha*. The following marine fish species were also recorded: *Eleutheronema tetradactylum*, *Bathygobius fuscus*, *Nemipterus bleekeri*, and *Otolithes ruber*.

In 2009, a Fisheries Frame survey was conducted at only one East Hammar sub-site (Al-Nagarah) and found 20 fishing boats there using floating gill nets (mesh size 1-3 cm), fixed gill nets (2-4 cm), seine nets (0.5 cm) as well as electro-fishing and poisons. The daily catch estimate was 16 kg/boat.

Conservation Issues: Two very high threats facing the East Hammar area are the lack of water (due to upstream dams and water resource management) and the expansion of the oil industry. The lack of water is causing an increasing tidal effect over the area and a rise in salinity in the eastern parts of the marsh. The oil industry also affects the western side of East Hammar due to its location between two of the largest oil fields in Iraq (West Qurna and Rumaila). In the western parts construction of embankments have cut large areas from the original marsh and added them to the oilfield. Two additional very high threats were unsustainable hunting and human disturbance caused by reed-cutting and fishing especially during the bird breeding season. Other high threats come from development that is affecting some areas around this KBA. Pollution from the oil industry affects mainly the western parts of this area, also includes the occasional use of poison for fishing.

Recommendations: In 2015, both East and West Hammar were jointly designated a Ramsar Wetland Site of International Importance and then in 2016, along with other wetland areas, East Hammar was listed as a UNESCO World Heritage Site. Developing and implementing a comprehensive management plan will be an important next step for this area.

As with West Hammar, this site needs a more stable supply of water, which will largely require a stronger commitment by the Iraqi government. It is important that water levels in the Shileichiya area of East Hammar are more stringently controlled to ensure a stable habitat. It is also clear that water levels in this area have decreased, most probably due to the closure of the regulator on the MOD downstream of the marsh. A local stated, "Yes, the water is saline here but it's better than nothing or we would have to leave!" The most urgent priority at the Slein sub-site is to stop the construction of embankments and check the expansion of oil fields. Based on the high number of waterfowl observed here it would also be advisable to designate an ecologically protected area in cooperation with the local police and sheikhs of the region. The National Protected Area Committee (NPAC) identified this site in 2014 as a proposed protected area.

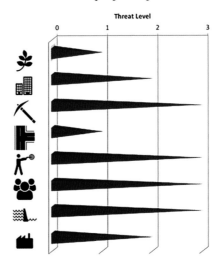

Salman (IQ078)

Muthanna - 30.416660°N 44.499999°E

IBA Criteria: **A1**
IPA Criteria: **Not assessed**

Area: **14959 ha** - Altitude: **210-330 m**
Ecoregion: **Arabian desert and east Sahero-Arabian Xeric Shrublands (PA1303)**
Status: **Unprotected**

©2010 MA SALIM/NI

Site Description: This site is located in the heart of the southwestern desert, southwest of the town of Salman. It consists mainly of dry habitats that include large seasonal watercourses (wadis) running northeastwards that carry water during the rainy season into Al-Suwaibaat depression (IQ074) west of the city of Nasiria. The area has scattered and low-growing vegetation throughout, but plant cover is denser in the wadis. A different large depression, in the southern part of the area, fills with water during the rainy season, and attracts various species.

Important Bird Area Criteria	Notes	
A1. Globally threatened species	**Breeding**	**Wintering/ Passage**
Macqueen's Bustard *Chlamydotis macqueenii* (Winter visitor)		Reported frequently by locals and hunters.

Additional Important Bird Observations: Only a few bird species were oberved during the 2010 winter and summer surveys, but the area appears to provide good habitat for resident and migrant desert birds because of the relatively good plant cover in the wadis. According to local reports, large numbers of raptors also utilize the area. The flat, non-rocky parts provide good habitat for Macqueen's Bustard *Chlamydotis macqueenii* (Vulnerable), which is hunted extensively here each season. The site held breeding populations of seven Sahara-Sindian Desert biome-restricted species but these did not trigger inclusion under criterion A3.

Other Important Fauna: According to frequent reports of locals and hunters, Honey Badger *Mellivora capensis*, Striped Hyena *Hyaena hyaena* (Near Threatened), and Grey

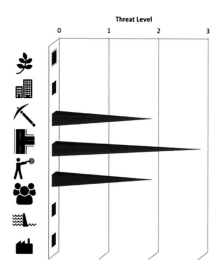

Wolf *Canis lupus* occur along with other common species. Gazelles used to be found in the western parts of this area, but it is not clear whether they still occur. Reptiles include the Desert Monitor *Varanus griseus* and Egyptian Spiny–tailed Lizard *Uromastyx aegyptia,* the latter is Vulnerable and listed on CITES Appendix II.

Conservation Issues: Salman provide a good example of the Shamiya, the southern Iraqi desert, which is poorly-known in regards to its wildlife and ecology. A very high threat is hunting and falconry, especially represented by the caravans of falconers, many from the Gulf States, that visit the area in winter with the aid and cooperation of local hunters as well as some governmental bodies (the Ministry of Interior provides them with assistance because police and police

vehicles were often seen accompanying these groups). Locals frequently report that these groups do not leave the area until it is "clear" of any birds, mammals or large lizards. Other impacts are human intrusion due to grazing and hunting and the traffic on the paved road from Salman to the Saudi border that runs along the northern and western parts of this area.

Recommendations: As the site was only visited in 2010, more surveys are recommended to fully characterize its biodiversity during different seasons. Conservation actions such as awareness-raising and outreach should focus on the town of Salman. It is also recommended to liaise with the local police and border police to ensure compliance with hunting laws, especially in regard to falconers visiting this area.

Lehais (IQ079)

Thi Qar & Basrah - 30.60583°N 6.52917°E

KBA Criteria: **Not Assessed**
IBA Criteria: **Not Assessed**
IPA Criteria: **B1**

Area: **797 ha** - Altitude: **20-29 m**
Ecoregion: **Arabian Desert and East Sahero-Arabian Xeric Shrublands (PA1303)**
Status: **Unprotected**

©2010 MA SALIM/NI

Site Description: This is a desert area with sparsely growing shrubs such as *Astragalus spinosus* and *Cornulaca aucheri.* The geology of this area is Lower Faris Series: marls, siltstone, gypsum/anhydrite, and limestone bands; the soil type is sandy. Subba & Lehais Oilfield is located to the southwest and a gas separation station is located close to the south end of the site. A small, abandoned airstrip exists near the northwestern extension and there is an area with vegetable farms to the east. Al-Lehais town is also located to the east of the site. A paved road leading to the gas station runs close to the area (from the south). The old Basrah-Nasiria main road runs close to the north side. The area is scattered with the remains of Iraqi military remains from the first Gulf War. Surveys for birds or other fauna were not conducted at Lehais in the KBA Project.

Conservation Issues: The Subba & Lehais Oilfield, which has been contracted to an international oil company for

development, and the gas separation causes some air and soil pollution. There is also a pipeline network in the area. Overgrazing and farming are significant impacts. Human disturbance is caused by nomadic movement and vehicle traffic on the Basrah-Nasiria main road. Soil pollution due to old war remains is also a potential concern.

Recommendations: This site is a good example of desert habitat and it is relatively rich in desert plant species, particularly annuals that grow after the rainy season. Therefore, it is recommended to find ways to regulate the oil industry, grazing, and disturbance from vehicle traffic to reduce the impact of these activities especially near the more pristine parts of the site, which are concentrated in the northwestern portion of Lehais.

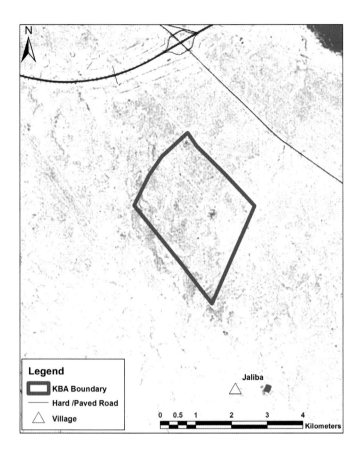

Important Plant Area Criteria

B1. The site is a particularly species-rich example of a defined habitat type

Herbaceous vegetation- Sparsely vegetated land habitat type

Jabal Senam (IQ080)

Basrah - 30.12444°N 47.62722°E

IPA Criteria: **B1**

Area: **2918 ha** - Altitude:**25–145 m**
Ecoregion: **Gulf Desert and Semi-Desert (PA1323)**
Status: **Unprotected**

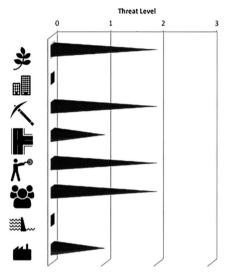

Site Description: This is an isolated mountain in the desert of southern Iraq surrounded by steppe. After the rains, the land is covered by annual herbs. It lies adjacent to the Iraqi-Kuwaiti border and there is a military base and border police office at the top of the mountain. There is extensive grazing and the area is surrounded by vegetable farms especially on the eastern side. The geology is sedimentary, with a sandy-clay soil type and the habitat was sparsely vegetated herbaceous vegetation.

©2008 MA SALIM/NI

Important Plant Area Criteria

B1. The site is a particularly species-rich example of a defined habitat type

Herbaceous vegetation- Sparsely vegetated land

Additional Important Bird Observations: Although this site does not qualify as an IBA (based on only one-year survey), one Eastern Imperial Eagle *Aquila heliaca* (Vulnerable) was recorded in winter and, according to locals, large numbers of Macqueen's Bustards *Chlamydotis macqueenii* (Vulnerable) are hunted in winter. Additionally, the site supported seven Sahara-Sindian Desert biome-restricted species but this did not triggar inclusion under the A3 criterion.

Other Important Fauna: According to frequent reports from locals and hunters, a considerable list of mammals (mainly desert species) have been seen and/or hunted in and around this area, including Rueppell's Fox *Vulpes ruepellii*, Gray Wolf *Canis lupus* and Honey Badger *Mellivora capensis*.

Conservation Issues: Jabal Senam is a unique geological formation, a rocky hill with height of more than 50m in a flat desert area. This creates special habitats for plants in particular in the valleys around the hill where the rainwater accumulates. This site is located close to Rumaila and Al-Zubayr Oilfields and it may be included in their future expansion plans. There is a small oil well on the Kuwaiti side of the border that releases smoke, affecting air quality. The site is also affected by the smoke produced by the Rumaila Oilfield to the west. There is significant human intrusion due to oil development, the military base on the summit of Jabal Senam and farming activities. Agricultural intensification, Settlement expansion, overgrazing, birds hunting, and noise from drilling and other oilfield activities, as well as pollution from agricultural chemicals are all issues here.

Recommendations: No important bird species were found around the mountain, but the site was surveyed only in one year and might be important for the migration of raptors, passerines and Houbara Bustards. It also represents a distinctive geological formation in the region that is home to a unique assemblage of plants and thus requires further survey work (though this may be partially hindered by the presence of the military base on the summit). This should also be taken into consideration when planning any oil or agricultural expansion in the area. Coordination with security forces for future surveys is necessary to factilitate access to the entire site and learn more about the species present. Coordination with Kuwaiti authorities and experts to plan a cross-border conservation program at this area is recommended.

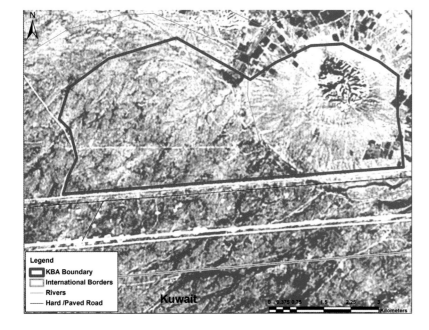

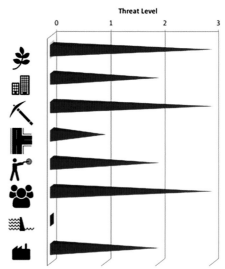

Khor Az-Zubayr (IQ081)

Basrah - 30.146166°N 47.842907°E

IBA Criteria: **A2**
IPA Criteria: **B1**

Area: **31854 ha** - Altitude: **less than 7 m**
Ecoregion(s): **South Iran Nubo-sindian Desert and semi-desert (PA1328); Arabian Desert and East Sahero-Arabian xeric shrublands (PA1303), and Gulf Desert and semi-desert (PA1323)**
Status: **Unprotected**

©2006 MA SALIM/NI

Site Description: Evans (1994) describes this area as a tidal inlet at the head of the Gulf with intertidal mudflats (IBA041). There are water channels that tidal water runs through, which become soft mud as the tide ebbs. The site can be reached from the city of Umm-Qasr just to the north of Umm-Qasr Port by crossing the canal from west to east. An unpaved road connects the site to the main road (paved), which in turn connects Umm-Qasr seaport with Basrah and the express highway. Khor Az-Zubayr Canal is the main shipping route used to reach Umm-Qasr Port and there is a small island in the canal called Hacham Island that supports halophytic vegetation. Khor Abdullah (IBA042), which forms the upper part of the Gulf and the border between Iraq and Kuwait, is located to the south of Khor Az-Zubayr. No bird hunting was observed as border forces protect the site.

The geology of the area is gravel, sand, silt and alluvium and about 60% of the site was unvegetated but habitats at the site include salt pioneer sward vegetation and desert shrublands. Six sub-sites were surveyed in this area.

Important Bird Area Criteria	All observations made 2005-2010.	
A2. Restricted-range species	**Breeding**	**Wintering/ Passage**
Iraq Babbler *Turdoides altirostris* (Resident)	50 pairs	
Important Plant Area Criteria		
B1. The site is a particularly species-rich example of a defined habitat type		
Salt Pioneer Swards vegetation habitat type and Desert-Shrub vegetation habitat type		

Additional Important Bird Observations: During the surveys 53 bird species were observed in the area. Greater Spotted Eagle *Aquila clanga* (Vulnerable) was also recorded in winter but in numbers below IBA threshold criteria. The site also held breeding populations of two Sahara-Sindian Desert biome-restricted species but this did not trigger inclusion under A3 criterion. Khor Az-Zubayr is the only KBA site reported to regularly hold Crab Plover *Dromas ardeola*.

Other Important Fauna: Fish data were collected for this site from 2005 to 2007, and again in 2009. A total of 18 species were recorded at Khor Az-Zubayr. Significant freshwater species, according to Coad (2010), were *Acanthopagrus cf. arabicus, Carassius auratus, Liza abu, L. klunzingeri, L. subviridis, Tenualosa ilisha* and *Silurus triostegus*.

The following marine fish species observed at the site are not mentioned in Coad (2010): *Caranx malabaricus, Chirocentrus dorab, Cynoglossus arel,* and *Eleutheronema tetradactylum, Epinephelus* spp. and *Liza* spp. Also, the following species seen are mentioned in Coad (2010) but there is little information about them: *Brachirus orientalis, Johnius belangerii, Otolithes ruber, Pampus argenteus,* and *Sillago sihama.* In addition, a species of Mudskipper was observed, which could be found everywhere in the soft mud within the tidal areas and is unique to this area of Iraq.

In 2009, a Fisheries Frame Survey was conducted a one of the sub-sites (located 100 meters east of Khor Az-Zubayr) and found that there were 15 ships (over 25 meters in length) and over 500 smaller craft (around 15 meters in length) fishing in the area using seine and trawling nets. Catch was estimated for the large ships at 8 tons/ship per week.

Conservation Issues: The upper part of Khor Az-Zubayr Canal and all the areas along the canal lie within an important complex of large factories and other industrial facilities including Umm-Qasr Port. Navigation and the loading/ unloading of ships in Umm-Qasr Port create significant amounts of water pollution and human disturbance. Road building and the construction of oil pipelines are also issues at the site. Khor Az-Zubayr is heavily affected by oil

pollution due to pipelines and shipping tankers, causing repeated impacts to biodiversity throughout the area. There are a number of shipwrecks either partially or completely submerged in the main channel and in 2010, there was still evidence of pollution from an oil tanker that sank in Khor Az-Zubayr Canal four years previously. Large numbers of fishing boats and passenger ships are also based in this area, which also contribute to impacts on water quality and fishing likely represents a high pressure in the area.

Recommendations: This site is one of two coastal KBA sites in Iraq, and is very important for coastal vegetation. Local and governmental initiatives are urgently required to clean up the area and prevent further pollution and environmental damage from the oil shipping industry. It is recommended that the noise from loading/unloading processes in Umm-Qasr Port be reduced through the use of improved equipment and methods. It is also recommended to regulate the use and volume of horns by the ships passing within Khor Az-Zubayr Canal. Stronger emergency response training and procedures as well as equipment are urgently needed to handle oil spills and other emergency response situation and derelict ships need to be removed.

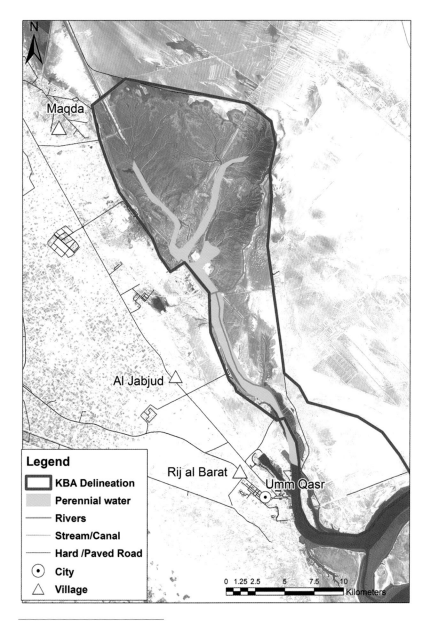

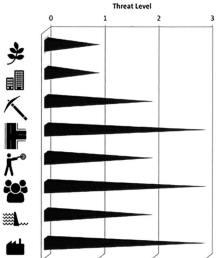

Fao (IQ082)

Basrah - 29.928888°N 48.602500°E

IBA Criteria: **A1 and A2**
IPA Criteria: **B1**

Area: **16909 ha** - Altitude: **less than 5 m**
Ecoregion: **South Iran Nubo-Sindian desert and semi-desert (PA1328)**
Status: **Unprotected**

©2010 MA SALIM/NI

Site Description: This site is located at the Iraq-Iran border, which divides the Shatt Al-Arab River between Iraq (western half) and Iran (eastern) and is considered part of two original Important Bird Areas (IBA040 & IBA042) both described by Evans (1994). The border is only political and no physical barriers divide the river. This site extends from Sayhan (northwest) to Ras Al-Beesha (southeast). It includes marshes along the Shatt Al-Arab that are affected by tides and saltwater of the Gulf. The geology of the area is gravels, sand, silt and alluvium, mainly sand. During the 1980s prior to the Iraq/Iran war this was an important region for date palm production, but the trees have since been destroyed. The habitats at the site are salt pioneer swards and marsh vegetation with both reed beds and rooted submerged vegetation, with approx. 30% unvegetated.

To the north of this site, the Karun River enters from Iran, providing freshwater to the Shatt Al-Arab. The area known as Ras Al-Beesha is located along the coast in the southernmost part of Iraq near the end of the Shatt Al-Arab. To the west of Ras Al-Beesha, there are drainage channels and mudflats, but in 2007 dredging work began and by the winter 2009 survey much of the previously inundated area was dry or drying (the depth of the drainage channels increased and no water overflowed from them to flood the surrounding areas). As a result, during the winter survey (2009), there was little water in these tidal wetlands and some of the former wetland had dried out with negative impacts on fish and wildlife populations. Locals at the site indicate that this is being done for development and investment projects.

Ras Al-Beesha is covered with thick and inaccessible reedbeds, 2-3 m high or more in some areas. Some new embankments and roads have been constructed, including paving the new network of roads. A number of people hunting birds were found within the reedbeds and there is some low-scale grazing in the site. There were many crabs, mudskippers (called locally Abu-Shlamboo) and small fish in the mudflats and channels in the tidal area. Two sub-sites were visited in the survey.

Important Bird Area Criteria	Observations made 2005-2010	
A1. Globally threatened species	**Breeding**	**Wintering/ Passage**
Marbled Duck *Marmaronetta angustirostris* (Resident)	80 pairs (counts)	66 (count)
Basra Reed Warbler *Acrocephalus griseldis* (Summer visitor)	100 pairs	
A2. Restricted-range species	**Breeding**	**Wintering/ Passage**
Acrocephalus griseldis	100 pairs	
Iraq Babbler *Turdoides altirostris* (Resident)	50 pairs	Present
Important Plant Area Criteria		
B1. The site is a particularly species-rich example of a defined habitat type		
Salt pioneer swards vegetation habitat type and Inland Standing Water- Aquatic communities- Rooted Submerged vegetation habitat type		

Additional Important Bird Observations: During the surveys, 52 bird species were observed. In addition to those listed in the table above, Greater Spotted Eagle *Aquila clanga* and Dalmatian Pelican *Pelecanus crispus* (both Vulnerable) were found in winter as well as flocks of Eurasian Curlew *Numenius arquata* (Near Threatened). The site also held four Sahara-Sindian breeding species but these did not trigger inclusion under A3 criterion. The endemic race of

Little Grebe *Tachybaptus ruficollis iraquensis* also occurs and probably breeds.

Other Important Fauna: Data were collected from 2007 to 2008. Golden Jackal *Canis aureus* was observed, as were Wild Boar *Sus scrofa*.

Fish: Data were collected from 2005 to 2007 and also in 2009, during which 36 species were found. Significant freshwater species at this site, according to Coad (2010) were: *Acanthopagrus cf. arabicus, Arabibarbus grypus* (Vulnerable), *Carasobarbus luteus, Carassius auratus, Cyprinus carpio* (Vulnerable), *Liza abu, L. klunzingeri, L. subviridis, Luciobarbus xanthopterus* (Vulnerable), *Tenualosa ilisha, Thryssa hamiltonii,* and *Silurus triostegus.*

The following, primarily marine species, were also found: *Ablennes hians, Acanthopagrus berda, Argyrosomus aeneus, Bothus pantherinus, Chirocentrus dorab, Cynoglossus arel, Eleutheronemate tradactylum, Epinephelus tauvina, Ilisha elongata, Johnius belangerii, Lutjanus rivulatus, Mugil cephalus, Nemipterus bleekeri, Otolithes ruber, Pampus argenteus, Platycephalus indicus, Pseudorhombus arsius, Sarda orientalis, Sciaena dussumieri, Scomberomorous commerson* (Near Threatened), *S. guttatus, Siganus canaliculatus,* and *Sillago sihama.*

Conservation Issues: Urban and rural development activities in towns and cities along Shatt Al-Arab River such as Fao and Seeba on the Iraqi side and Abadan on the Iranian side are a very high threat. The most challenging threat to this area is the Iraqi government's efforts to build a seaport covering a large area of the coastal zone of Iraq (including the southern part of this site). A related threat is the fishing port and ships berthing in Fao city. The dredging of the area west of Ras Al-Beesha, which noticeably affected bird populations due to the drying of some mudflats and the recent management of the freshwater sources in both Iranian and Iraqi sides is also a major threat. The site was heavily impacted when the Karun River was closed for temporary periods by the Iranian government, and when the Iraqi government reduced the freshwater coming to the Shatt Al-Arab from the Tigris and Euphrates.

High impact threats are caused by widespread pollution from oil refineries and shipping. In 2011 when Iran closed the Karun River, pollution of the area by waste from the Abadan refinery grew severe. Another high threat is road construction especially for border police activities, where the new network of roads has disfigured the natural habitat in the Fao wetlands and turned many scattered ponds and marshes and even reedbeds into dry areas. Hunting of birds and animals and extensive fishing in the Shatt Al-Arab and northern part of the Gulf are also high threats, as is human disturbance from picnic activities at Seeba, ship movements on the Shatt Al-Arab and road traffic along the river. The Iraqi government's plans to construct two power plants and oil export pipelines in this area as well.

Recommendations: This site is one of only two coastal KBA sites in Iraq and is very important for its coastal vegetation. Much of the area is far from the city of Fao and favorable for coastal birds and other wildlife. It also provides important spawning and nursery areas for fish. The extensive changes going on this region requires more integration of biodiversity protection into the development planning process for Fao. Future projects need environmental impact assessments, especially for the Great Fao Seaport. Ensuring an adequate flow of freshwater from the Shatt Al-Arab and Karun Rivers is also important, which will require negotiations with Iran. Fishing is a key activity in the area and requires stronger oversight and controls. Ticehurst (1922) reported that great numbers of Steppe Buzzards were observed migrating over Fao, which suggests that the site could be an important bottleneck for soaring birds so a survey is recommended.

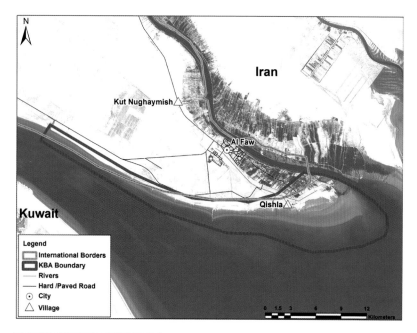

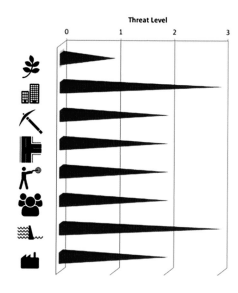

7.C FORMER IBA SITES THAT COULD NOT BE VISITED OR ADEQUATELY ASSESSED

Seven sites remain on the Important Bird & Biodiversity Areas list for Iraq but are not provided with detailed assessments. Originally listed by Evans (1994), many of these sites could not be visited due to security or logistical reasons. Others may have been partially surveyed or visited on only one occasion but this was not deemed adequate to determine the status of the biological diversity of these sites. In all cases, these sites that are listed below should be priority areas to visit and assess in the future.

1. Anah & Rawa (Anbar – 34.475278°N, 41.883889°E): This site was an original IBA site (IBA006 by Evans). The western edge near Anah features elevated cliffs declining gradually westward into arid steppes filled with halophytic vegetation. The water is shallow along the muddy river-banks, which are covered in grassy vegetation as well as date palm, citrus and other fruit orchards. Small, flat islands rise sporadically in the middle of the river, which are important roosting spots for many resident and migrant waders and shorebirds. This site may no longer found meet IBA Criteria but more site visits are recommended in this area.

2. Attariya Plains (Diyala – 33.528056°N 44.763056°E): Attariya Plains was listed as an Important Bird Area (IBA013) by Evans. It is a combination of grassy, arid steppes and uncultivated open land dotted with shrubs and seasonal pools bordered by reeds and Typha vegetation. There are a small number of farms and orchards throughout, consisting mainly of citrus and date-palms irrigated from the many irrigation canals linked to the Diyala River. Towards the southern and eastern ends of the site the landscape changes to open steppe featuring halophytic and xerophytic plants. The area may be an important winter habitats for Houbara in the region (Al-Sheikhly, 2011). It was not considered to meet the KBA criteria though unstable security conditions limited the survey effort, therefore it likely warrants more field work.

3. Augla (Anbar – 33.916667°, 41.033333°E): Evans described this site (IBA010) as "Part of a limestone and sand desert plateau c.30 km west-south-west of Haditha and c.200 km west-north- west of Baghdad. The area contains large, deep wadis surrounded by shallow depressions which are wet from November to March and carry a rich vegetation of *Haloxylon, Artemisia* and *Astragalus.*" The KBA Team was not able to visit this site due to poor security in this area.

4. Baquba Wetland (Diyala – 33.916667°N 44.833333°E): This site, originally designated an Important Bird Area (IBA011) by Evans, consists of a group of scattered marshlands in the Diyala River valley along the road to Shahraban township, most of which have been dried and developed for human settlements and military bases. Agriculture, based on wheat and barley are the main practices in this area but certain areas of the site contains scattered seasonal pools of rainwater. Semi-desert and arid lands also appear where the road turns north toward Kafri and Kalar in southern Sulaymaniyah governorate. A single survey was not adequate to document the biological diversity at the site and it is highly recommended that more detailed surveys are conducted.

5. Huweija Marshes (Kirkuk – 34.976667°N, 44.002222°E): Evans described this site (IBA005) as "an area of marshes and lakes c.65 km west-south-west of Kirkuk, along the pipeline road." Under the KBA program this site, which was called "Huweija Marshes and Beagi" referring to an area near that referred to by Evans, were surveyed along with Shayat Wetlands, with which it is hydrologically connected. Access to the site was extremely limited and few observations could be made but the team noted that the large strategic oil pipeline transporting crude oil from Kirkuk to Turkey, which traverses this area suffers from many leaks in various places within the site. This produces chronic environmental impacts to wildlife and their habitat. Further surveys are recommended when security improves.

6. Khor (Khawr) Abdallah (Basrah – 29.55°N, 48.32°E): Evans listed this site as IBA042 and described is as an "enormous areas of swampy grassflats (c.90,000 ha) and intertidal mudflats (c.36,000 ha) backed by a belt of date palms and then by silt desert, immediately west of the point where the Shatt Al Arab enters the Gulf, and of the town of Fao." This area could not be surveyed by the KBA Team.

7. Shubaicha Marsh (Wasit & Diyala – 33.250000°N 45.300000°E): Hor Shubaicha is a seasonal marsh that was listed as IBA017 by Evans. It gets water mainly from the eastern uplands in Iran and the agricultural drainage network east of the Tigris River. It was known, as a part of the eastern Tigris seasonal marshes and was a stopover and feeding site for large numbers of waterfowl and waders. During the only visit in winter under the KBA Program, the site, which supposedly fills with water during the winter rainy season, was almost dry except for some water patches, mainly formed from the agricultural fields that are scattered throughout the area. The security conditions did not allow the team to survey the area as required, so not all of it was covered and additional surveys are recommended.

253

7.D FORMER IBA SITES THAT NO LONGER MEET KBA CRITERIA

Thirteen sites that were formally listed as Important Bird Areas by Evans (1994) were visited in the KBA Program but were found to be destroyed or so altered that the biodiversity of these sites no longer met any of the criteria systems applied in this effort (KBA for non-avian fauna, IBA or IPA Criteria). These sites are effectively delisted but could, in some cases, be potential sites for restoration activities. The list of these 13 sites is provided below with a short description each.

1. **Abu Dalaf and Shari Depression** (Salah Ad-Din – 34.358889°N 43.857500°E): Evans described a rain-fed and saline lake (Shari Lake) between 5000 to 8000 hectares in this site (IBA009). The area is now composed of farmlands, marshes and open sandy steppes. The ancient "Wall of Shnass," which dates from the Abbasid dynasty, is found northwest of the site. The ancient Abbasid mosque of Abu Dalaf, is one of the many traces of ancient civilizations that once flowered here. Several date-palm orchards extend eastward near Samara. This site was not found to meet KBA Criteria.

2. **Abu Habba** (Baghdad – 33.20°N, 44.20°E): Evans described this site (IBA014) as lying "12 km west of Mahmudiya and was originally a flat, silt desert with sparse herbs, but much of it has long been under cultivation, irrigated by water from the Mahmudiya channel of the River Tigris." This site is now fully converted to farms and agricultural use and was not included in the KBA Program Surveys.

3. **Gasr Muhaiwir** (Anbar – 33.543611°N 41.003889°E): Gasr Muhaiwir was originally listed as an IBA site (IBA012) by Evans. It is an ancient site of biological and historical importance located in the Western Iraqi desert. It is known for the castle-like ruins, from which the site name is derived, located on the eastern edge of Wadi Horan, the main valley that leads towards the western desert and the Euphrates River flood plain. The area was also used as a field base by British forces participating in the Arabian Revolution of 1916 against the Ottoman Empire and the tracks of the military caravans remain to the present day. The site is dominated by desert and semi-desert steppes, with a variety of typical desert vegetation. In the distance, the rocky valleys and limestone slopes of the Wadi Horan can be clearly seen, a landmark used by Bedouins as a guidepost during desert navigation. Seasonal pools of ground and rainwater are scattered throughout the site,

which are difficult to locate during the winter and are a popular resting point for migrant birds. Despite three site visits, the team found that it did not meet KBA criteria.

4. **Hor (Haur) Al-Abjiya and Haur Umm Al-Baram** (Wasit – 32.28N, 46.04E): Evans has little information on these two seasonal freshwater marshes or lakes that make up IBA022 but indicates that they lie south of the Tigris River. "Haur Umm Al-Baram (32°32 'N 46°07 'E, 5,000 ha) is c.25 km east of Kut, whilst Haur Al Abjiya (32°25 'N 46°03 'E, 5,000 ha) is 20 km southeast of Kut. The two hors are connected by the Dujaila Canal." These seasonal marshes no longer exist and the region is arid with some agriculture. It was not included in the KBA Site List.

5. **Hor (Haur) Al-Hachcham and Haur Maraiba Hor** (Wasit – 32.05N, 46.12E): Evans described IBA027 as "two small haurs largely overgrown with reeds Phragmites, on the plains southeast of Al-Hayy. The haurs are fed by the Shatt Al-Gharraf." These marshes no longer exist and the region is arid with some agriculture. It was not included in the KBA Site List.

6. **Hawijat Albu Dheab and Al-Ramadi Marshes** (Anbar – 33.47527°N, 43.268056°E): Evans listed this as IBA016. Albu Dheab is the name of the tribe who inhabit this area. The word "Hawija" means marshland, with the unique species and habitat of this site in Western Iraq being similar to the large marsh ecosystems in the south of the country. There are also a small number of wheat and cornfields, interspersed with uncultivated areas of thick halophytic vegetation. The site also holds many suitable spots for resident and migrant bush warblers and songbirds. Though visited, the site did not appear to meet KBA Criteria but more surveys are recommended.

7. **Hor Lafta** (Missan – 31.401111°N 45.511389°E): His-torically, this site was listed as a wetland of international importance by Carp (1980), and was considered by Carp & Scott (1979) to be possibly of great importance for wintering waterbirds, but there is no specific ornithological information available. Evans described this area (IBA037) as "an isolated haur c.5 km north of the River Euphrates and c.80 km west north-west of Nasiriya, fed by floodwater from the Euphrates." Currently, the site is dry and most of the area is arid with scattered low shrubs. There are a number of canals that transport irrigation water to neighboring fields. The surveys showed that the site no longer qualified as an IBA; nor is there evidence that the site qualified as a KBA for other, non-avian fauna or as an IPA.

8. **Hor Sadiya** (Missan – 32.193056°N 46.639722°E): Hor Al-Sadiya was listed by Scott (1995) as a large wetland of international importance. In Evans, the site (IBA025) was considered important for several wintering, threatened species. Other bird species were considered to have 1% or more of their regional population wintering at the site. But by 2005, most of the marsh area had been turned into agricultural fields, and the rest was dry and thus no longer qualifying as an IBA (nor is there as yet evidence that it meets KBA Criteria for non-bird fauna or IPA Criteria for plants). Some of these agricultural fields were still exploited as wheat farms at the time of the survey, but the majority of them were abandoned perhaps because of high soil salinity.

9. **Hor Uwaina** (Thi Qar – 31.408333°N 46.200000°E): Evans described this area (IBA035) as "a complex of large haurs and associated marshes on the plains to the east of the Shatt Al-Gharraf waterway, east and southeast of the village of Shatra. The two principal marshes are Hor Abu Ajul in the north, and Hor Chamuqa (Ghamuga) in the south" Both fed mainly from the Shatt Al-Gharraf. Evans documents extensive drainage of the area in the 1970s but as of the mid-1990s stated "much of this remaining marsh is still in excellent condition with extensive Typha beds and large areas of open water". Unfortunately during the survey of winter 2005, the team found most of the area turned into agricultural farms with a network of irrigation and drainage canals and dry areas with very few, small patches of wetlands among them. The site was not included in the KBA Site List.

10. **Jadriyah and Umm Al-Khanazeer Island** (Baghdad – 33.275278°N, 44.376667°E): The University of Baghdad, which contains part of this site (Jadriyah), has made the area familiar to many Iraqis. Prior to the construction of the University in the 1960s, the habitat was date palm orchards and farmland with dense thickets. Today it is predominantly urban with little of the original habitat left, although pockets are still found on the southern and western edges of the campus. But a zone of cultivated land extends in an arc along the Tigris River where there is year-round agricultural activity and many vegetable fields. Dense orchards of date palms are found here, although they are more common in the Doura district of Baghdad across the river to the south (outside the survey area). Umm Al-Khanazeer Island, which is not really an island, was named for the large wild boar *Sus scrofa* population that once lived there. It has similar habitat to the site across the river but now has restricted access. Evans listed both these areas as IBA015. These areas were not included in the KBA Site List.

11. **Rayan** (Missan – 31.578333°N 47.033333°E): Evans described Rayan (IBA033) as an important complex of shallow lagoons and reedbeds and sedge marsh between the villages of Maymoona and Salam supporting many threatened species. Unfortunately today this marsh is a dry agricultural area, which the local people do not want re-flooded and it no longer qualifies as an IBA nor is there yet evidence that it meets KBA (non-avian) or IPA criteria.

12. **Saniya** (Missan – 31.919444°N 46.763889°E): Evans listed Saniya as a chain of large marshes connected hydrologically to Hor Sadiya (CM21) in the north. Though there was no information available in the mid-1990s on the site when Evans wrote his account, Saniya (IBA031) was still considered a marsh "of major importance given its size and proximity to wetlands of known importance." Unfortunately, this area is now almost completely dry except for a small water channel used for irrigating the extensive agricultural lands in the region. Most of the unused and arid areas are covered with scattered Tamarix. It no longer qualifies as an IBA and there is no evidence that it meets KBA (non-bird) or IPA criteria.

13. **Shatt Al-Arab Marshes** (Basrah – 30.583056°N, 47.771667° to 29.928889°N 48.6025°): Evans listed this as IBA040 and stated that "these marshes lie along the c.165 km of the Shatt Al Arab waterway from 31°00 'N 47°25 'E to 29°55 'N 48°30 'E, mainly between Qurnah and Basrah (c.65 km), and include Haur Al Shaibah (15 km north of Basrah), Qarmat Ali, Khamisiyah and Shafi." During the KBA Program, the team visited several sites along the Shatt Al-Arab from the Euphrates & Tigris Junction at Qurnah to Ras Al-Beesh near Fao (the latter is now considered part of the Fao KBA site IQ028). The Shatt Al-Arab is heavily impacted by sewage, garbage, pollution from oil and other industries. Areas along the waterway are highly developed and the waterway has extensive boat traffic. While there are scattered wetlands and riparian habitat along this waterway, they are much degraded and it was not included in the KBA Site List.

7.E ADDITIONAL SURVEY SITES IN THE KBA PROGRAM

An additional 12 sites were included in the survey. They were not found to meet any of the criteria systems applied in the KBA Program but some may warrant further study. These sites are listed below with a short description.

1. **Aski-Kalak & Safea** (Erbil – 36.272778°N 43.642778°E): This site is an open area bordered by hills on the eastern and southwestern sides, located along the Greater Zab River west of Erbil. The Zab is wide at this point, flowing

southwest towards the Tigris River. The predominant habitat here is steppe grasslands and the site is located in the Moist Steppe Zone. There is also some Riverine Forest in the Plains (Al-Ahrash) habitat. This site was not found to meet KBA Criteria.

2. **Erbaidh** (Salah Ad-Din – 34.376944°N 43.735556°E): The site consists of dry steppes with halophytic vegetation and scattered sand dunes in the southeast extend south towards the northern edge of Tharthar Lake. Al-Sahra Airfield and military base is nearby. The site is a key site for migrant raptors and soaring birds. It was used as a base for many hunters to trap migrant species, including the Common Crane *Grus grus*, Saker Falcon *Falco cherrug* and Barbary Falcon *Falco pelegrinoides*, which pass through after nesting in the nearby Makhool foothills. Hunting of Houbara or McQueen's Bustard *Chlamydotis macqueenii*, especially in the area near Al-Sukariya, was reported by locals. Police authorities cooperated with the team in 2009-2010 when a large contingent of falconers from Arabian Gulf countries were based in the area and increased hunting pressure on local species (Nature Iraq, 2009). The site may provide suitable habitat for Arabian Sand gazelle *Gazella subgutturosa marica* (Vulnerable). More surveys are warranted.

3. **Habbariyah** (Anbar – 32.331053°N 42.221119°E): Habbariyah is one of Iraq's main faidhatt, a natural depression that collects rain water flowing from the surrounding hills and provides a key grazing area that is important for nomadic tribes. Faidhatt Al-Habbariya is situated on the western plateau of Iraq in the Al-Nekheab Desert. It measures 20-25 km in length and 2-3 km in width and includes Wadi Itbel, one of the main wadis in western Iraq. The site has a general desert landscape with xerophytic vegetation surrounded by a few scattered rocky hills. Local hunting reports indicate the presence of Arabian Sand Gazelle G*azella marica* (Vulnerable) though they are considered rare. Egyptian Spiny-tailed Lizard *Uromastyx aegyptia* is a common reptile in the area and is targeted by locals and Bedouin. Further surveys are recommended here.

4. **Khazar & Kalakchi** (Dohuk - 36.650278°N 43.481663°E): Khazar and Kalakchi is a flat area bisected by the Khazar Stream flowing south-southwest, which is formed from a few streams that are also used for irrigation. Two main habitat types in this area were moist steppe zone (grassland) and riverine grass and herbs. The site is located in the Moist Steppe Zone, where the geology is sandstones, clay, and sandy gravels, and the soil type is sandy clay. The northern part of the site (the Khazar area), located north of Kalakchi village, is a steppe region with agricultural land. The survey sites are covered by *Phragmites australi*s. South of this area is Kalakchi, which contains agricultural land with herbaceous plants as well as species such as *Typha* sp., *Juncus* sp., *Mentha* sp. and *Rumex* sp. These area were not found to meet KBA Criteria yet.

5. **Kteibaan** (Basrah – 30.708333°N 48.027222°E): Plant cover at Kteibaan is poor as the soil has high salinity levels. Although KBA and IBA criteria were not met, this site may hold waterbird populations during migration and winter due to extensive shallow water and mudflats present at that time. There is an embankment on the border beyond which is a large body of water extending into Iran. Added to this factor, its location on the Iraq-Iran border gives the site added protection as no one is allowed to enter from either sides for security reasons.

6. **Qasir Al-Khubbaz** (Anbar – 34.845706°N 42.387719°E): The site is situated on the eastern edge of Wadi Horan one of the main wadis in the western desert of Iraq. It contains typical desert vegetation and some elevated ground extending to the north and northwest edge. During the wet seasons the site is used for grazing especially by people from Hit and Kubaissa. Only one visit was conducted so further field surveys are recommended.

7. **Rutba and Al Massad Gazelle Reserve** (Anbar – 32.912914°N, 40.223994°E): Al-Massad enclosure is c.125 km long x 5 km wide with a BRC fence c.3m high and is surrounded by high hills. It was established in 1974 and began with a captive breeding group of 11 gazelles, consisting of seven individuals trapped in the Al-Nikhaeb and Ara'ar area in southwest Iraq, three that were brought from Northern Iraq, and one individual from Falluja. At the time of the first survey the Al Massad enclosure held 800 gazelles consisting mainly of adult males. A second enclosure, Faidhat Al-Dhaba'a, is c.35 km east of Rutba and formerly held 120 individuals brought from Al-Massad. Poor knowledge of wildlife management and lack of veterinary care has caused a marked decline of the Al-Massad population between 2005 and 2006. The animals appear to be Arabian Sand Gazelle G. [s.] *marica*. Two main habitats were observed within the survey area, desert shrublands and herbaceous, steppe vegetation. This site was not included in the KBA Site List.

8. **Sabkhat Albu Garis** (Anbar – 34.698333°N 41.219167°E): From December to February, rain and floodwater accumulate at Sabkhat Albu Garis (located in the Al-Qae'em sub-district), forming a shallow seasonal pool extending over an area shared by Iraq and Syria. During April-May it forms one of the largest steppe areas in

Western Iraq, known as Sheeb Albu Garis. The site has few if any accessible roads. A hunter reported small, scattered groups of Arabian Sand Gazelle *Gazella s. marica* dwelling in the border area between Iraq and Syria (though not at levels that indicate yet if the site would meet KBA Vulnerability Criterion). Only the margins of the site were surveyed due to security concerns so more comprehensive surveying is recommended when security improves.

9. **Shayat Wetlands** (Salah Ad-Din & Kirkuk – 34.79233°N, 44.315526°E): The site is one of the main wetlands in Salah Ad-Din Governorate. It includes several narrow tributaries (the largest is the Zegaton River) of the Tigris that flows from south of Kirkuk Governorate toward Al-Shayat Lake on the eastern edge of Salah Ad-Din Governorate. Al-Shayat Lake is formed by the Al-Udheim Dam on the border between Diyala and Salah Ad-Din and is surrounded by permanent wetland. This is in turn is surrounded by a chain of elevated ground and hills on both sides. Arid land and grassy open steppes are extensive. Zegaton River drains in from the north and is one of the main features of the site that attract birds and other wildlife as well as hunters. The Himreen Hills border the site on its southwest side and extend into the site as the rocky hills flatten into marshland. A small number of irrigated wheat and cornfields are located close to the road. Although the security situation of the site is still critical, hunters and falcon trappers now regularly visit it from Tikrit and Baquba as well as nearby townships, primarily in winter. The surrounding area attracts hunters who are after See-see Partridge and Black Francolin and gazelle. More surveys are needed here as this site may meet KBA Criteria.

10. **Tell Al-Lahm** (Thi Qar – 30.067607°N 46.390720°E): This is a desert area located west of the Basrah-Baghdad express highway and north of a strategic oil pipeline. There is a paved road in the south. Many species of grasses and herbs grow in depressions along the road, while the surrounding area has sparsely growing shrub desert vegetation.

11. **Turaq Steppe** (Erbil - 36.021389°N 43.941111°E): This is a flat agricultural area with about 2 km² of remaining steppe habitat. The village of Turaq is located on the north side of the surveyed site. This land is used for farming field crops (wheat and barley) and grazing. It was not found to meet KBA Criteria.

12. **Wadi Horan - Al Hussayniyah** (Anbar – 33.419167°N, 41.021389°E): The site consists of two large desert valleys (or wadis) in Western Iraq and the area between them. Wadi Horan, the larger of the two, extends west towards Saudi Arabia and Wadi Amij extends northeast toward Kubasa Township near Hit. No botanical survey was conducted but the dominant habitats are arid desert and semi-desert with halophytic vegetation and limestone hillsides with occasional cliff faces. There is a seasonal pool, called the Wahatt Horan or Hussaninyatt Wadi Horan. Hussaninyatt is a dam constructed by the Iraq government during the 1990s to collect runoff from the surrounding hills during the rainy season for grazing cattle herds. The ancient highlands of Wadi Horan were formed by erosion in a previous geological age and represent a typical Iraqi desert ecosystem. Dry streambeds extend into the valley toward the survey site, forming water pathways during the spring and harboring grassy vegetation in the middle of the desert. More surveys are recommended here.

Abdulhasan, N.A. (2009). *Habitat Mapping Project of the Proposed Iraqi Marshlands National Park Area.* In: Krupp, F., Musselman, L.J., Kotb, M.M.A., & Weidig, I. (Eds). Environmental, Biodiversity and Conservation in the Middle East. Proceedings of Middle Eastern Biodiversity Congress, Aqaba, Jordan, 20-23 October 2008. *Biorisk* 3:55-68.

Abdulhasan, N.A., Salim, M.A., Al-Obaidi, G.S., Ali, H.J., Al-Saffar, M.A., Abd, I.M., & Minjil, M.Sh. (January 2009). *Habitat Project – Classification and Description of Southern Iraqi Marshlands* (No. NI-0109-01). Unpublished document, Nature Iraq.

Abed, S.A., Altaey, M.M., Salim, M.A. (2013). Status and conservation of ducks in Dalmaj wetlands, Southren Iraq. International Journal of Advanced Research. Volume 2, Issue 8, 931-937. ISSN 2320-5407.

Addax Petroleum. (2012). *Kurdistan Region of Iraq: Country overview & License Area.* Retrieved from www.addaxpetroleum.com/operations/middle-east on 21 June 2012.

Ahmad, S.A. (Abdulrahman, S.). (2013a). *Vascular plants of Hawraman region in Kurdistan Iraq.* (Unpublished doctoral dissertation). Sulaimani University, Sulaimani, Iraq.

Ahmad, S.A. (Abdulrahman, S.). (2013b). Eighteen species new to the flora of Iraq. *Feddes Repert, 124:* 65–68. Wiley-Vch Verlag.

Ahmad, S.A. (Abdulrahman, S.). (2013c). *Ferula Shehbaziana (Apiaceae), A new species from Kurdistan, Iraq. Harvard Papers in Botany, Vol. 18*(2), 2013, pp. 99–100. President and Fellows of Harvard College.

Ahmad, S.A. (Abdulrahman, S.). (2014). *Petrorhagia sarbaghiae* (Caryophyllaceae), a new species from Kurdistan, Iraq. *Willdenowia, 44 (1):* 35 – 38. 6 March 2014. – Version of record published online ahead of inclusion in April 2014 issue; ISSN 1868-6397. BGBM Berlin-Dahlem.

Al-Abbasi, T.M., Al-Farhan, A., Al-Khulaidi, A.W., Hall, M., Llewellyn, O.A., Miller, A.G. & Patzelt, A. (2010). Important Plant Areas in the Arabian Peninsula. *Edinburgh Journal of Botany, 67*(1): 25-35.

Al-Rawi A. (1964). *Wild plants of Iraq with Their Distribution.* Baghdad: Government Press.

Al-Rawi, A. & Chakravarty, H.L. (1964). *Medicinal Plants of Iraq* (2nd ed.). Baghdad, Iraq: Al-Yagatha Print House.

Al-Sheikhly, O.F. (2012). The hunting of endangered mammals in Iraq. *Wildlife Middle East 6* (2&3): 10.

Al-Sheikhly, O.F. (2011). A Survey Report on the Raptors Trapping and Trade in Iraq. *Wildlife Middle East 6*(1).

Al-Sheikhly, O.F. (2012). *Some Ecological observations on Lesser Kestrel Falco naumanni in north and northwest of Iraq.* (Unpublished Masters thesis). Baghdad University, Baghdad, Iraq.

Al-Sheikhly, O.F. & Mallon, D. (2013) The Small Asian Mongoose *Herpestes javanicus* and the Indian Gray Mongoose *Herpestes edwardsi* in Iraq (Mammalia: Carnivora: Herpestidae). *Zoology in the Middle East 59:* 173-175.

Al-Sheikhly, O.F. and Nader, I.A. (2013). The status of the Iraqi Smooth-coated Otter *Lutrogale perspicillata,* Hayman 1956 and Eurasian Otter *Lutra lutra,* Linnaeus 1758 in Iraq. *IUCN Otter SG Bulletin 30*(1): 18-30.

Alessandrini, D. (2011). Venomous Snakes in Iraq. *Reptiles Magazine.* Retrieved from www.reptilechannel.com/snakes/venomous-snakes/venomous-snakes-iraq.aspx on March 1, 2011.

Allouse B. (1953). The avifauna of Iraq. *Iraq Nat. Hist. Mus.* Publication No. 3 Baghdad: Nat. Hist. Mus.

Allouse, B. (1960-62). *Birds of Iraq* (Vols. 1-3). (in Arabic). Baghdad: Al-Rabita Press.

Alwan, A.R.A. (2006). Past and present status of the aquatic plants of the marshlands of Iraq. *Marsh Bulletin 1*(2): 160-172.

Amr, Z. (February 2009a). *Mammals of Iraq* (No. NI-0209-002). Unpublished document, Nature Iraq.

Amr, Z. (February 2009b). *Reptiles and Amphibians of Iraq* (No. NI-0209-003). Unpublished document, Nature Iraq.

Anderson, S.C. (1999). The Lizards of Iran. *Contributions to Herpetology, 15.* Ithaca, NY: Society for the Study of Amphibians and Reptiles.

Ararat, K., Abid, I.M. & Abdulrahman, S. (2008). *Key Biodiversity Areas Survey of Kurdistan, Northern Iraq: Site Review for Birds, Plants & Fisheries – Winter & Summer 08 Survey* (No. NI-1208-001). Unpublished report, Nature Iraq.

Armesto, M.J., Boehm, C. & Bowden, C. (Compilers). (2006). *International Single Species Action Plan for the Conservation of the Northern Bald Ibis Geronticus eremita.* AEWA Technical Series No. 10. Bonn, Germany: AEWA.

Austrian Institute of Technology (AIT) (2012). *Towards Halgurd-Sakran National Park Kurdistan, Iraq.* Vienna, Austria: AIT. Retrieved from www.ait.ac.at.

Bachmann, A. (June 2009). *Nature Iraq KBA Report for Sakran & Barzan Areas* (No. NI-0609-01). Unpublished document, Nature Iraq.

Banister, K.E. & Bunni, N.K. (1980). A new blind cyprinid fish from Iraq. *Bulletin of the British Museum (Natural History) Zoology 38*(3): 151-158.

Batanouny, K.H. (2001). *Plants in the deserts of the Middle East.* Berlin: Springer.

Bermani, A.K. (1981). *Systematic Study of The Genus Verbascum (Scrophulariaceae) as it is occurs in Iraq.* Baghdad, Iraq: Baghdad University.

BirdLife International. (2000) *Threatened Birds of the World.* (1st ed.). Cambridge, UK: BirdLife International.

BirdLife International. (2006). *Monitoring Important Bird Areas: a global framework.* Version 1.2. Compiled by Bennun, L., Burfield, I., Fishpool, L., Nagy, S., & Stattersfield, A. Cambridge: BirdLife International.

BirdLife International. (2010). *Regional Checklist.* Cambridge, UK: BirdLife International.

BirdLife International. (2013a). *Global IBA Criteria.* Retrieved from www.birdlife.org/datazone/info/ibacritglob on 15 March 2013.

BirdLife International. (2013b). *IUCN Red List for birds.* Retrieved from www.birdlife.org on 15 March 2013.

BirdLife International (2013c). *State of the world's birds.* Retrieved from www.birdlife.org/datazone/sowb/ SOWB2013 on 17 September 2013.

BirdLife International (2014) *Endemic Bird Area factsheet: Mesopotamian marshes.* Retrieved from www.birdlife.org on 18 May 2014.

BBC. (2013). Saddam's Iraq Key Events: Chemical Warfare (UN experts confirmed in 1986 that Iraq had contravened the Geneva Convention by using chemical weapons against Iran). *BBC.* Retrieved from news.bbc.co.uk/2/shared/spl/ hi/middle_east/02/iraq_events/html/chemical_warfare. stm on 14 November 2014.

Carp, E. (1980). Directory of Western Palaearctic Wetlands. Switzerland: IUCN.

Carp, E. & Scott, D.A. (1979). The Wetlands and Waterfowl of Iraq: Report on the joint expedition of the International Waterfowl Research Bureau and the University of Basrah, Iraq. 10 January to 3 February, 1979. Unpublished transcript.

Central Intelligence Agency (CIA). (2012, May 11). World Factbook – Iraq. Retrieved from www.cia.gov/library/ publications/the-world-factbook/geos/iz.html on 11 May 2012.

Coad, B.W. (2010). *Freshwater Fishes of Iraq.* Pensoft Series Faunistica #93. Sofia-Moscow: Pensoft Publishers.

Coad, B.W. (2013). *Freshwater Fishes of Iran.* Retrieved from www.briancoad.com/Species%20AccountsContents% 20new.htm on 19 August 2013.

Convention on Biological Diversity (CBD) (2014). *2020 Aichi Biodiversity Targets.* CBD Secretariat. Retrieved from www.cbd.int/sp/targets/ on 13 November 2014.

Darwall, W., Carrizo, S., Numa, C., Barrios, V., Freyhof, J. and Smith, K. (2014). Freshwater Key Biodiversity Areas in the Mediterranean Basin Hotspot: Informing species conservation and development planning in freshwater ecosystems. Cambridge, UK and Malaga, Spain: IUCN. x + 86pp.

Davis, P.H. (1978). *Flora of Turkey.* (Vol. 6). Edinburgh: University of Edinburgh, University Press.

Davis, P.H. (1982). *Flora of Turkey and the East Aegean Islands* (Vol. 7). Edinburgh: University of Edinburgh, University Press.

Davis, P. (1984). *Flora of Turkey and the East Aegean Islands* (Vol. 8). Edinburgh: University of Edinburgh, University Press.

Delany, S. and Scott, D. (2002 & 2006). *Waterbird Population Estimates* (3rd ed. (2002) & 4th ed. (2006)). Wageningen: Wetlands International.

Esmaeili, H.R., Sayyadzadeh, G., Ozulug, M., Geiger, M. and Freyhof, J. (2014b). Three new species of *Turcinoemacheilus* from Iran and Turkey (Teleostei: Nemacheilidae) *Ichthyol. Explor. Freshwaters.* München, Germany: Verlag Dr. Friedrich Pfeil. March 2014. 24(3), pp. 257-273.

Evans, M. (1994). *Important Bird Areas of the Middle East.* BirdLife Conservation Series No. 2. Cambridge: BirdLife International.

Fazaa, N.A. and Whittingham, M. (2014). Management of Animal Ecology and Adaptation to Climate Change in the Iraqi Marshlands. (Unpublished dissertation in prep.). UK: School of Biology, Newcastle University.

Freyhof, J., Esmaeili, H.R., Sayyadzadeh, G., and Geiger, M. (2014). Review of the crested loaches of the genus *Paracobitis* from Iran and Iraq with the description of four new species (Teleostei: Nemacheilidae). *Ichthyol. Explor. Freshwaters.* München, Germany: Verlag Dr. Friedrich Pfeil. August 2014. 25(1), pp. 11-38.

Freyhof, J. (2012) *Fieldwork in Iraqi Kurdistan in June 2012.* Retrieved from http://joerg-freyhof.de/fieldwork/278-fieldwork-in-iraqi-kurdistan-in-june-2012 on 31 Oct 2013.

Gari, L. (2006). A History of the Hima Conservation System. *Environment and History 12:* 213–228. Cambridge, UK: The White Horse Press.

Garshelis, D. & McLellan, B. (2011). What's So Special about the Syrian Bear? *International Bear News.* Quarterly Newsletter of the International Association for Bear Research and Management (IBA) and the IUCN/SSC Bear Specialist Group 20(1). Retrieved from www.bearbiology.com/fileadmin/tpl/Downloads/IBN_Newsletters/IBN_Low_February_2011.pdf on 13 May 2011.

George, P.V. & Mahdi, A. (1969). Systematic list of Iraqi vertebrates - Aves. *Iraq Natural History Museum Publication 26:* 34–63.

Ghahraman, A. (1983). *Flora Iranica* (Vol. 4). Tehran University, Tehran, Iran. Published and Distributed by RIFR, POB.

Ghahraman, A. (1987). *Flora Iranica* (Vol. 9). Tehran University, Tehran, Iran. Published and Distributed by RIFR, POB.

Ghahraman, A. (1999). *Flora Iranica* (Vol. 18). Tehran: Tehran University, Tehran, Iran. Published and Distributed by RIFR, POB.

Ghahraman, A. (2001). *Flora Iranica* (Vol. 22). Tehran University, Tehran, Iran. Published and Distributed by RIFR, POB.

Ghahraman, A. (2003). *Flora Iranica* (Vol. 24). Tehran University, Tehran, Iran. Published and Distributed by RIFR, POB.

Ghazanfar, S.A. & Edmonson, J.R. (2013. *Flora of Iraq: Lythraceae to Campanulaceae* (Vol. 5, Part 2). London: Royal Botanic Gardens, Kew.

Guest, E. (1960). *Flora of Iraq* (Vol. 1). Baghdad: Ministry of Agriculture of the Republic of Iraq.

Hamilton, A.M. (1930). *Road through Kurdistan: travels in Northern Iraq.* Retrieved from https://goo.gl/SKNFKx on 18 June 2012.

Harrison, D.L. (1964-1972). *The Mammals of Arabia* (Vols. 1-3). London: Ernest Benn Limited.

Harrison, D.L. & Bates, P.J.J. (1991). *The Mammals of Arabia* (2nd ed.). Tonbridge, UK: Harrison Zoological Museum.

Hatt, R.T. (1959). *The Mammals of Iraq.* Museum of Zoology, Univ.of Michigan.

Helmen, C. (2007). "Trouble is My Business." *Forbes.* Retrieved from www.forbes.com/forbes/2007/1015/099.html on 21 June 2012.

Houri, A. & Houri, N.M. (2001a). *Photographic Guide of Wild Flowers of Lebanon* (Vol. 1). Beirut, Lebanon: Chemaly and Chemaly.

Houri A. & Houri N.M. (2001b). *Photographic Guide of Wild Flowers of Lebanon* (Vol. 2). Beirut, Lebanon: Dots Dar El Kotob s.a.l.

Human Rights Watch (HRW). (2003). *The Iraqi Government Assault on the Marsh Arabs.* Human Rights Watch Briefing Paper. January 2003. Retrieved from www.hrw.org/legacy/backgrounder/mena/marsharabs1.htm on 14 November 2014.

Hussain D.A. & Alwan A.A. (2008). *Evaluation of Aquatic macrophytes vegetation after restoration in East Hammar marsh, Iraq.* Marsh Bulletin 3(1): pp. 32-44.

Iraq & United States Army Corps of Engineers (2003). Iraqi Dam Assessments. Retrieved from www.envirozan.info/EZ_Docs/Dams/D_Iraqi%20Dam%20Assessments.pdf on 2 July 2012.

Iraq Ministry of Health & Environment (IMoHE) (2010). *National Report on Biodiversity in Iraq.* Fourth national report to the Convention on Biological Diversity (CBD), July 2010. Available at www.cbd.int/doc/world/iq/iq-nr-04-en.doc

Iraq Ministry of Health & Environment (IMoHE) (2014). Programme of Work for Protected Areas (PoWPA). Retrieved from www.cbd.int/protected/implementation/actionplans/country/?country=iq on 10 January 2014.

Iraq Ministry of Health & Environment (IMoHE) (2014). *National Report on Biodiversity in Iraq.* Fifth national report to the Convention on Biological Diversity (CBD), March 2014. Available at www.cbd.int/doc/world/iq/iq-nr-05-en.pdf

IUCN. (2010). The IUCN Red List of Threatened Species. Version 2010. www.iucnredlist.org>. Retrieved from from: www.birdlife.org/datazone/userfiles/file/Species/Taxonomy/BirdLife_Checklist_Version_3.zip.

IUCN. (2012). IUCN Red List Categories and Criteria: Version 3.1 (2nd ed.). Gland, Switzerland and Cambridge, UK: IUCN. iv + 32pp.

IUCN (2014a). Key Biodivesity Areas. Retrieved from www.iucn.org/about/union/secretariat/offices/iucnmed/iucn_med_programme/species/key_biodiversity_areas/ on 20 March 2014.

IUCN (2014b). IUCN Protected Areas Categories System. Retrieved from www.iucn.org/about/work/programmes/gpap_home/gpap_quality/gpap_pacategories/ on 22 January 2014.

IUCN (2014c). Threats Classification Scheme ver. 3.2 Retrieved from www.iucnredlist.org/documents/Dec_2012_Guidance_Threats_Classification_Scheme.pdf on 13 November 2014.

Kamangar B.B., Prokofiev A.P., Ghaderi E., and Nalbant T.T., (2014). Stone loaches of Choman River system, Kurdistan, Iran (Teleostei: Cypriniformes: Nemacheilidae). ZOOTAXA. Magnolia Press. 3755(1): pp. 33–61.

Knees, S., Zantout, N., Gardner, M., Neale, S. & Miller, A. (2009). *Flora of Iraq, preliminary checklist, field version.* Edinburgh: Royal Botanic Garden Edinburgh.

Kubba, S. and Salim, M. (2011). 'The Wetlands Wildlife and Ecosystem' in S. A. A. Kubba (ed. 2011). The Iraqi Marshlands and Marsh Arabs: The Ma'dan, Their Culture and the Environment. Reading: ITHACA Press (2011).

Kurdistan Environmental Protection and Improvement Commission (KEPIC) (2014a). The Law of Environmental Protection and Improvement Board in Iraqi Kurdistan Region, Law Number 3 for the year 2010. Retrieved from http://zhenga.net/laws_en.aspx on 22 July 2014.

Kurdistan Environmental Protection and Improvement Commission (KEPIC) (2014b). The Law of Environmental Protection and Improvement in Iraqi Kurdistan Region, Law Number 8 for the year 2008. Retrieved from http://zhenga.net/laws_en.aspx on 22 July 2014.

Langhammer P.F., Bakarr M.I., Bennun L.A., Brooks T.M., Clay R.P., Darwall W., ... and Tordoff A.W. (2007). *Identification and Gap Analysis of Key Biodiversity Areas: Targets for Comprehensive Protected Area Systems.* Gland, Switzerland: IUCN.

Latifi, M. (1991). *The Snakes of Iran.* USA: Society for the Study of Amphibians and Reptiles.

Leviton, A.E., Anderson, S.C., Kraig, A., & Minton, S.A. (1992). *Handbook to Middle East Amphibians and Reptiles.* USA: Society for the study of Amphibians and Reptiles

Maahzide, A.M. (2003). *A Comparative Systematic Study of Prunus L. (Rosaceae) in Iraq.* (Unpublished doctoral dissertation). Mosul: University of Mosul.

Mabberley, D.J. (2008). *Mabberley's Plant-book: A Portable Dictionary of Plants, their Classifications, and Uses* (3rd ed.). Cambridge University Press.

Mahir, A.M., Radhi, A.G., Falih, H.A., Al-Obaidi, G.S. and Al-Saffar, M.A.T. (2009). *Key Biodiversity Survey of Kurdistan, Northern Iraq: Water Quality Review – Winter & Summer 08 Survey* (No. NI-0209-004). Unpublished document, Nature Iraq.

Mashhadani, A.N. (1992). *A Comparative Systematic Study of Onosma L. Spp. (Boraginaceae) in Iraq.* (Unpublished doctoral dissertation). Baghdad: University of Baghdad.

Mullarney, K., Svensson, L., Zetterström, D., & Grant, P. (2001). *The Complete Guide to the Birds of Europe.* Princeton University Press.

Murie, O.J. and Elbroch, M. (2005). *Peterson Field Guide to Animal Tracks* (3rd ed.) Houghton Mifflin Harcourt.

Nature Iraq. (2008a). *Management Plan for the Hawizeh Marsh Ramsar Site of Iraq.* (Vol. 1: Background, Vision, Principles and Annexes). A Report prepared for the Iraq National Marshes and Wetlands Committee. December 2008. C.D.A. Rubec (Ed.). Retrieved from www.natureiraq.org/site/sites/default/files/Hawizeh%20Plan%202nd%20Draft%20Volume%201%20Dec%201%202008.pdf on 16 September 2012.

Nature Iraq. (2008b). *Management Plan for the Hawizeh Marsh Ramsar Site of Iraq.* (Vol. 2: Management Issues and Recommendation). A Report prepared for the Iraq National Marshes and Wetlands Committee. December 2008. C.D.A. Rubec (Ed.). Retrieved from http://www.natureiraq.org/site/sites/default/files/Hawizeh%20Plan%202nd%20Draft%20Volume%201%20Dec%201%202008.pdf on 16 September 2012.

Nature Iraq & Twin Rivers Institute. (2009). *State of the Environment Report – Darbandikhan Lake Basin.* Prepared for the Kurdistan Ministry of Environment. June 2009. Sulaymaniyah, Iraq: Nature Iraq & TRI.

New Eden Group. (2006). *The New Eden Master Plan for Integrated Water Resource Management in the Marshland Areas.* Italy: New Eden Group.

New Eden Group (2010). *Mesopotamia Marshland National Park Management Plan: Site Description (Vol. 1) and Strategies and Objectives* (Vol. 2). Italy: New Eden Group

Omer, S., Wronski, T., Alwash, A., Elamin, M.H., Mohammed, O., & Lerp H. (2012). Evidence for persistence and a major range extension of the Arabian Smooth-coated Otter, *Lutrogale perspicillata maxwelli* (Mustellidae, Carnivora) in Iraq. Folia Zool. 61(2): 172–176.

Paley, A.R. (2007, October 30). "Iraqi Dam Seen In Danger of Deadly Collapse." The Washington Post. Retrieved from www.washingtonpost.com/wp-dyn/content/article/2007/10/29/AR2007102902193.html?hpid=topnews on 2 May 2012.

Plantlife International. (2008). IPA Criteria & Methodology. Retrieved from www.plantlife.org.uk/international/ wild_plants/IPA/ipa_criteria_and_methodology/ on 25 October 2014.

Pohl, T., Al-Muqdadi, S.W., Ali, M.H., Fawzi, N. Al-M., Ehrlich, H., & Merkel, B. (2014, March 6). Discovery of a living coral reef in the coastal waters of Iraq. Sci. Rep. 4. Art. No. 4250. Macmillan Publishers Limited. Retrieved from www.nature.com/srep/2014/140306/srep04250/full/ srep04250.html on 10 March 2014.

Porter, R.F. (2014). The mystery of the 'Syrian' Serins wintering in northern Iraq is solved. *Sandgrouse.* 36: 58 – 60.

Porter, R.F. and Aspinal, S. (2010). *Birds of the Middle East* (2nd ed.). Princeton University Press.

Porter, R.F., Christensen, S. & Schiermacker-Hensen, P. (1996). *Field Guide to the Birds of the Middle East.* Princeton University Press.

Ramsar (2011). *National Report on the Implementation of the Ramsar convention on Wetlands. National Reports to be submitted to the 11th Meeting of the Conference of the Contracting Parties, Romania, June 2012.* Retrieved from www.ramsar.org/pdf/cop11/nr/cop11-nr-iraq.pdf on 16 September 2012.

Ramsar (2014). *About the Ramsar Convention.* Retrieved from www.ramsar.org/ on 24 January 2014.

Raza, H.A., Ahmed, S.A., Hassan, N.A., Ararat, K., Qadir, M, & Ali, L. (2012). First photographic record of the Persian leopard in Kurdistan, northern Iraq. *Cat News* 56, 34-35.

Raza, T.J. and Dawd, W.H. (1983). *Geographical distribution of Wild Vascular Plants.* Baghdad: Ministry of Agriculture and Agrarian Reform Republic.

Rechinger, K.H. (Ed.). (1963-in progress). *Flora Iranica: Flora des Iranischen.* Graz, Austria: Hochlandes und der umrahmenden Gebirge. 172 Fascicles.

Rechinger, KH (1964). *Flora of Lowland Iraq.* Stuttgart, Germany: J. Cramer Publisher In Weintheim.

Rosti, N. (2011, January 16). "Zarze and Sara caves: Neanderthal Shelters." Kurdish Globe. No. 288. nadr_ rwsty@yahoo.com

Rubec C. (2013). A Wetland Future for Iraq? *Marsh Bulletin* 8(2): 114-130.

Salim M.A. (2002). The first records, including breeding, of the Black-winged Kite *Eleanus caeruleus* in Iraq. *Sandgrouse* 24: 136–138.

Salim, M.A. (2004). Field Observation on Birds in "Abu-Zarag" and "Kirmashiyah" Wetlands. 30 Jun – 4 Jul, 2004 in Southern Iraq. Iraq Foundation. Unpublished report.

Salim M.A. (2008a). *Natural Overview of Ar-Razzaza Lake.* Unpublished internal report, Nature Iraq.

Salim, M.A. (2008b). The first Namaqua Dove *Oena capenses* in Iraq. *Sandgrouse* 30: 100–101.

Salim, M.A. (2009a). Biodiversity of the Middle Euphrates: Current Status and Potentials for Conservation Action. Unpublished report.

Salim, M.A. (2009b). Birds in Iraq: Executive Summary - 2012 - Unpublished report.

Salim, M.A. (2010). First record of White-crowned Black Wheatear *Oenanthe leucopyga* for Iraq. *Sandgrouse* 32: 149–150.

Salim, M.A. (2011). [*The illegal hunting and trade increase the threat on Macqueen's Bustard Chlamydotis macqueenii in Iraq.*] Nature Iraq Technical Publications NI-1011-02. www.natureiraq.org/site/ar/node/284. [In Arabic]

Salim M.A., Al-Sheikhly O.F., Majeed K.A. & Porter R.F. (2012). An annotated checklist of the birds of Iraq. *Sandgrouse* 34: 4-43.

Salim, M., Porter, R.F., & Rubec, C. (2009). A summary of birds recorded in the marshes of southern Iraq, 2005–2008. Krupp, F., Musselman, L.J., Kotb, M.M.A., Weidig, I. (Eds). Environment, Biodiversity and Conservation in the Middle East. Proceedings of the First Middle Eastern Biodiversity Congress, Aqaba, Jordan, 20–23 October 2008. *BioRisk* 3: 205–219. doi: 10.3897/biorisk.3.14.

Salim, M.A, Porter, R.F., Schiermacker-Hansen, P., Christensen, S., & Al-Jbour, S. (2006). [*Field guide to the birds of Iraq*], (In Arabic). Amman, Jordan: Nature Iraq/ BirdLife International.

Sardar, A.S. (2003). *Morphological-Systematic Study of the Genus Scabiosa L. (Dipsacaceae) in Kurdistan of Iraq.* Erbil: University of Salahadin.

Schwartz, M.W. (1999. Choosing the appropriate scale of reserves for conservation. *Annual Review of Ecology and Systematics* 30: pp. 83–108.

Scott, D.A. (1995). *Directory of Middle East Wetlands.* Wetlands International.

Shell Global (2013, April 30). "Shell Signs Historic MOU with IUCN to Conserve Iraq's Environmental Biodiversity." Retrieved from www.shell.com/irq/en/aboutshell/media-centre/news-and-media-releases/2013/mou.html on 4 Sep, 2013.

Sinan, O. (2007, June 24). Iraqi Kurds Await Chemical Ali Verdict. Associated Press. Retrieved from www. washingtonpost.com/wp-dyn/content/article/2007/ 06/23/AR2007062300673.html on 14 November 2014.

Smith, K.G., Barrios, V., Darwall, W.R.T. and Numa, C. (Editors). (2014). *The Status and Distribution of Freshwater Biodiversity in the Eastern Mediterranean.* Cambridge, UK, Malaga, Spain and Gland, Switzerland: IUCN. xiv+132pp.

Stokes, D. & Stokes, L. (1987). *Stokes Guide to Animal Tracking and Behavior.* Stokes Nature Guides. New York: Little, Brown and Co.

Szczerbak, N.N. & Golubev, M.L. (1986) *Gecko Fauna of the USSR and contiguous regions.* Akademia Nauk Ukrainskoj SSR Zoologicheskij Institutim. I. I. Schmalbgauzena. Kieve Naukova Dumka.

Ticehurst, C.B., Buxton, P.A., & Cheesman, R.E. (1922). The Birds of Mesopotamia. *Journal of the Bombay Natural History Society* 28: 210-250, 381-427, 650-674, & 937-056.

Tohme, G. and Tome, H. (2002). *A Thousand and One Flowers of Lebanon.* Beirut: Lebanese University.

Townsend, C.C. and Guest, E. (1966). *Flora of Iraq* (Vol. 2): Introductory Taxonomic Material. Baghdad: Ministry of Agriculture of the Republic of Iraq.

Townsend, C.C. and Guest, E. (1968). *Flora of Iraq* (Vol. 9): Gramineae. Baghdad: Ministry of Agriculture of the Republic of Iraq.

Townsend, C.C. and Guest, E. (1974). *Flora of Iraq* (Vol. 3): Leguminales. Baghdad: Ministry of Agriculture of the Republic of Iraq.

Townsend, C.C. and Guest, E. (1980a). *Flora of Iraq* (Vol. 4, Part 1): Cornaceae - Rubiaceae. Baghdad: Ministry of Agriculture of the Republic of Iraq.

Townsend, C.C. and Guest, E. (1980b). *Flora of Iraq* (Vol. 4, Part 2): Bignoniaceae to Resedaceae. Baghdad: Ministry of Agriculture of the Republic of Iraq.

Townsend, C.C. and Guest, E. (1985). *Flora of Iraq* (Vol. 8): Monocotyledones. Tonbridge, United Kingdom: The Whitefriars Press Ltd.

Trewavas, E. (1955). A blind fish from Iraq, related to *Garra. Ann. Mag. Nat. Hist.,* 8(91) (Series 12, Article 67): 551-555.

United Nations Education, Scientific and Cultural Organization (UNESCO). (2014). *Criteria for Selection.* Retrieved from http://whc.unesco.org/en/criteria on 22 January 2014.

United Nations Education, Scientific and Cultural Organization (UNESCO). (2016). *The Ahwar of Southern Iraq: Refuge of Biodiversity and the Relict Landscape of the Mesopotamian Cities.* Retrieved from http://whc.unesco.org/en/list/1481 on 19 August 2016.

United Nations Environment Programme (UNEP) (2003). *Desk Study on the Environment in Iraq.* Switzerland: UNEP. Retrieved from http://postconflict.unep.ch/publications/Iraq_DS.pdf on 26 October 2014.

United Nations Environment Programme (UNEP) (2006). *Global Desert Outlook.* Nairobi Kenya: UNEP. Retrieved from www.unep.org/geo/gdoutlook/ on 8 July 2014.

Wikipedia. (2012, May 11). *Cheekha Dar.* Retrieved from en.wikipedia.org/w/index.php?title=Cheekha_Dar&oldid=490631565 on 11 May 2012.

Wilson, D.E. & Reeder, D.M. (Eds.). (2005). *Mammal Species of the World. A Taxonomic and Geographic Reference* (3rd ed). USA: Johns Hopkins University Press. See also www.vertebrates.si.edu/msw/mswcfapp/msw/index.cfm

World Wide Fund for Nature (WWF). (2014). *Ecoregions.* Retrieved from worldwildlife.org/biome-categories/terrestrial-ecoregions on 22 January 2014.

ADDITIONAL KBA PAPERS

Ararat, K. (2009). Key Biodiversity Areas: Rapid assessment of birds in Kurdistan, northern Iraq. In: Krupp, F., Musselman, L.J., Kotb, M.M.A., Weidig, I. (Eds). Environment, Biodiversity and Conservation in the Middle East. Proceedings of the First Middle Eastern Biodiversity Congress, Aqaba, Jordan, 20–23 October 2008. *BioRisk 3:* 187–203. doi: 10.3897/biorisk.3.21

Abd, I.M., Rubec, C.D.A., & Coad, B.W. (2009). Key Biodiversity Areas: Rapid assessment of fish fauna in southern Iraq. Krupp, F., Musselman, L.J., Kotb, M.M.A., Weidig, I. (Eds). Environment, Biodiversity and Conservation in the Middle East. Proceedings of the First Middle Eastern Biodiversity Congress, Aqaba, Jordan, 20–23 October 2008. *BioRisk 3*: 161–171. doi: 10.3897/biorisk.3.15.

Al-Obaidi, G.S., Salman, S.K., & Rubec, C.D.A. (2009). Key Biodiversity Areas: Rapid assessment of phytoplankton in the Mesopotamian Marshlands of southern Iraq. Krupp, F., Musselman, L.J., Kotb, M.M.A., Weidig, I. (Eds). Environment, Biodiversity and Conservation in the Middle East. Proceedings of the First Middle Eastern Biodiversity Congress, Aqaba, Jordan, 20–23 October 2008. *BioRisk 3:* 111–126. doi: 10.3897/biorisk.3.20.

Rubec, C., Alwash, A., & Bachmann, A. (2009). The Key Biodiversity Areas Project in Iraq: Objectives and Scope 2004–2008. Krupp, F., Musselman, L.J., Kotb, M.M.A., Weidig, I. (Eds). Environment, Biodiversity and Conservation in the Middle East. Proceedings of the First Middle Eastern Biodiversity Congress, Aqaba, Jordan, 20–23 October 2008. BioRisk 3: 39–53. doi: 10.3897/biorisk.3.12.

WEB RESOURCES

Amphibia Web: This website provides information on amphibian declines, natural history, conservation, and taxonomy. www.amphibiaweb.org

Angiosperm Phylogeny: The focus of this site is on angiosperm families, although treatments of gymnosperm groups were added in 2005. Emphasis is placed on plant families because they are the groups — admittedly partly arbitrary as to circumscription, but now monophyletic (including all and only the known species of a commonn ancestor) — around which many of us organize our understanding of plant diversity. Attention to groupings of families has been given because much progress is being made in sorting them out, while infrafamilial groups in families like Poaceae, Apocynaceae, Malvaceae and Ericaceae are being added as studies become available. In larger families there is a tendency to focus on literature that deals with clades with fifty or more taxa, in smaller families the coverage is more detailed. www.mobot.org/MOBOT/research/APweb

BirdLife Data Zone: Provides extensive data on species, Important Bird & Biodiversity Areas (IBAs), Country Profiles, State of the World's Birds, and Threatened Bird Forums. It is also the 'public face' of BirdLife's World Birds / Biodiversity Database (WBDB). The WBDB is used by BirdLife Partners to collect, manage and share their national IBA information. The BirdLife Secretariat uses the WBDB to validate IBA data prior to publishing the official list of confirmed IBAs on the Data Zone. Hosted by BirdLife International. www.birdlife.org/datazone/home

BirdLife International: BirdLife International is the world's largest nature conservation Partnership with 120 BirdLife Partners worldwide and many partner designate and affiliate groups. BirdLife is widely recognized as the world leader in bird conservation. Rigorous science informed by practical feedback from projects on the ground in important sites and habitats enables us to implement successful conservation programmes for birds and all nature. Nature Iraq is a BirdLife Affiliate in Iraq. www.birdlife.org

British Birds (BB): British Birds is a monthly journal for all keen birdwatchers. They publish articles on a wide variety of topics, including behaviour, conservation, distribution, identification, status and taxonomy. Their contributors include both professional and amateur ornithologists, and content is always abreast of current ideas and thinking, yet written in a clear and simple style which is easy to interpret. www.britishbirds.co.uk

Catalog of Fishes: Online database edited by W.N. Eschmeyer on all fish genera, species, and references hosted by the California Academy of Sciences. researcharchive.calacademy.org/research/ichthyology/catalog/fishcatmain.asp

Centre for Middle Eastern Plants (CMEP): The Centre for Middle Eastern Plants (CMEP) is an authority on the Middle Eastern environment. They work with local partners to tackle contemporary environmental challenges including sustainable development, biodiversity conservation, water conservation and climate change. CMEP projects leave pragmatic and environmentally sustainable legacies. www.cmep.org.uk.

Convention on Biological Diversity: The Convention on Biological Diversity, which went into force in 1993, was inspired by the world community's growing commitment to sustainable development. It represents a dramatic step forward in the conservation of biological diversity, the sustainable use of its components, and the fair and equitable sharing of benefits arising from the use of genetic resources. This website provides all convention documents including national reports from convention parties. www.cbd.int

FishBase: FishBase is a relational database with information to cater to different professionals such as research scientists, fisheries managers, zoologists and many more. FishBase on the web contains practically all fish species known to science. www.fishbase.org

Ford Motor Company Conservation and Environmental Grants Program: Since its inception in 2000, the Ford Grants has supported at least 150 projects with more than $1.3 million granted to date. The program was created to empower individuals and non-profit groups donating their time and effort to preserve the environmental well-being of their communities. www.me.ford.com/en/omn/about/environment/overview

Freshwater Fishes of Iraq: Brian W. Coad's Personal Website. Brian is an ichthyologist (student of fishes), working for the Canadian Museum of Nature in Ottawa, Canada. His main research interests are on Canadian freshwater and marine fishes and Southwest Asian, principally Iranian, freshwater fishes. He also has an inordinate fondness for words. www.briancoad.com

Integrated Biodiversity Assessment Tool (IBAT): IBAT for Research and Conservation Planning is an innovative tool designed to facilitate access to a range of global and national data layers, such as protected area boundaries, biological information about habitat and species diversity indices, and key areas for biodiversity, which can be useful for research and conservation planning purposes. www.ibat-alliance.org/ibat-conservation/login

Iraqi Ministry of Health & Environment (IMoHE), Department of Environment: www.moen.gov.iq (in Arabic only)

IUCN Red List: The IUCN Global Species Programme working with the IUCN Species Survival Commission (SSC) has been assessing the conservation status of species, subspecies, varieties, and even selected subpopulations on a global scale for the past 50 years in order to highlight taxa threatened with extinction, and thereby promote their conservation. The IUCN Red List of Threatened Species™ provides taxonomic, conservation status and distribution information on plants, fungi and animals that have been globally evaluated using the IUCN Red List Categories and Criteria. This system is designed to determine the relative risk of extinction, and the main purpose of the IUCN Red List is to catalogue and highlight those plants and animals that are facing a higher risk of global extinction (i.e. those listed as Critically Endangered, Endangered and Vulnerable). www.iucnredlist.org

Kurdistan Environmental Protection and Improvement Commission (KEPIC): www.zhenga.net (in Kurdish & English)

Nature Iraq: The largest Iraqi conservation, non-governmental organization in Iraq. Nature Iraq is an affiliate of BirdLife International. www.natureiraq.org

Ornithological Society of the Middle East (OSME): The Ornithological Society of the Middle East, the Caucasus and Central Asia is a registered charity (no 282938) and exists to collect, collate and publish data on all aspects of the Ornithology of the Middle East, the Caucasus and Central Asia region. Publishes the journal Sandgrouse. www.osme.org. OSME also maintains the regional list of birds. www.osme.org/orl

Ramsar Convention: The Convention on Wetlands of International Importance, called the Ramsar Convention, is the intergovernmental treaty that provides the framework for the conservation and wise use of wetlands and their resources. The Convention was adopted in the Iranian city of Ramsar in 1971 and came into force in 1975. Since then, almost 90% of UN member states, from all the world's geographic regions, have acceded to become "Contracting Parties". Website contains all convention documents including national reports from convention parties. www.ramsar.org

Royal Society for the Protection of Bird (RSPB): The RSPB is the UK's largest nature conservation charity, inspiring everyone to give nature a home. They also play a leading role in a worldwide partnership of nature conservation organizations. www.rspb.org.uk

The Reptile Database: This database, edited by P. Uetz, provides a catalogue of all living reptile species and their classification. The database covers all living snakes, lizards, turtles, amphibians, tuataras, and crocodiles. Currently there are about 10,000 species including another 2,800 subspecies. The database focuses on taxonomic data, i.e. names and synonyms, distribution and type data and literature references. www.reptile-database.org

World Wide Fund for Nature (WWF): One of the world's leading conservation organizations, WWF works in 100 countries and is supported by close to 5 million members globally. WWF's mission is to conserve nature and reduce the most pressing threats to the diversity of life on Earth. worldwildlife.org

WorldBirds Database: An Internet-based spatial database about birds, their science and their distribution, that allows users from anywhere in the world to input and query the bird data. The system provides the facility to download data into Excel and analyze them for personal requirements. A portion of this database is specific to Middle East Birds. Recently the World Birds database has been amalgamated to become part of another personalized bird data recording system called Bird Track. See app.bto.org/birdtrack2/main/data-home.jsp

Pars Herpetologists Institute: Website for the Iranian Herpetology Institute. www.pars-herp.com

World Birds/Biodiversity Database (WBDB): see BirdLife Data Zone above.

APPENDICES

APPENDIX 1: CRITERIA USED IN SITE SELECTION

KBA Criteria and thresholds used for other fauna (Langhammer et al., 2007):

Criterion	Sub-criteria	Provisional threshold for triggering KBA Status
V. Vulnerablity Regular occurrence of a globally threatened species (according to the IUCN Red List) at the site		The presence of a single individual of either a Critically Endangered (CR) or Endangered (EN) species or the presence of 30 individuals or 10 pairs of a Vulnerable (VU) species.
I. Irreplaceability Site holds X% of a species' global population at any stage of the species' lifecycle	a. Restricted-range species.	Species with a global range less than 50,000 km2 or 5% of global population at site
	b. Species with large but clumped distribution.	5% of global population at site
	c. Globally significant congregations.	1% of global population seasonally at the site
	d. Globally significant source populations.	Site is responsible for maintaining 1% of global population
	e. Bioregionally- restricted assemblages	To be defined

Important Bird & Biodiversity Area Criteria (BirdLife International, 2013a):

Criterion	Sub-criteria	Provisional threshold for triggering KBA Status
A1. Globally threatened species - The site is known or thought regularly to hold significant numbers of a globally threatened species, or other species of global conservation concern.		The presence of a single individual of either a Critically Endangered (CR) or Endangered (EN) species or the presence of 30 individuals or 10 pairs of a Vulnerable (VU) species.
A2. Restricted-range species - The site is known or thought to hold a significant component of a group of species whose breeding distributions define an Endemic Bird Area (EBA) or Secondary Area (SA).		
A3. Biome-restricted species - The site is known or thought to hold a significant component of the group of species whose distributions are largely or wholly confined to one biome.		
A4. Congregations. A site may qualify on any one or more of the four criterion listed	i. Site known or thought to hold, on a regular basis, 1% of a biogeographic population of a congregatory waterbird species.	≥ 1% of a biogeographic population of a congregatory waterbird species.
	ii. Site known or thought to hold, on a regular basis, 1% of the global population of a congregatory seabird or terrestrial species.	≥ 1% of the global population of a congregatory seabird or terrestrial species.

Criterion	Sub-criteria	Provisional threshold for triggering KBA Status
	iii. Site known or thought to hold, on a regular basis, 20,000 waterbirds or 10,000 pairs of seabirds of one or more species.	≥ 20,000 waterbirds or ≥ 10,000 pairs of seabird of one or more species.
	iv. Site known or thought to exceed thresholds set for migratory species at bottleneck sites.	At least 20,000 storks (Ciconiidae), raptors (Accipitriformes and Falconiformes) or cranes (Gruidae) regularly pass during spring or autumn migration.

Important Plant Area Criteria (Al-Abbasi et al., 2010):

Criterion	Sub-criteria
A. Threatened Species.	A1. Site contains globally threatened species or infraspecific taxa (i.e. subspecies and varieties)
	A2. Site contains regionally (Arabian) threatened species or infraspecific taxa
	A3. Site contains national threatened species or infraspecific taxa
	A4. Site contains national endemic, near-endemic, regional endemic and/or regional range-restricted species or infraspecific taxa
	A5. Site contains species of special interest.
B. Species Richness. Site contains high number of species of special interest.	B1. The site is a particularly species-rich example of a defined habitat type in Arabia
	B2. The site is a refuge for:
	a) elements of one biogeographic zone that fall within another (an important aspects of the biogeography of the Arabian Peninsula)
	b) biogeographically and bioclimatically restricted plants to 'retreat to' in the face of global climatic change.
C. Threatened habitat type - Site is identified as an outstanding example of a globally or regionally (Arabian) threatened habitat type.	

APPENDIX 2: LIST OF SURVEY SITES, SURVEY PERIODS, SUB-SITES, & THEIR GPS LOCATIONS

Site	Site Code	Survey Period	Sub-Site Name	Sub-Site Code	N	E
Kurdistan Region Sites						
Ahmed Awa	IQ042	W '07 & S '07-'10.	Ahmed Awa	S4A	35.299722°	46.078056°
Altun Kopri Marsh	IQ033	W '07, '08 & '10 & S '07-'09.	Altun Kopri Marsh	E3	35.715833°	44.119444°
Not KBA		W '07-'08 & S '07.	Aski-Kalak & Safea	E10	36.272778°	43.642778°
Amedi & Sulav	IQ008	S & W '07 & '10.	Ser Amadiya	D2A	37.108056°	43.366111°
			Sulav Resort	D2B	37.108056°	43.480833°
Assos Mountain	IQ024	W & S '10	Assos Mountain, north face	S32A	36.065556°	45.25°
			Assos Mountain, south face	S32B	35.988611°	45.213889°
Atrush & Bania Area	IQ012	W '07 & S '07-'08.	Atrush A	D3A	36.867222°	43.232778°
			Atrush B	D3B	36.865833°	43.233333°
			Bania Area	D3C	36.926389°	43.203333°
Bahrka	IQ020	W '07-'10 & S '07-'08.	Bahrka	E11	36.453611°	43.810278°
Bakhma & Bradost Mountain	IQ014	W '07 & '10 & S '07-'10.	Bakhma	E7	36.703889°	44.278056°
			Bradost	E18	36.701944°	44.381667°
Barzan Area & Gali Balnda	IQ004	W '07 & '09-'11 & S '07-'11; S '09 (Gali B.)	Barzan	E8	36.943611°	44.195556°
			Shanidar Cave	E8B	36.833056°	44.218333°
			Gali Balnda	D14	37.044722°	43.848333°
Benavi & Sararu	IQ002	W & S '07 & S '08-'09; Sararu S '08.	Benavi	D6	37.234444°	43.406944°
			Sararu	D13	37.264173°	43.405827°
Chamchamal	IQ035	S '07 & '08 & W '09.	Chamchamal Area	S9	35.421389°	44.618333°
Chami Razan	IQ026	W & S '07, S '08-'10 & Spr '09-'12.	Chami Razan Area A	S10A	35.809167°	45.020556°
			Chami Razan Area B	S10B	35.801389°	44.9775°
Darbandikhan Lake	IQ040	S '07-'09 & W '07-'10.	Darbandikhan Lake	S1	35.144722°	45.755°
De Lezha	IQ036	S '09 & '10.	De Lezha	S23	35.460278°	45.194444°
Not KBA		W & S '07	Turaq Steppe	E4	36.021389°	43.941111°
Dohuk Lake	IQ010	W & S '07 & '08.	Dohuk Lake	D9	36.885833°	43.0075°
Doli Plngan	IQ022	S '09.	Doli Plngan	S28	36.405833°	44.751944°
Doli Smaquli & Ashab	IQ021	W '07 & S '07-'10.	Ashab Valley	E5B	36.292778°	44.378333°
			Doli Smaquli	E5A	36.363611°	44.322778°
Dukan Lake	IQ023	W & S '07-'10.	Dukan Lake	S2	36.0925°	44.935833°
Dure	IQ003	S '09-'10.	Dure	D16	37.225556°	43.509444°
Fishkhaboor	IQ001	W '08-'09 & S '08 & '10.	Fishkhaboor - Syrian Border	D11A	37.046389°	42.376667°
			Fishkhaboor - Turkish Border	D11B	37.111944°	42.383330°
Gali Zanta & Garbeesh	IQ013	W & S '07, S '08 & '09.	Gali Zanta	D1A	36.741111°	43.972222°
			Garbeesh Mountain	D1B	36.813889°	43.963333°

Site	Site Code	Survey Period	Sub-Site Name	Sub-Site Code	N	E
Gara Mountain & Garagu	IQ007	W '07 & S '07'-10	Gara Mountain	D4	37.010833°	43.365°
			Garagu	D5	37.030833°	43.392778°
Hagi Omran	IQ018	W & S '07, & S '08 & '10.	Hagi Omran	E1	36.666944°	45.050000°
Halgurd Mountain	IQ017	S '09	Halgurd Mountain	E13	36.729722°	44.884444°
Hawraman Area	IQ043	W '07 & S '07-'10.	Awesar	S4B	35.215833°	46.186111°
Hazardmerd	IQ037	S '10.	Hazardmerd	S34	35.497500°	45.311663°
Kalar	IQ044	W '07-'09 & S '07-'08.	Kalar A	S3	34.555000°	45.285275°
			Kalar B	S3C	34.587778°	45.301389°
Not KBA		W & S '07-'08.	Kalakchi	D12A	36.566389°	43.531667°
Not KBA		W & S '07-'08.	Khazar	D12B	36.650278°	43.481667°
Kherazook	IQ005	W '07-'08 & S '07-'09.	Kherazook	E9	36.959167°	44.324722°
Maidan Area	IQ045	W '09 & '10.	Maidan Area	S22	34.655833°	45.680278°
Mangesh	IQ006	S '07-'09.	Mangesh	D8	37.030278°	43.088333°
			Mangesh Area	D8B	36.954167°	43.110833°
			Mangesh Valley	D8A	37.03°	43.071389°
Mawat	IQ029	S '07	Mawat	S36	35.939167°	45.382500°
			Mawat	S8	35.959444°	45.386667°
Mosul Lake	IQ009	S '07-'10 & W '08-'10	Mosul Lake	D10	36.762222°	42.749722°
			Mosul Lake A	D10A	36.741111°	42.786111°
			Mosul Lake B	D10B	36.762222°	42.749444°
			Steppe Area north of Mosul Lake	D17	36.899167°	42.675278°
Parazan	IQ034	S '09 & '10	Parazan	S26	35.626944°	45.738611°
Penjween	IQ032	W '07 & S '07 & '09	Penjween	S5	35.755556°	45.943889°
Peramagroon Mountain	IQ027	S '07-'12, W '07 & Sprs '09-'11	Homer Qowm & Shadala Valley	S24	35.785°	45.2525°
			Peramagroon	S6	35.76°	45.241389°
Qara Dagh	IQ039	W & S '07 & S '08-'10; Spr train'g '08-'10 & Wild Goat survey '11.	Qara Dagh Area	S11	35.331111°	45.290278°
Rawanduz Gorge	IQ015	W & S of '07 & S of '08.	Jundyan	E6	36.626389°	44.594167°
			Gali Ali Beg	E12	36.631392°	44.446108°
Sakran Mountain	IQ019	S '09 & '10.	Sakran Mountain	E14	36.573889°	44.986111°
Sangaw	IQ038	S '08 & W '09-'10.	Sangaw	S14	35.290278°	45.218889°
Sargalu	IQ028	W & S '07, & S '08 & '10-'11.	Sargalu	S7	35.875278°	45.165278°
Sari Hassan Bag Mountain	IQ016	S '09	Sari Hassan Bag Mountain	E15	36.71917°	44.64583°
Sharbazher Area	IQ031	S '10	Gmo Mountain	S33	35.912778°	45.550275°
			Sharbazher Area	S13	35.950281°	45.563611°
Taq Taq	IQ025	W '07-'09 & S '07-'08	Taq Taq	E2	35.892778°	44.623333°

Site	Site Code	Survey Period	Sub-Site Name	Sub-Site Code	N	E
Kuradawe & Waraz	IQ030	S of '09	Kuradawe	S27	35.840833°	45.499444°
			Waraz	S25	35.791667°	45.514167°
Zalm River	IQ041	S '08 & W '08-'09	Zalm	S12	35.306111°	45.970003°
Zawita	IQ011	S '07 & '08	Zawita	D7	36.896667°	43.146667°
Central Iraq Sites						
Ajeel Himreen Hills	IQ048	Summer '09, winter, summer, & fall '10, spring '11-'12.	Jallet Albu Ageel	SD5	34.692403°	43.871158°
Not KBA		Summers '09-'11, winter '10	Anah & Rawa	AN3	34.475278°	41.883889°
Not KBA		Winter & summer '10, spring '11	Sabkhat Albu Garis	AN11	34.698333°	41.219167°
Not KBA			Augla	AN5	33.916667°	41.033333°
Not KBA		Summer '09, winter & summer '10	Gasr Muhaiwir	AN6	33.543611°	41.003889°
Not KBA			Abu Habba	BG2	33.333333°	44.333333°
Not KBA		Spring '11	Baquba Wetl&s	DY2	33.916667°	44.833333°
Not KBA		Summer & winter '10, Spring '11	Attariya Plains	DY3	33.528056	44.763056°
Not KBA			Huweija Marshes & Beagi	KK1	34.976667°	44.002222°
Not KBA		Summers '09-'10 & winter '10	Abu Dalaf & Shari Depression	SD4	34.358889°	43.8575°
Ga'ara	IQ055	Spring '12	Ga'ara	AN17	37.010833°	43.365000°
Habbaniya Lake	IQ056	Winters & summers '09-'10; summer & autumn '11	Habbaniya Lake	AN1	33.196667°	43.460556°
Haditha Wetlands & Baghdadi	IQ050	Winter, summer, & autumn '10; spring & autumn '11, spring '12.	Haditha Wetl&s & Baghdadi	AN2	33.905833°	42.532778°
Not KBA		Winters & summers '09-'10; Summer '11	Hawijat Albu Dheab & Al Ramadi Marshes	AN8	33.475278°	43.268056°
Himreen lake	IQ053	Winter '09, summer '10 & spring '11	Himreen lake	DY1	34.193056°	45.003056°
Jabal Makhool	IQ046	Spring '11	Jabal Makhool	SD7	35.240660°	43.332073°
Not KBA		Summer '09, winter & summer '10, & spring '11	Jadriyah & Umm Al Khanazeer Isl&	BG1	33.275278°	44.376667°
Mahzam & Al-Alam Area	IQ047	Summer '09, winter '10, & spring '11	Mahzam & Al-Alam Area	SD3	34.715208°	43.678417°
Mandli	IQ054	Summers '09-'10.	Mandli	DY4	33.764606°	45.552758°
Qadissiya Lake	IQ049	Summer '09 & winter & summer '10	Al Qadissiya or Haditha Dam	AN7	34.3575°	42.073333°

Site	Site Code	Survey Period	Sub-Site Name	Sub-Site Code	N	E
Not KBA		Summer '10 & spring '12	Rutba & Al Massad Gazelle Reserve	AN12	32.912914°	40.223994°
Samara Wet-lands	IQ052	Winters & sum-mers '09-'10	Samarra dam & Wetl&s	SD1	34.1925°	43.852222°
Not KBA		Summer & fall '09, summer '10, & spring '11	Shayat Wetlands	SD10	34.376944°	43.735556°
harthar Lake & Dhebaeji Fields	IQ051	Winters & sum-mers '09-'10, fall '10, spring '11-'12	The Western Edge of Al Thar-thar Lake	AN9	33.698889°	43.304722°
Tharthar Lake & Dhebaeji Fields			Tharthaar Lake & Dhebaeji Field	SD2	34.283889°	43.183056°
Not KBA		Summer '09-'10	Wadi Horan & Al-Hussayniyah	AN4	33.419167°	41.021389°
Southern Iraq Sites						
Not KBA			Msandag Marsh, South	CM19	32.412222°	46.475278°
Not KBA			Msandag	CM20	32.438611°	46.445833°
Not KBA			Haffaar Open-ing 3	HA20	30.939444°	46.946944°
Not KBA			Suweibaat Wet-lands	MT2	30.908889°	45.7325°
Not KBA			Euphrates & Tigris Junction	SA1	30.583056°	47.771667°
Not KBA			Umm Ar-Risaas	SA2	30.448889°	48.091944°
Not KBA			Sayhan	SA3	30.314167°	48.217778°
Not KBA			Chabbaasi	SA5	30.549722°	47.824722°
Auda Marsh	IQ072	W & S '05-'09	Auda Marsh	CM23	31.641389°	46.851389°
Central Marsh	IQ075	W & S '05-'10	Baghdadiya, South	CM1	31.024444°	47.015833°
			Fuhood, North	CM10,HAB_AZ_3	30.986111°	46.725556°
			Abu Zirig	CM16	31.149167°	46.621111°
			Muwayjid Area	CM17	31.083333°	46.634167°
			Abu Zirig, South near Al-Fuhood	CM18	30.993056°	46.711111°
			Hmaara Al-Kabi-ra	CM2	31.133611°	46.989722°
			Hmaara Al-Kabi-ra, East	CM3	31.145556°	46.999722°
			Baghdadiya, West	CM4	31.048611°	47.018056°
			Zichri	CM5	31.055278°	47.221944°
			Subaytiya	CM6	30.993056°	47.220556°
			Khnayziiri	CM7	30.968611°	47.1375°
			Abu An-Narsi	CM8	30.968333°	47.060278°
			Hammaar Town Area	CM9	30.9875°	46.823889°

Site	Site Code	Survey Period	Sub-Site Name	Sub-Site Code	N	E
Dalmaj	IQ064	W & S '05-'11; Add'l material gathered since 1990	Dalmaj Marsh, South	ME10	32.125°	45.451944°
			Dalmaj Marsh, East	ME11	32.174167°	45.643611°
			Dalmaj Marsh, North	ME12	32.3575°	45.258889°
			Basroogiya	ME13	31.985694°	45.592972°
			Dalmaj Marsh, South B	ME2	32.107222°	45.583333°
			Dalmaj Marsh, West	ME3	32.156389°	45.386389°
Not KBA		W & S '10	Kteibaan	BR2	30.708333°	48.027222°
Not KBA		W '05	Hor Uwaina	CM15	31.408333°	46.2°
Not KBA		W & S '05-'06	Hor Sadiya	CM21	32.193056°	46.639722°
Not KBA		W & S '05-'06	Saniya	CM22	31.919444°	46.763889°
Not KBA		W '05	Rayan	CM24	31.578333°	47.033333°
Not KBA		W & S '05	Lafta Marsh	ME1	31.401111°	45.511389°
Not KBA		W '09	Shubaicha Marsh	SM6	33.25°	45.3°
Not KBA		Spr '10	Tell Al-Lahm	TQ2	30.067607°	46.39072°
Fao	IQ082	W & S '05-'10	Ras Al-Beesha	SA4	29.928889°	48.6025°
Gharraf River	IQ065	W & S '05-'06	Gharraf East, near An-Nassir	CM11	31.526111°	46.13°
			Gharraf West, near An-Nassir	CM12	31.537222°	46.16°
			Gharraf, Rifaa'ii Area	CM14	31.660278°	46.108333°
			Gharraf, between An-Nassir & Ar-Rifaa'ii	CM13	31.606944°	46.514167°
Hammar (East)	IQ077	W & S '05-'10 & S '12	Al-Mas'hab	HA14	30.668056°	47.640556°
			Al-Sallal	HA15	30.678889°	47.616389°
			Al-Nagarah	HA16	30.6875°	47.601667°
			Al-Shileichiya	HA17	30.625556°	47.625556°
			Slein (Ghatra)	HA21	30.688056°	47.471111°
			Shaafi	HA26	30.825556°	47.446667°
Hammar (West)	IQ076	W & S '05-'10	Teena, Northern	HA1	30.888611°	46.906667°
			Sinaaf Marsh, Western	HA10	30.731667°	46.585278°
			Shuwaya'riya Area	HA11	30.7825°	46.624444°
			Sinaaf Marsh, Eastern	HA12	30.738056°	46.726389°
			Ibn Maajid Lake	HA13	30.730833°	46.729167°
			Haffaar Opening 1	HA18	30.951389°	46.926111°
			Haffaar Opening 2	HA19	30.936111°	46.970278°
			Teena, Southern	HA2	30.899722°	46.866389°
			Abu Hedeeda	HA22	30.802778°	46.813611°
			Abu-'Ajaj	HA23	30.871667°	46.803056°
			Nuwashi	HA24	30.86°	46.453333°
			Rashid Lake	HA25	30.682778°	46.631111°
			Abu-Ajaj, East	HA27	30.835278°	46.88°

Site	Site Code	Survey Period	Sub-Site Name	Sub-Site Code	N	E
			Buhaira Al-Hilwa	HA3	30.781667°	47.050278°
			Umm At Tiyaar near Al-Buhaira	HA4	30.899722°	46.866389°
			Umm At Tiyaar near Al-Buhaira	HA4	30.899722°	46.866389°
			5th Irrigation Channel/Al-Irwaai' Al-Khaamis	HA5	30.871667°	47.085833°
			Umm Nakhla	HA6	30.821111°	46.642222°
			Khwaysa Area in Al-Kermaashiya Marsh	HA7	30.778056°	46.6575°
			Kermashiya Marsh	HA8	30.798889°	46.623611°
			Hammar Marshes, Southern	HA9	30.738889°	46.537778°
			Ghabishiya	HA28	30.678056°	46.884167°
			Umm An-Ni'aaj	HZ1	31.593056°	47.582222°
			Udheim	HZ2	31.686944°	47.748889°
			Sewaalif	HZ3	31.695556°	47.715278°
			E'jayrda	HZ4	31.331944°	47.630833°
			E'jayrda, East	HZ5	31.327222°	47.630556°
Hawizeh	IQ073	W & S '05-'10	E'jayrda Border Station	HZ6	31.286111°	47.612778°
			E'jayrda, North	HZ7	31.289444°	47.455833°
			Majnoon	HZ8	31.094722°	47.577222°
			Umm Al-Ward Bushes	HZ9	31.568056°	47.501111°
Hindiya Barrage	IQ060	W & S '09-'10	Hindiya Barrage	ME7	32.733889°	44.263889°
Hoshiya & Saaroot	IQ066	W & S '05-'07	Hoshiya	SM1	32.108056°	46.915556°
			Saaroot	SM2	32.347778°	46.798889°
			Saaroot, Northern	SM3	32.478611°	46.731667°
Ibn Najm	IQ063	W & S '05-'10	Ibn Najm	ME4	32.149167°	44.641944°
Jabal Senam	IQ080	W & S '10 (fauna) & Spr '10 (plants)	Jabal Senam	BR1	30.124444°	47.627222°
Jazman (Zurbatia)	IQ057	W & S '09-'10	Jazman (Zurbatia)	WT1	33.147222°	46.0775°
			Khuwaysaat	KZ1	30.309444°	47.828889°
			Khor Az-Zubayr Canal	KZ2	30.088611°	47.951667°
Khor Az-Zubayr	IQ081	W & S '05-'10	Khor Az-Zubayr Canal-100 meters east	KZ3	30.090833°	47.953611°
			Hachaam Area	KZ4	30.041667°	47.964167°
			Khor Az-Zubayr, West	KZ5	30.306944°	47.823611°
			Umm Qasr Port	KZ6	30.062222°	47.939444°

Site	Site Code	Survey Period	Sub-Site Name	Sub-Site Code	N	E
Lehais	IQ079	Spr '10 (plants)	Lehais	BR4	30.605833°	46.529167°
Musayab	IQ059	W '09	Musayab	ME6	32.808056°	44.275556°
North Ibn Najm	IQ062	W & S '09-'10	North Ibn Najm	ME8	32.315278°	44.406944°
Razzaza	IQ058	W & S '08-'10; Spr '10 (plants)	Al-Rahaliya	AN10	32.773889°	43.451667°
			Al-Taar	KR1	32.481944°	43.736667°
			Ein Al-Tamr	KR2	32.549167°	43.503056°
			Razzaza, East	ME5	32.614795°	43.805724°
Salman	IQ078	W & S '10	Salman	MT3	30.416667°	44.5°
Sawa Lake	IQ071	W & S '10	Sawa Lake	MT1	31.313889°	45.003611°
Shuweicha Marsh	IQ061	W & S '09-'10, '10 (Plants)	Shuweicha Marsh	SM7	32.709167°	45.808889°
Sinnaf Seasonal Wetlands	IQ069	W & S '05-'10	Sinnaf Area, Eastern	SM0	31.841000°	47.637222°
			Sinnaf area, Central-south (called Haur Chubaisah by Evans)	SM4	31.871667°	47.324167°
			Sinnaaf Area, Western	SM5	31.880833°	47.215556°
Suwaibaat	IQ074	W & S (fauna); & Spr (plants) '10	Suwaibaat (or Sleibaat)	TQ1	30.806111°	45.966389°
Teeb Oasis & Zubaidaat	IQ067	W & S '09 & '10 (fauna), & S '09/ Spr '10 (plants)	Teeb oasis	MN1	32.388611°	47.341667°
			Zubaidaat	MN2	32.394444°	47.390833°
Teeb Seasonal Wetlands	IQ068	S & W '09-'10 (fauna), & S '09/ Spr '10 (plants)	Teeb Seasonal Wetlands	SM8	32.166663°	47.38333°
Wadi Al-W'eir & Sh'eeb Abu-Talha	IQ070	W & S '10 (fauna), & Spr '10 (plants)	Wadi Al-W'eir	NJ1	31.683889°	44.2925°
			Sh'eeb Abu-Talha	NJ2	31.076389°	44.021944°

APPENDIX 3: BREEDING CODES FOR BIRD OBSERVATIONS DURING THE SUMMER SURVEY

Code	Description
Non-Breeding	
F	Flying over
M	Species observed but suspected to be still on Migration
U	Species observed but suspected to be sUmmering non-breeder
Possible breeder	
H	Species observed in breeding season in suitable nesting Habitat
S	Singing male present in breeding season in suitable breeding habitat
Probable breeding	
P	Pair observed in suitable nesting habitat in breeding season
T	Permanent Territory presumed through registration of territorial behaviour (song etc) on at least two different days a week or more at the same place or many individuals on one day
D	Courtship and Display (judged to be in or near potential breeding habitat; be cautious with wildfowl)
N	Visiting probable Nest site
A	Agitated behavior or anxiety calls from adults, suggesting probable presence of nest or young nearby
I	Brood patch on adult examined in the hand, suggesting Incubation

APPENDIX 4: NON-AVIAN FAUNA REPORTED IN THE KBA ASSESSMENTS

The Iraq checklists currently have nine threatened mammals (2 Endangered, 5 Vulnerable and 2 Near Threatened); four reptiles (1 Endangered, 2 Vulnerable, 1 Near Threatened), and three amphibians (1 Critically Endangered, 1 Vulnerable, 1 Near Threatened). For fish see Appendix 6.

Most of the Iraqi non-avian fauna checklists are not yet complete and the lists provided below only include species seen during the KBA surveys (these are marked with an asterisk (*)) or on other surveys (including camera trapping). Also included in the list are species that were reliably reported as well as unconfirmed reports by locals. Species that were observed and used in the assessments (indicating that a site meets KBA Criteria) are marked with a caret (^). Conservation Concern species are listed with their IUCN Red List status in parenthesis.

KBA MAMMALS LIST

1. *Canis aureus* Golden Jackal. Camera trapped at Qara Dagh in 2011Kurdistan region- 10 sites, Center- five sites, and South- eight sites.

2. *Canis lupus* Grey Wolf. Local reported attacking herds every year. Camera trapped in Hawizeh. Kurdistan region- nine sites, Center- five sites, and South- 12 sites.

3. *Canis lupus arabs* Arabian Wolf. Local reports and based on suitable habitat. Center- one site.

4. *^*Capra aegagrus* Wild Goat. (VU) Kurdistan region- eight sites. Local forestry police reported the sighting of 30 individuals in Bakhma & Bradost Mountain. Commonly seen at Barzan.

5. *Capreolus capreolus* Roe Deer. Seen and photographed by locals. Kurdistan region- two sites.

6. *Caracal caracal* Caracal. Reported killed. Center-one site and South-one site.

7. *Dama mesopotamica* Persian Fallow Deer. (EN). Unconfirmed local report from Ser Amadiya, Kurdistan region.

8. *Felis chaus* Jungle Cat. Sometimes confused with *Felis silvestris*. Kurdistan region- two sites, Center- two sites, and South- nine sites.

9. *Felis silvestris* Wild Cat. Sometimes confused with *Felis chaus*. Camera trapped at Qara Dagh in 2011 and in Hawizeh. Kurdistan- two sites, Center- three sites, and South nine sites.

10. ^*Gazella subgutturosa* Goitered Gazelle. (VU) Local report. Observation was made in 2011 in the Kalar Area. Gazelles used to be found in the lower parts of Wadi Al-W'eir & Sh'eeb Abu-Talha until the last couple of decades but it is not clear whether this population still exists. During the KBA surveys it was found in Kurdistan- four sites, in Center- four sites, and in South- seven sites.

11. **Herpestes edwardsii* Indian Grey Mongoose. Kurdistan- two sites, Center- one site, and South- two sites.

12. **Herpestes javanicus* Small Asian Mongoose. South- one site.

13. *Hyaena hyaena* Striped Hyena. (NT) Kurdistan region- four sites, Center- five sites, and South- eight sites.

14. **Hystrix indica* Indian Crested Porcupine. Camera trapped by team at Qara Dagh and Peramagroon Mountain in 2011. Kurdistan region - four sites, Center- two sites, and South- one site

15. **Lutra lutra* Eurasian Otter. (NT). Local report. Sighted with flashlights at night in Barzan. Tracks found at Fishkhaboor. Kurdistan- seven sites, Center- two sites, and South- two sites

16. ^*Lutrogale perspicillata* Smooth-coated Otter. (VU). Specimen from Taq Taq and Hawizeh DNA tested. The isolated population in the Lower Mesopotamian marshlands is regarded as a separate subspecies (L. p. *maxwelli*) but its status and distribution has been unclear due to confusion with the Eurasian Otter *Lutra lutra*. Recent surveys (Omer et al., 2012, Al-Sheikhly and Nader, 2013) have confirmed the presence of smooth-coated otter in parts of the southern marshes for the first time since the 1950s-1960s. Kurdistan region- one site and South- five sites.

17. *Lynx lynx* Eurasian Lynx. Seen and photographed by a local in Barzan. Local report in Darbandikhan.

18. *Meles canescens* Southwest Asian Badger. Reliable local reports at Peramagroon, Qara Dagh, and Darbandikhan area and photographs of the animal hunted were confiscated from hunters in these areas.

19. *Mellivora capensis* Honey Badger. Center- one site and South- 11 sites.

20. ^*Panthera pardus saxicolor* Persian Leopard. (EN). Local report. Camera trapped at Qara Dagh in 2011. Kurdistan region- six sites.

21. **Sciurus anomalus* Persian Squirrel. Heavily persecuted for the pet trade in Iraq. Kurdistan region- eight sites.

22. **Sus scrofa* Wild Boar. Common; Kurdistan region six sites, central and west Iraq four, and southern Iraq almost all sites.

23. *Ursus arctos syriacus* Syrian Brown Bear. Local reported sighting in 2009 in Assos and forestry police reported the sighting of 2 in Bakhma & Bradost Mountain. Camera trapped by team in 2011. Kurdistan region- nine sites.

24. *Vulpes vulpes* Red Fox. Camera trapped by team at Qara Dagh in 2011. Kurdistan region- two sites, Center- three sites, and South- seven sites.

25. **Vulpes rueppelli* Rüppell's Fox. Center- three sites and South- 11 sites

KBA REPTILES LIST

1. *Acanthodactylus opheodurus*. Arnold's Fringe-fingered Lizard. At Jabal Makhool

2. *Apathya cappadocica urmiana*. Urmia Rock Lizard. At Doli Smaquli, Gali Zanta & Garbeesh, Gara Mountain & Garagun, Peramagroon Mountain.

3. *Asaccus griseonotus* Peramagroon Mountain. Observed on a reptile survey in 2013

4. *Cerastes gasperettii*. Arabian Horned Viper. At Ga'ara,

5. *Dolichophis jugularis*. Large Whip Snake. At Jabal Makhool.

6. *Eirenis collaris*. Collared Dwarf Snake. At Jabal Makhool.

7. *Eumeces schneideri princeps*. Schneideri's Red-marked Skink. At Chami Razan.

8. *Hemidactylus turcicus*. Turkish Gecko. At Tharthar Lake and Al-Dhebaeji Fields.

9. *Lacerta media.* Three-lined Lizard. At Doli Smaquili and Gara Mountain & Garagu.

10. *Paralaudakia caucasia.* At Peramagroon Mountain. Observed on a reptile survey in 2013

11. *Laudakia nupta.* Large-scaled Rock Agama. At Chami Razan, Dukan Lake, Gara Mountain & Garagu, Peramagroon Mountain, and Qara Dagh. A common species in the mountains

12. *Mediodactylus heteropholis.* Iraqi Gecko. Photograph taken during KBA, which was not identified until later hence site of occurrence is not clear.

13. *Mesalina brevirostris.* Blanford's Short-nosed Desert Lizard. At Ga'ara.

14. *Macrovipera lebetina.* Levantine Viper. At Parazan.

15. *Montivipera raddei kurdistanica.* Kurdistan Viper. (NT). At Sakran Mountain, restricted to the mountains of Kurdistan region.

16. *Natrix tessellata.* Dice Snake. A Common Snake

17. *Platyceps najadum.* Dahls Wipe Snake. At Mawat Area.

18. *Ophisops elegans.* Snake Eyed Lizard. At Halgurd Mountain, Mangesh.

19. *Pseudopus apodus.* European Glass Lizard. At Qara Dagh.

20. ^*Rafetus euphraticus.* Euphrates softshell Turtle. At Darbandikhan, Dukan Lake, Central Marshes, Gharraf River.

21. *Saara loricata.* Iraqi Mastigure. At Himreen Lake. CITES Appendix II Species.

22. *Spalerosophis diadema.* Diadem Snake. At Haji Omran, Doli Plngan.

23. ^*Testudo graeca.* Spur-thighed Tortoise. (VU). At Chami Razan, Halgurd Mountain. Would likely meet KBA Vulnerability criteria; requires more study.

24. *Timon princeps kurdistanica* Zagrosian Lizard. At Barzan, Benavi & Sararu, and Dure.

25. *Trachylepis aurata.* Levant Skink. At Jabal Makhool.

26. *Trachylepis aurata transcaucasica.* At Peramagroon Mountain. Observed on a reptile survey in 2013

27. *Trapelus lessonae.* At Peramagroon Mountain. Observed on a reptile survey in 2013

28. *Uromastyx aegyptia.* Egyptian Spiny-tailed Lizard. At Ga'ara, Habbaniyah Lake, Tharthar Lake and Al-Dhebaeji Fields, Salman, Teeb Oasis & Zubaidaat, Wadi Al-W'eir and Sh'eeb Abu-Talha. Would likely meet KBA Vulnerability criteria; requires more study. Also a CITES Appendix II species because of potential declines due to hunting for food and trade.

29. *Varanus griseus.* Desert Monitor. At Dalmaj, Salman, Teeb Oasis & Zubaidaat, Wadi Al-W'eir and Sh'eeb Abu-Talha.

30. *Walterinnesia morgani* Desert Cobra. At Habbaniyah Lake and Jabal Makhool.

KBA AMPHIBIANS LIST

1. ^*Neurergus crocatus* Azerbaijan Newt. (VU). Kurdistan- Gara Mountain & Garagu, Halgurd Mountain & Sakran Mountain.

2. ^*Neurergus derjugini* Kurdistan Newt. (CE). Kurdistan- Ahmad Awa, Hawraman Area, Mawat Area & Penjween.

3. *Salamandra infraimmaculata* Spotted Belly Salamander. (NT). Kurdistan region- Hawraman Area

4. A number of toads and frogs were often heard or directly observed at KBA sites but rarely identified, the exceptions were *Bufotes variabilis* Variable Toad (1 site in Central Iraq), *Hyla savignyi* Lemon-yellow Tree Frog (1 site in Kurdistan region) and *Pelophylax ridibundus* Marsh Frog (a common species, 2 sites in the south).

APPENDIX 5: THE BIRD SPECIES OF IRAQ REPORTED IN THE IBA ASSESSMENTS

For the full checklist of Iraq Birds please consult Salim et al. (2012)

1. **See-see Partridge** *Ammoperdix griseogularis*. Fairly widespread breeding resident in hills and mountains in northern and eastern Iraq. This Irano-Turanian biome-restricted species was recorded breeding at seven sites in northern Iraq during the KBA surveys.

2. **Greater White-fronted Goose** *Anser albifrons*. Winter visitor to northern wetlands and agricultural land, also southern marshes. Recorded at one site during the KBA surveys in numbers that exceeded 1% of the biogeographic population, thus qualifying for inclusion under A4 criteria.

3. **Lesser White-fronted Goose** *Anser erythropus* Vulnerable. Rather local winter visitor in northern Iraq, rare in southern marshes, Recorded at six sites in winter during the KBA surveys.

4. **Red-breasted Goose** *Branta ruficollis*. Endangered. Rare winter visitor to wetlands in northern and southern Iraq. Recorded at one site in northern Iraq and one site in southern Iraq in winter during the KBA surveys.

5. **Common Shelduck** *Tadorna tadorna*. Winter visitor to wetlands throughout Iraq. Recorded at one site in northern Iraq during the KBA surveys in numbers that exceeded 1% of the biogeographic population, thus qualifying for inclusion under A4 criteria.

6. **Ruddy Shelduck** *Tadorna ferruginea*. Local breeding resident in wetlands in northwest and central Iraq; fairly widespread passage migrant and winter visitor. Recorded at one site in northern Iraq during the KBA surveys in numbers that exceeded 1% of the biogeographic population, thus qualifying for inclusion under A4 criteria.

7. **Marbled Duck** *Marmaronetta angustirostris*. Vulnerable. Local breeding resident in wetlands in central Iraq, more widespread in the southern marshes, where wintering population is probably largest in the world. Recorded breeding at 11 sites and 15 wintering sites during the KBA surveys.

8. **Red-crested Pochard** *Netta rufina*. Very local breeding resident in the southern marshes; fairly widespread but generally uncommon passage migrant and winter visitor. Recorded at one site in southern Iraq during the KBA surveys in numbers that exceeded 1% of the biogeographic population, thus qualifying for inclusion under A4 criteria.

9. **White-headed Duck** *Oxyura leucocephala*. Endangered. Rare or uncommon winter visitor to central and southern marshes, Recorded at two sites in winter during the KBA surveys.

10. **Greater Flamingo** *Phoenicopterus roseus*. Passage migrant and winter visitor, especially to southern marshes, where also summers; formerly bred in the southern marshes but apparently not in recent years. Recorded at one site in southern Iraq during the KBA surveys in numbers that exceeded 1% of the biogeographic population, thus qualifying for inclusion under A4 criteria.

11. **Great Cormorant** *Phalacrocorax carbo*. Fairly widespread passage migrant and winter visitor. Recorded at one site during the KBA surveys in numbers that exceeded 1% of the biogeographic population, thus qualifying for inclusion under A4 criteria.

12. **Dalmatian Pelican** *Pelecanus crispus* Vulnerable. Uncommon and probably irregular winter visitor. Recorded at one site in winter during the KBA surveys, but did not trigger inclusion under criterion A1.

13. **Egyptian Vulture** *Neophron percnopterus*. Endangered. Breeding summer visitor to the mountains and rocky hills in northern and western Iraq; also a passage migrant Recorded breeding at 33 sites during the KBA surveys, most in northern Iraq.

14. **Greater Spotted Eagle** *Aquila clanga* Vulnerable. Rather uncommon passage migrant and winter visitor, widespread but most frequent in the southern marshes. Recorded in large numbers at one site in winter in southern Iraq during the KBA surveys, also recorded in 10 sites but did not trigger the inclusion under criterion A1, as the numbers were low. Found below threshold levels at 2 sites.

15. **Eastern Imperial Eagle** *Aquila heliaca* Vulnerable. Uncommon, but fairly widespread passage migrant and winter visitor. Recorded at 15 sites in winter during the KBA surveys, but the numbers seen did not trigger inclusion under criterion A1.

16. **Saker Falcon** *Falco cherrug* Endangered. Apparently a rare resident in northwest Iraq, otherwise an uncommon or rare winter visitor. A single bird recorded in winter at one site during the KBA surveys and thus did not trigger inclusion under criterion A1.

17. **Macqueen's Bustard** *Chlamydotis macqueenii.* Vulnerable. Winter visitor mainly in the dry grasslands and deserts of western and southern Iraq; breeding frequently reported by hunters in southern and western Iraq. Recorded at nine sites in winter in southern Iraq during the KBA surveys.

18. **Eurasian Coot** *Fulica atra.* Breeding resident in very small numbers mainly in the south; widespread passage migrant and winter visitor, especially in southern marshes. Recorded at one site in northern Iraq during the KBA surveys in numbers that exceeded 1% of the biogeographic population, thus qualifying for inclusion under A4 criteria.

19. **Sociable Lapwing** *Vanellus gregarius* Critical Endangered. Rare passage migrant, some may winter, formerly common. Recorded at two sites during the KBA surveys: one bird by satellite signal from a tagged bird in March, the other reported at Al-Asad camp near Baghdadi.

20. **White-tailed Lapwing** *Vanellus leucurus.* Local breeding resident in the wetlands of southern, central and western Iraq; more widespread passage migrant, including to northern Iraq where also found in summer and may breed, This Sahara-Sindian Desert biome-restricted species was recorded breeding at 22 sites in southern and central Iraq during the KBA surveys. Ten of these sites qualified for inclusion under A3 for this biome.

21. **Kentish Plover** *Charadrius alexandrinus.* Breeding resident in central and southern Iraq; passage migrant and winter visitor. Recorded at two sites in southern Iraq during the KBA surveys in numbers that exceeded 1% of the biogeographic population, thus qualifying for inclusion under A4 criteria.

22. **Cream-coloured Courser** *Cursorius cursor.* Breeding resident in the desert and semi-desert areas of northwest, central and southern. This Sahara-Sindian Desert biome-restricted species was recorded breeding at four sites in southern and central Iraq during the KBA surveys, most in southern Iraq.

23. **Collared Pratincole** *Glareola pratincola.* Fairly widespread breeding summer visitor to dry grassland areas near wetlands; passage migrant. Recorded breeding at one site in northern Iraq during the KBA surveys. in numbers that exceeded 1% of the biogeographic population, thus qualifying for inclusion under A4 criteria.

24. **Slender-billed Gull** *Chroicocephalus genei.* Breeding resident in central and southern Iraq; breeding resident or summer visitor to the north; passage migrant and winter visitor. Recorded breeding at six sites in northern and southern Iraq during the KBA surveys, most in northern Iraq in numbers that exceeded 1% of the biogeographic population, thus qualifying for inclusion under A4 criteria.

25. **Whiskered Tern** *Chlidonias hybrida.* Resident and breeding summer visitor in central and southern Iraq, also possibly northeast; fairly widespread passage migrant; winter visitor, but not in north. Recorded breeding at one site in southern Iraq during the KBA surveys in numbers that exceeded 1% of the biogeographic population, thus qualifying for inclusion under A4 criteria.

26. **Spotted Sandgrouse** *Pterocles senegallus.* Very local breeding resident in deserts and semi-deserts areas, This Sahara-Sindian Desert biome-restricted species was recorded breeding at seven sites in southern and western Iraq during the KBA surveys.

27. **Pallid Scops Owl** *Otus brucei.* Local resident or breeding summer visitor. This Sahara-Sindian Desert biome-restricted species was recorded breeding at two sites in southern Iraq during the KBA surveys.

28. **Egyptian Nightjar** *Caprimulgus aegyptius.* Breeding summer visitor to semi-deserts and arid areas of southern and central Iraq, and possibly northeast; passage migrant in south and central Iraq. This Sahara-Sindian Desert biome-restricted species was recorded breeding at six sites in southern Iraq during the KBA surveys.

29. **Masked Shrike** *Lanius nubicus.* Breeding summer visitor to north and parts of central Iraq; passage migrant. This Mediterranean biome-restricted species was recorded breeding at 11 sites in nothern Iraq during the KBA surveys.

30. **Yellow-billed Chough** *Pyrrhocorax graculus.* Very local breeding resident in the high mountains of north Iraq. This Eurasian High-Montane biome-restricted species was recorded breeding at one site in northern Iraq during the KBA surveys.

31. **Brown-necked Raven** *Corvus ruficollis*. Uncommon breeding resident in desert and arid grasslands of southern and western Iraq. This Sahara-Sindian Desert biome-restricted species was recorded at three sites in southern and western Iraq during the KBA surveys.

32. **Hypocolius** *Hypocolius ampelinus*. Breeding summer visitor to southern and central Iraq where especially found in oases, date orchards, *Tamarix* and thorny woodlands, usually near water, This Sahara-Sindian Desert biome-restricted species was recorded breeding at seven sites in southern and central Iraq during the KBA surveys, most in southern Iraq.

33. **Sombre Tit** *Poecile lugubris*. Widespread and not uncommon breeding resident in the northern woodlands, especially *Quercus*. This Mediterranean biome-restricted species was recorded breeding at 11sites in northern Iraq during the KBA surveys.

34. **Greater Hoopoe-Lark** *Alaemon alaudipes*. Widespread breeding resident in the western, central and southern deserts. This Sahara-Sindian Desert biome-restricted species was recorded breeding at six sites in western and southern Iraq during the KBA surveys, most in southern Iraq.

35. **Bar-tailed Lark** *Ammomanes cinctura*. Breeding resident in western deserts. This Sahara-Sindian Desert biome-restricted species was recorded breeding at two sites in southern Iraq during the KBA surveys

36. **Desert Lark** *Ammomanes deserti*. Fairly widespread breeding resident but absent from the extreme north. This Sahara-Sindian Desert biome-restricted species was recorded breeding at six sites in southern and central Iraq during the KBA surveys, most in southern Iraq.

37. **Temminck's Lark** *Eremophila bilopha*. Breeding resident in western and southern deserts and arid grasslands. This Sahara-Sindian Desert biome-restricted species was recorded breeding at three sites in western and southern Iraq during the KBA surveys.

38. **White-eared Bulbul** *Pycnonotus leucotis*. Widespread breeding resident in woodland groves, especially palm, in central, western, southern and northeast Iraq; appears to be spreading north as now found in northern areas, This Sahara-Sindian Desert biome-restricted species was recorded breeding at seven sites in southern and western Iraq during the KBA surveys.

39. **Pale Crag Martin** *Ptyonoprogne obsoleta*. Breeding recorded at two mountain sites in north Iraq; otherwise status uncertain, passage bird recorded in the south, This Sahara-Sindian Desert biome-restricted species was recorded at one site in southern Iraq during the KBA surveys.

40. **Plain Leaf Warbler** *Phylloscopus neglectus*. This Irano-Turanian biome-restricted species, which is a summer visitor to Iraq, was recorded at one site (its only known breeding site) in northern Iraq during the KBA surveys.

41. **Basra Reed Warbler** *Acrocephalus griseldis*.Endangered. Breeding summer visitor to the extensive reedbeds of the southern marshes and recently discovered further north in the marshes of central Iraq and at one site in western Iraq. Endemic, though now recorded breeding outside Iraq in Kuwait and Israel. Recorded breeding at 11 sites during the KBA surveys.

42. **Upcher's Warbler** *Hippolais languida*. Fairly widespread breeding summer visitor to scrub and wooded hills in northern Iraq. This Irano-Turanian biome-restricted species was recorded breeding at three sites in northern Iraq during the KBA surveys

43. **Iraq Babbler** *Turdoides altirostris*. Breeding resident in reedbeds, mainly along the Tigris and Euphrates, and extending its range northwards along the latter. Endemic, though now recorded along the Euphrates in Syria and southern Turkey. Recorded breeding at 21 sites during the KBA surveys. But only 11 sites triggered inclusion under A2 for this restricted-range species.

44. **Ménétries's Warbler** *Sylvia mystacea*. Fairly widespread breeding summer visitor to scrub, especially riverine, in northern Iraq, and locally elsewhere; passage migrant and winter visitor, This Irano- Turanian biome-restricted species was recorded breeding at nine sites in northern and southern Iraq during the KBA surveys, most in northern Iraq.

45. **Western Rock Nuthatch** *Sitta neumayer*. Widespread breeding resident in the mountains and foothills, often with woodland, of northern Iraq. This Mediterranean biome-restricted species was recorded breeding at 11 sites in northern Iraq during the KBA surveys.

46. **Eastern Rock Nuthatch** *Sitta tephronota*. Widespread breeding resident of the rocky hills with open woodland of northern Iraq. This Irano-Turanian biome-restricted species was recorded breeding at ten sites in northern Iraq during the KBA surveys.

47. **White-throated Robin** *Irania gutturalis*. Fairly local breeding summer visitor to open woodlands in hills and mountains in northern Iraq; passage migrant. This Irano-Turanian biome-restricted species was recorded breeding at 10 sites in northern Iraq during the KBA surveys.

48. **Kurdistan Wheatear** *Oenanthe xanthoprymna*. Very local breeding summer visitor to mountain slopes in northeast Iraq; uncommon passage migrant, mainly in north and central Iraq. This Irano-Turanian biome-restricted species was recorded breeding at four sites in northern Iraq during the KBA surveys.

49. **Eastern Black-eared Wheatear** *Oenanthe melanoleuca*. Breeding summer visitor to northern hilly country; widespread passage migrant. This Mediterranean biome-restricted species was recorded breeding at 11 sites in northern Iraq during the KBA surveys

50. **Finsch's Wheatear** *Oenanthe finschii*. Resident breeder in northern foothills winter visitor. This Irano-Turanian biome-restricted species was recorded breeding at nine sites in northern Iraq during the KBA surveys

51. **Hume's Wheatear** *Oenanthe albonigra*. Rare resident in rocky valleys in hills of extreme east. . This Sahara-Sindian Desert biome-restricted species was recorded breeding at one site in southern Iraq during the KBA surveys.

52. **Dead Sea Sparrow** *Passer moabiticus*. Local breeding resident found especially along major watercourses; widespread in winter, This Sahara-Sindian Desert biome-restricted species was recorded breeding at six sites during the KBA surveys, most in southern Iraq.

53. **Pale Rock Sparrow** *Carpospiza brachydactyla*. Local breeding summer visitor in the northern hills; passage migrant. This Irano-Turanian biome-restricted species was recorded breeding at two sites in northern Iraq during the KBA surveys.

54. **White-winged Snowfinch** *Montifringilla nivalis*. Very local breeding resident in high mountains of northern Iraq; winter visitor. This Eurasian High-Montane biome-restricted species was recorded breeding at one site in northern Iraq during the KBA surveys.

55. **Red-fronted Serin** *Serinus pusillus*. Very local breeding resident in the northern mountains; winter visitor to northern Iraq. This Eurasian High-Montane biome-restricted species was recorded breeding at one site in northern Iraq during the KBA surveys.

56. **Desert Finch** *Rhodospiza obsoleta*. Fairly widespread winter visitor; breeding recently proven. This Sahara-Sindian Desert biome-restricted species was recorded at two sites in southern Iraq during the KBA surveys

57. **Grey-necked Bunting** *Emberiza buchanani*. Rare summer visitor to mountains in northeast Iraq, where breeding recorded at one site in 2011. This Irano-Turanian biome-restricted species was recorded at two sites in northern Iraq during the KBA surveys.

58. **Black-headed Bunting** *Emberiza melanocephala*. Widespread breeding summer visitor to open woodlands in northern Iraq; passage migrant. This Mediterranean biome-restricted species was recorded breeding at 11 sites in northern Iraq during the KBA surveys.

APPENDIX 6: LIST OF CONSERVATION CONCERN FRESHWATER & MARINE FISH SPECIES MENTIONED IN THE KBA ASSESSMENTS

Eighty-six fish species (60 freshwater and 26 marine species) were seen in the KBA surveys and by subsequent surveys including those conducted by Dr. Jörg Freyhof of the Leibniz-Institute of Freshwater Ecology and Inland Fisheries in Germany, which included ten threatened fish species (3 Critically Endangered (CR), 5 Vulnerable (VU), and 1 Near Threatened (NT) freshwater and 1 Near Threatened (NT) marine species). Additional effort is needed to complete the Iraqi freshwater and marine fish checklists and recent work indicates that Iraq has 100 freshwater and marine fish species with 4 CR, 2 Endangered (EN), 11 VU, and 10 NT species. In the lists below, species that were observed during the KBA surveys are marked with an asterix (*). Each species is listed with its IUCN Red List status in parenthesis, if known (note that many fish have not been assessed by the IUCN).

FRESHWATER SPECIES

Alien species

1. *Carassius auratus Goldfish (LC). Kurdistan region- 10 sites, Center- two sites, and South- 10 sites
2. *Ctenopharyngodon idella Grass Carp. South- five sites
3. *Cyprinus carpio Wild Common Carp (VU). Kurdistan region- nine sites, Center- two sites and South- 10 sites
4. *Gambusia holbrooki Eastern Mosquitofish (LC). Kurdistan region- five sites and South- one site
5. *Heteropneustes fossilis Stinging Catfish (LC). Kurdistan region- Five sites
6. *Hypophthalmichthys molitrix (NT). Kurdistan region- Five sites
7. *Hypophthalmichthys nobilis Bighead Carp (DD). Kurdistan region- one site
8. *Hemiculter leucisculus Sharpbelly (LC). Found by Dr. Freyhof, 2012- Kurdistan region-Zalm area
9. *Pseudorasbora parva Topmouth Gudgeon (LC). Kurdistan region- Altun Kopri Marsh.
10. *Tilapia zillii (LC). This introduced species is the synonym for Coptodon zillii. It has become increasing reported by fishermen and was found in the south at five sites.

Native species

11. *Acanthobrama marmid Mesopotamian Bream (LC). Kurdistan region region- seven sites, Center- one site, and South- five sites
12. *Acanthopagrus arabicus. Arabian Yellowfin Seabream (LC). Stocked at the site in Razzaza by the government to support fishing in the lake. South-eight sites
13. Alburnoides velioglui Velioglu's Chub. Found by Dr. Freyhof and not mentioned in Coad.
14. *Alburnus caeruleus Black Spotted Bleak (LC). Kurdistan region- eight sites
15. *Alburnus mossulensis Mossul Bleak (LC). Kurdistan region- 19 sites, Center- one site, and South- six sites
16. *Arabibarbus grypus Shabout (VU). Kurdistan region- five sites and South four sites
17. *Leuciscus vorax Mesopotamian Asp (LC). Kurdistan region- five sites, Center- two sites, and South- nine sites
18. *Barbus lacerta Lizard barbel (LC). Kurdistan region- five sites
19. *Bathygobius fuscus Brown Frillfin (LC). Lack of information in Iraq about this species. South- three sites
20. Caecocypris basimi Haditha Cavefish (CR). Cave-dwelling fish Center- one site
21. *Capoeta trutta Longspine Scraper (LC). Kurdistan region- one site
22. *Capoeta damascina Levantine Scraper (LC). Kurdistan region- eight sites
23. Capoeta umbla Tigris Scraper (LC). This species is widespread in rivers
24. *Carasobarbus luteus Mesopotamian Himri (LC). Kurdistan region- four sites, Center- two sites, and South- 10 sites

25. *Chondrostoma regium* Mesopotamian Nase (LC). Kurdistan region- three sites, center- one site

26. *Cobitis taenia* Spined Loach (LC). Status in Iraq is unknown. The name for this species is to be updated to *Cobitis avicennae*. Kurdistan region- one site

27. *Cyprinion kais* Smallmouth Lotak (LC). Kurdistan region- four sites, Center- one site, and South- one site

28. *Cyprinion macrostomum* Largemouth Lotak (LC). Kurdistan region- 14 sites, South- one site

29. *Eidinemacheilus proudlovei*. Kurdistan region – Described from subterranean waters in the Little Zab River Drainage (Freyhof et al, 2016)

30. *Garra elegans* Mesopotamian Garra (LC). Kurdistan region- one site.

31. *Garra rufa* Red Garra (LC). Kurdistan region- 14 sites

32. *Garra widdowsoni* Haditha Cave Garra (CR). A cave-dwelling fish. South- one site

33. *Glyptothorax kurdistanicus* Mesopotamian Sucking Catfish (DD). Kurdistan region- one site

34. *Liza abu* Abu Mullet (LC). Kurdistan region- three sites, Center- three sites, and South- 10 sites. All Liza species will soon be updated to the genus *Planiliza*.

35. *Liza carinata* Keeled Mullet. South- one site. All Liza species will soon be updated to the genus *Planiliza*.

36. *Liza klunzingeri* Klunzinger's Mullet. South- four sites. All Liza species will soon be updated to the genus *Planiliza*.

37. *Liza subviridis*. South- three sites. All Liza species will soon be updated to the genus *Planiliza*.

38. *Liza* spp. Unidentified species- South- one site. All *Liza* species will soon be updated to the genus *Planiliza*.

39. *Luciobarbus esocinus* Pike Barbel (VU). Kurdistan region- seven sites, Center- one site, and South- one site

40. *Luciobarbus subquincunciatus* Leopard Barbel (CR). Kurdistan region- two sites

41. *Luciobarbus xanthopterus* Gattan (VU). Kurdistan region- eight sites, Center- three sites, and South- eight sites

42. *Mastacembelus mastacembelus* Mesopotamian Spiny Eel (LC). Kurdistan region- 11 sites and South- eight sites

43. *Mesopotamichthys sharpeyi* Binni (VU). While this species is present in eight southern sites, two in Central Iraq and one site in Iraqi Kurdistan region, it appears that the only healthy stock of this fish was found in Dalmaj and thus this site might be an important source for the re-introduction of this species into other sites within Iraq.

44. *Mystus pelusius* Zugzug Catfish (LC). Kurdistan region- five sites and South- two sites

45. *Oxynoemacheilus bergianus* Kura Sportive Loach (LC). New Species for Iraq by Jörg Freyhof, Germany. Kurdistan region- two sites

46. *Oxynoemacheilus frenatus* Tigris Loach (LC). Found by Dr. Freyhof. Kurdistan region-Penjween.

47. *Oxynoemacheilus argyrogramma* Two Spot Loach (LC). Found by Dr. Freyhof.

48. *Oxynoemacheilus chomanicus*. This species is not evaluated by IUCN but was found by Dr. Freyhof and is widespread in Lesser Zab River.

49. *Oxynoemacheilus kurdistanicus*. This species is not evaluated by IUCN but was found by Dr. Freyhof and is widespread in Lesser Zab River.

50. *Oxynoemacheilus zagrosensis*. This species is not evaluated by IUCN but was found by Dr. Freyhof and is widespread in Lesser Zab River.

51. *Paracobitis* sp. This species was seen at two sites in Kurdistan (Barzan & Zalm). The identification at Zalm has been recently clarified to be *Paracobitis molavii* and the identification at Barzan has been recently clarified to be *Paracobitis zabgawraensis*; Dr. Freyhof observed both.

52. *Silurus triostegus* Mesopotamian Catfish (LC). Appears to be increasing. This species is not fished commercially (scaleless) but is of ecological importance as a predator fish in the marshes and its conservation status in Iraq is not well understood. Kurdistan Region- 11 sites, Center- two sites, and South- 11 sites.

53. *Squalius berak* Mesopotamian Chub (LC). Found by Dr. Freyhof. Kurdistan region, Zalm River.

54. *Squalius cephalus* Chub (LC). Noted at two sites (Dukan & Penjween) within the Lesser Zab River Basin in Iraqi Kurdistan but a recent IUCN Red List assessment indicates that *S. cephalus* does not occur in the Euphrates/Tigris and therefore additional survey work will likely clarify if there was a potential misidentification at these sites.

55. *Squalius lepidus* Mesopotamian Pike Chub (LC). Kurdistan region- two sites and South- one site

56. *Tenualosa ilisha* Hilsa (LC). South- three sites

57. *Thryssa hamiltonii* Hamilton's Thrysaa. South- one site

Though not listed in the site assessments recent communication with Dr. Jörg Freyhof has alerted us to the addition of three species to the Iraqi checklist (Freyhof et al., 2014; Esmaeili et al., 2014, & Kamangar et al., 2014):

58. *Paracobitis molavii.* New species from the Sirwan and Lesser Zab drainages in the Iranian and Iraqi Tigris catchments.

59. *Paracobitis zabgawraensis.* New species from the Greater Zab drainage in the Iraqi Tigris catchment. Found in the Rezan River in the Barzan area.

60. *Turcinoemacheilus kosswigi* Zagroz Dwarf Loach (LC). Widespread in Iraqi headwaters, recorded from the Sirwan, Greater and Lesser Zab Rivers.

MARINE FISH SPECIES

1. *Ablennes hians* Flat Needlefish (LC). South- two sites

2. *Acanthopagrus berda* Goldsilk Seabream. South- one site

3. *Argyrosomus aeneus.* South- one site

4. *Bothus pantherinus* Leopard Flounder. South- one site

5. *Brachirus orientalis* Oriental Sole. Little information about this species. South- one site

6. *Caranx malabaricus* Malabar Trevalley. Not mentioned in Coad. South- one site

7. *Chirocentrus dorab* Dorab Wolf-herring. South- two sites

8. *Cynoglossus arel* Largescale Tonguesole. South- two sites

9. *Eleutheronema tetradactylum* Fourfinger Threadfin. There is a lack of information in Iraq about this species. South- five site

10. *Epinephelus tauvina* Greasy Grouper (DD). South- one site

11. *Epinephelus spp.* Unidentified species not mentioned in Coad. South- one site

12. *Ilisha elongata.* South- one site

13. *Johnius belangerii* Belanger's Croaker. South- two sites

14. *Lutjanus rivulatus* Blubberlip Snapper. South- one site

15. *Mugil cephalus* Flathead Mullet (LC). South- one site

16. *Nemipterus bleekeri.* There is a lack of information in Iraq on this species. South- four sites

17. *Otolithes ruber* Tigertooth Croaker. South- three sites

18. *Pampus argenteus* Silver Pomfret. South- two sites

19. *Platycephalus indicus* Bartail Flathead (DD). South- one site

20. *Pseudorhombus arsius* Largetooth Flounder. South- one site

21. *Sarda orientalis* Oriental Bonito (LC). South- one site

22. *Sciaenadus sumieri* Sin Croaker. South- one site

23. *Scomberomorous commerson* Narro-barred Spanish Macherel (NT). South- one site

24. *Scomberomorous guttatus* Indo-Pacific King Mackerel (DD). South- one site

25. *Siganus canaliculatus* White-spotted Spinefoot. South- one site

26. *Sillago sihama* Silver Sillago. South- two sites

APPENDIX 7: THE PLANT SPECIES OF IRAQ REPORTED IN THE IPA ASSESSMENTS

#	Species	Endemic	Near endemic	Nationally rare	Locations
1	Acantholimon blackelockii*		✓		Sakran Mountain; Peramagroon; Qara Dagh
2	Acantholimon petraeum*	✓			Peramagroon
3	Alcea arbelensis			✓	Ahmed Awa
4	Alcea sulphurea			✓	Darbandikhan
5	Alchemilla kurdica*	✓			Halgurd
6	Alkanna orientalis			✓	Zalm
7	Allium arlgirdense*		✓		Halgurd
8	Allium calocephalum*	✓			Barzan; Halgurd
9	Allium iranicum*		✓		Haji Omran Mt.
10	Allium vinicolor*		✓		Ga'ara; Haditha & Baghdadi
11	Alyssum penjwinense*	✓			Hawraman; Halgurd; Qara Dagh
12	Alyssum singarense*	✓			Barzan; Sakran Mountain
13	Anogramma leptophylla			✓	Gali Zanta & Garbeesh
14	Anthemis gillettii*		✓		Jazman
15	Anthemis micrantha*	✓			Peramagroon
16	Anthemis plebeia*	✓			Chamchamal
17	Aristolochia paecilantha			✓	De Lezha; Doli Smaquli & Ashab; Peramagroon; Sakran Mountain
18	Asperula asterocephala*	✓			Dohuk lake; Zawita
19	Asperula friabilis*	✓			Qara Dagh
20	Asperula insignis*		✓		Peramagroon
21	Astragalus carduchorum		✓		Hawraman
22	Astragalus caryolobus		✓		Hawraman
23	Astracantha crenophila*		✓		Gara Mountain & Garagu; Haji Omran Mt.; Peramagroon; Hawraman
24	Astragalus dendroproselius*	✓			Barzan
25	Astragalus gudrunensis		✓		Hawraman
26	Astragalus helgurdensis*	✓			Halgurd; Sakran Mountain; Sari Hassan Beg
27	Astragalus lobophorus var. pilosus*	✓			Peramagroon
28	Astragalus octopus		✓		Hawraman
29	Astracantha peristerea*	✓			Peramagroon
30	Astragalus porphyrodon*	✓			Halgurd
31	Astragalus sarae	✓			Hawraman
32	Astragalus tawilicus			✓	Hawraman
33	Astracantha zoharyi*	✓			Peramagroon
34	Astragauls globiflorus		✓		Hawraman
35	Asyneuma amplexicaule			✓	Gara Mountain & Garagu
36	Bellavalia kurdistanica		✓		Benavi; Haji Omran Mt.; Sargalu
37	Bellevalia parva	✓			Bakhma & Bradost
38	Briza minor			✓	Gara Mountain & Garagu; De Lezha

*Plant is recorded historically at the site.

#	Species	Endemic	Near endemic	Nationally rare	Locations
39	*Bromus brachystachys*			✓	Barzan; Gali Zanta & Garbeesh; Qara Dagh
40	*Buffonia calycina*			✓	Hawraman
41	*Bunium avromanum**	✓			Barzan; Hawraman
42	*Bunium cornigerum**		✓		Halgurd; Peramagroon; Qara Dagh; Sakran Mountain
43	*Bupleurum leucocladum**		✓		Jazman
44	*Camelinopsis kurdica**	✓			Peramagroon
45	*Campanula acutiloba**		✓		Peramagroon; Qara Dagh
46	*Campanula radula**		✓		Gara Mountain and Garagu
47	*Carex iraqensis**		✓		Sakran Mountain; Halgurd
48	*Centaurea elegantissima**	✓			Sakran Mountain
49	*Centaurea gigantea**		✓		Dukan Lake; Gara Mountain & Garagu; Qara Dagh; Sakran Mountain
50	*Centaurea gudrunensis**	✓			Peramagroon
51	*Centaurea imperialis**		✓		Hawraman
52	*Centaurea irritans**		✓		Qara Dagh
53	*Centaurea koeieana**		✓		Hawraman
54	*Centaurea longipedunculata**	✓			Dure; Gara Mountain and Garagu
55	*Centaurea urvillei subsp. Deinacantha**		✓		Barzan; Halgurd; Peramagroon
56	*Colpodium gillettii**		✓		Halgurd
57	*Cousinia acanthophysa**	✓			Barzan; Gali Zanta and Garbeesh
58	*Cousinia algurdina**		✓		Halgurd
59	*Cousinia carduchorum**	✓			Haji Omran Mt.; Halgurd; Sakran Mountain
60	*Cousinia cymbolepis**			✓	Gara Mountain & Garagu
61	*Cousinia haussknechtii**		✓		Peramagroon
62	*Cousinia inflata**		✓		Ahmed Awa; Hawraman; Peramagroon
63	*Cousinia kopi-karadaghensis**		✓		Darbandikhan; Qara Dagh
64	*Cousinia kurdica**	✓			Sakran Mountain
65	*Cousinia leptolepis**		✓		Sakran Mountain; Halgurd; Hawraman
66	*Cousinia macrolepis**	✓			Peramagroon
67	*Cousinia masu-shirinensis**	✓			Barzan; Peramagroon; Sari Hassan Beg
68	*Cousinia odontolepis*		✓		Peramagroon
69	*Cousinia straussii**		✓		Peramagroon
70	*Delphinium kurdicum*			✓	Mangesh
71	*Delphinium micranthum**	✓			Gara Mountain & Garagu; Halgurd; Sakran; Hawraman
72	*Delphinium pallidiflorum*			✓	Hawraman
73	*Dianthus asperula*			✓	Dure
74	*Dianthus bassianicus*		✓		Peramagroon
75	*Dianthus floribundus*			✓	Sangaw
76	*Dionysia bornmuelleri*			✓	Ahmed Awa; Rawanduz Gorge

#	Species	Endemic	Near endemic	Nationally rare	Locations
76	Dionysia bornmuelleri			✓	Ahmed Awa; Rawanduz Gorge
77	Dipsacus laciniatus			✓	Darbandikhan; Zalm
78	Echinops armatus var. papillosus*		✓		Jazman
79	Echinops inermis*		✓		Qara Dagh; Hawraman
80	Echinops nitens*	✓			Barzan; Rawanduz Gorge
81	Echinops parviflorus*		✓		Peramagroon
82	Echinops rectangularis*	✓			Haji Omran Mt.
83	Echinops tenuisectus*		✓		Jazman
84	Equisetum arvense			✓	Sharbazher
85	Eremurus rechingeri*	✓			Teeb Oasis & Zubaidat
86	Ergocarpon cryptanthum		✓		Mandali
87	Erysimum boissieri*	✓			Haji Omran Mt.; Qara Dagh; Zawita; Hawraman
88	Erysimum kurdicum		✓		Peramagroon
89	Ferula shehbaziana*	✓			Ahmed Awa
90	Ferulago bracteata*		✓		Hawraman; Qara Dagh
91	Fibigia suffruticose			✓	Awesar; Hawraman; Sakran Mountain
92	Fritillaria crassifolia subsp. Poluninii*	✓			Peramagroon
93	Galium hainesii*	✓			Qara Dagh
94	Galium kurdicum			✓	Hawraman
95	Galium qaradaghense*	✓			Qara Dagh
96	Gladiolus kotschyanus			✓	Dure
97	Glaucium corniculatum			✓	Chamchamal
98	Globularia sintenisii*		✓		Gara Mountain and Garagu
99	Grammosciadium cornutum*		✓		Sur Amadia & Sulav Resort
100	Gypsophila sarbaghiae*	✓			Ahmed Awa
101	Hesperis kurdica var unguiculata	✓			Assos Mountain
102	Hesperis novakii			✓	Awesar; Hawraman
103	Hesperis straussii			✓	Assos Mountain
104	Himantoglossum hircinum	✓			Peramagroon
105	Hymenocrater longiflorus*		✓		Awesar; Hawraman
106	Iris barnumae			✓	Haji Omran Mt.
107	Iris germanica			✓	Awesar; Hawraman
108	Juncus effuses			✓	Assos Mountain; Chamchamal; Darbandikhan; Doli Plngan; Dukan Lake; Penjween; Sharbazher; Zalm
109	Lactuca hispida			✓	Peramagroon
110	Leutea rechingeri*	✓			Peramagroon; Sakran Mountain; Halgurd
111	Linum mucronatum subsp. Pubifolium*	✓			Zawita
112	Linum velutinum*	✓			Dure; Gara Mountain & Garagu; Sur Amadia & Sulav Resort
113	Malabaila secacul subsp. Aucheri*		✓		Qara Dagh; Jazman; Hawraman

#	Species	Endemic	Near endemic	Nationally rare	Locations
114	*Minuartia sublineata*			✓	Hawraman
115	*Muscari tenuiflorum*			✓	Darbandikhan
116	*Nepeta elymaitica**		✓		Halgurd; Halgurd
117	*Nepeta wettsteinii**		✓		Sari Hassan Beg
118	*Onopordum canum*		✓		Haditha Wetland & Baghdadi
119	*Onosma albo-roseum var. macrocalycinum**	✓			Awesar; Chami Razan; Darbandikhan; Doli Smaquli & Ashab; Gara Mountain & Garagu; Hawraman; Mangesh; Qara Dagh; Sakran; Amedi & Sulav; Zawita
120	*Onosma cardiostegium**		✓		Ahmed Awa; Hawraman
121	*Onosma cornuta**	✓			Halgurd
122	*Onosma hawramanensis**	✓			Ahmed Awa; Hawraman
123	*Onosma macrophyllum var. angustifolium*			✓	Ahmed Awa
124	*Orchis collina*			✓	Awesar; Hawraman
125	*Orchis tridentate*			✓	Awesar; Hawraman
126	*Ornithogalum iraqense**	✓			Barzan; Darbandikhan; Dukan Lake; Haji Omran Mt.; Himreen Lake
127	*Ornithogalum kurdicum**	✓			Qara Dagh
128	*Pelargonium quercetorum**		✓		Barzan
129	*Phelypaea coccinea*			✓	Peramagroon
130	*Phlomis kurdica*			✓	Ahmed Awa; Hawraman
131	*Picris strigosa subsp. kurdica**		✓		Gara Mountain and Garagu; Hawraman
132	*Primula auriculata*			✓	Halgurd
133	*Quercus macranthera*			✓	Assos Mountain; Dure; Gara Mountain & Garagu; Mangesh; Peramagroon
134	*Ranunculus bulbilliferus*			✓	Halgurd
135	*Ranunculus sphaerospermus*			✓	Penjween
136	*Rhus coriaria*			✓	Assos Mountain
137	*Rhynochorys elephas subsp. carduchorum*	✓			Halgurd
138	*Rosularia rechingeri**		✓		Halgurd; Sakran
139	*Rubus caesius*			✓	Gara Mountain & Garagu
140	*Salix babylonica*			✓	Dukan Lake
141	*Salvia kurdica**		✓		Amedi & Sulav
142	*Satureja metastashiantha**	✓			Gara Mountain & Garagu; Halgurd
143	*Korshinskia assyriaca**		✓		Barzan; Qara Dagh
144	*Scilla kurdistanica**	✓			Qara Dagh; Sakran
145	*Scilla siberica*			✓	Halgurd
146	*Scorzonera luristanica**		✓		Haji Omran Mt.
147	*Scrophularia atroglandulosa**	✓			Haji Omran Mt.; Halgurd
148	*Scrophularia gracilis**	✓			Halgurd
149	*Scrophularia kurdica* subsp. kurdica*	✓			Peramagroon; Hawraman
150	*Scrophularia sulaimanica**	✓			Hawraman

#	Species	Endemic	Near endemic	Nationally rare	Locations
151	Serratula grandifolia*		✓		Qara Dagh
152	Sideritis libanotica			✓	Atrush and Bania
153	Silene araratica			✓	Ahmed Awa; Hawraman
154	Silene avramana			✓	Ahmed Awa; Hawraman
155	Sparganium erectum			✓	Darbandikhan; Zalm
156	Stachys kotschyii*		✓		Gara Mountain & Garagu; Peramagroon; Qara Dagh; Amedi & Sulav
157	Stachys kurdica			✓	Ahmed Awa; Hawraman
158	Stipa kurdistanica			✓	Hawraman
159	Tamarix brachystachys			✓	Penjween
160	Teucrium melissoides			✓	Ahmed Awa; Hawraman; Zalm
161	Thymus neurophyllus*	✓			Chamchamal
162	Trachomitum venetum subsp. sarmatiense			✓	Atrush and Bania
163	Tragopogon bornmuelleri*		✓		Halgurd; Sakran; Hawraman
164	Tragopogon rechingeri*		✓		Haji Omran Mt.
165	Trigonosciadium viscidulum*		✓		Barzan; Peramagroon; Sakran
166	Tulipa buhseana			✓	Halgurd
167	Tulipa kurdica*	✓			Haji Omran Mt.; Halgurd; Sakran; Sari Hassan Beg
168	Turgenia lisaeoides*	✓			Hawraman
169	Typha lugdunensis			✓	Fishkhaboor
170	Verbascum alceoides*		✓		Dukan Lake; Peramagroon; Qara Dagh
171	Verbascum froedinii*		✓		Barzan
172	Verbascum phyllostachyum*		✓		Hawraman
173	Veronica beccabunga			✓	Halgurd
174	Veronica davisii*		✓		Halgurd; Sakran
175	Veronica macrostachya var. schizostegia*		✓		Barzan; Peramagroon; Hawraman
176	Viola pachyrrhiza*		✓		Gara Mountain and Garagu
177	Vitis hissarica subsp. Rechingeri*	✓			Hawraman; Halgurd
178	Zeugandra iranica			✓	Assos Mountain; Dukan Lake
179	Ziziphora clinopodioides subsp. Kurdica*		✓		Gara Mountain & Garagu; Peramagroon, Sakran; Hawraman
180	Zoegea crinita Boiss. subsp. crinite		✓		Hawraman
Total		58	66	56	

APPENDIX 8: THREAT ASSESSMENT TABLE

The following table provides the scoring at each site and each threat based on Timing (Tm), Scope (Sc) and Severity (Sv). Each of these was scored on a scale from 0 to 3 and added together to obtain an Impact Score (see Methodology). The higher the impact score, the higher the threat level.

Impact Score (0 to 9) = Tm Score + Sc Score + Sv Score

Where: Tm = Timing of threat (0 to 3)
Sc = Scope of threat (0 to 3)
Sv = Severity of threat (0 to 3)

Site #	1. Agriculture			2. Residential & comm. develop.			3. Energy Production & mining			4. Trans. & service corridors			5. Over-exploitation, persecution & control			6. Human intrusions & disturbance			7. Natural systems mod.			8. Pollution		
#	Tm	Sc	Sv	Tm	Sc	Sv	Tm	Sc	Sv	Tm	Sc	Sv	Tm	Sc	Sv	Tm	Sc	Sv	Tm	Sc	Sv	Tm	Sc	Sv
1	3	1	1	3	1	1	3	2	3	3	2	3	3	2	1	3	2	2	3	2	3	3	1	2
2	3	2	2	3	0	0	3	0	0	3	1	2	3	1	1	3	1	2	3	0	0	3	0	0
3	3	1	1	3	0	1	3	0	1	3	0	1	3	1	1	3	2	2	3	0	0	3	1	1
4	3	1	2	3	1	2	3	2	3	3	2	2	3	1	1	3	2	2	3	2	2	3	1	2
5	3	2	2	3	1	1	3	0	0	3	0	0	3	1	1	3	2	3	3	0	0	3	0	0
6	3	2	2	3	1	1	0	0	0	3	1	1	3	1	1	3	1	2	3	2	2	3	1	1
7	3	1	1	3	1	1	3	0	0	3	1	2	3	1	2	3	2	3	3	1	2	3	2	2
8	3	1	1	3	1	1	0	0	0	3	2	2	3	1	1	3	3	3	3	2	2	3	2	2
9	3	2	2	3	1	1	3	2	2	3	2	1	3	1	2	3	1	2	3	2	3	3	1	2
10	3	1	2	3	0	0	3	0	0	3	1	1	3	0	0	3	2	3	3	2	2	3	2	3
11	3	1	1	3	1	1	3	0	0	3	1	1	3	1	1	3	1	2	3	0	0	3	1	2
12	3	1	2	3	1	1	2	0	0	3	1	1	3	1	1	3	1	1	3	2	2	3	1	2
13	3	2	2	3	1	1	3	1	1	3	1	1	3	1	1	3	1	2	3	2	2	3	1	2
14	3	1	2	3	0	0	3	1	2	3	1	1	3	1	1	3	1	1	3	2	3	3	0	0
15	3	0	0	3	3	3	3	0	0	3	3	3	3	0	0	3	3	3	3	3	2	3	2	3
16	3	1	2	3	0	0	3	0	0	3	0	0	3	1	1	3	1	2	3	0	0	3	1	2
17	3	0	0	3	0	0	0	0	0	3	0	0	3	1	2	3	2	3	2	0	0	2	0	0
18	3	1	2	3	1	2	3	1	1	3	1	2	3	1	2	3	1	2	3	1	1	3	1	2
19	3	1	1	3	1	1	0	0	0	2	0	1	3	1	2	3	2	2	3	1	1	3	1	1
20	3	2	3	3	1	1	3	2	3	3	1	2	3	1	1	3	1	2	3	1	2	3	1	2
21	3	1	2	3	1	2	3	1	1	3	1	2	3	1	1	3	1	2	3	1	1	3	1	1
22	3	1	2	3	1	1	3	1	1	3	1	1	3	1	2	3	1	2	3	1	1	3	1	2
23	3	3	3	3	2	3	3	2	2	3	2	2	3	2	2	3	3	3	3	2	3	3	3	3
24	3	1	1	3	1	2	3	2	3	3	1	2	3	2	3	3	1	2	3	2	3	3	1	1
25	3	1	2	3	1	1	3	2	3	3	0	0	3	0	0	3	0	0	3	0	0	3	1	1
26	3	1	1	3	1	2	0	0	0	3	1	1	3	1	1	3	1	2	3	2	3	3	2	3
27	3	1	1	3	2	3	3	2	2	3	2	2	3	3	3	3	2	2	3	2	2	3	2	2
28	3	1	1	3	2	3	3	1	1	3	2	3	3	2	2	3	2	3	3	0	0	3	1	1
29	3	1	1	3	0	0	3	0	0	3	1	2	3	1	2	3	2	3	3	0	0	3	1	2
30	3	2	3	3	1	1	3	1	1	3	1	1	3	1	2	3	2	3	3	1	1	3	1	1
31	3	2	2	3	0	0	3	2	3	3	1	2	3	1	1	3	1	2	3	0	0	3	1	2

Site #	1. Agriculture			2. Residential & comm. develop.			3. Energy Production & mining			4. Trans. & service corridors			5. Over-exploitation, persecution & control			6. Human intrusions & disturbance			7. Natural systems mod.			8. Pollution		
#	Tm	Sc	Sv	Tm	Sc	Sv	Tm	Sc	Sv	Tm	Sc	Sv	Tm	Sc	Sv	Tm	Sc	Sv	Tm	Sc	Sv	Tm	Sc	Sv
32	3	2	3	3	2	2	3	2	3	3	1	2	3	1	2	3	1	2	3	1	2	3	1	1
33	3	2	2	3	0	0	3	1	1	3	2	2	3	2	2	3	1	2	0	0	0	3	1	2
34	3	1	2	3	1	2	3	1	1	3	1	1	3	1	1	3	2	3	0	0	0	3	1	2
35	3	2	3	3	1	1	3	0	0	3	1	2	3	2	3	3	1	2	3	1	1	3	1	2
36	3	1	1	3	1	0	3	0	0	3	1	2	3	2	3	3	1	1	3	1	1	3	1	2
37	3	1	2	3	1	1	3	1	1	3	1	2	3	2	3	3	2	3	3	0	0	3	2	3
38	3	2	2	3	1	1	3	2	3	3	1	2	3	2	3	3	1	1	3	0	0	3	1	1
39	3	1	2	3	1	1	3	2	2	3	1	1	3	1	2	3	1	2	3	2	2	3	2	2
40	3	2	3	3	1	1	3	3	3	3	1	2	3	3	3	3	3	3	3	2	3	3	3	3
41	3	2	2	3	0	0	3	0	0	3	1	1	3	1	1	3	1	1	3	0	0	3	2	2
42	3	1	1	3	1	2	0	0	0	3	2	3	3	1	2	3	2	3	3	1	2	3	1	2
43	3	0	0	3	2	3	0	0	0	3	1	2	3	1	2	3	1	2	3	1	1	3	1	2
44	3	1	2	3	1	2	3	2	3	3	1	1	3	1	2	3	1	2	3	1	2	3	1	2
45	3	1	2	3	1	1	3	2	3	3	1	2	3	2	3	3	1	2	3	1	1	3	1	1
46	3	1	1	0	0	0	3	2	2	3	1	1	3	2	3	3	1	2	3	1	1	3	2	3
47	3	1	1	3	2	2	3	2	2	3	2	2	3	2	3	3	2	2	3	1	1	2	3	3
48	3	2	3	3	1	1	3	2	3	3	1	1	3	3	3	3	1	1	0	0	0	3	1	1
49	0	0	0	0	0	0	0	0	0	3	1	1	3	1	1	1	1	1	3	2	2	3	2	2
50	3	1	1	3	1	1	0	0	0	3	1	1	3	1	2	3	1	2	0	0	0	3	1	1
51	3	3	3	0	0	0	0	0	0	0	0	0	3	3	3	0	0	0	2	3	3	2	2	3
52	3	2	2	3	1	1	3	1	1	3	1	1	3	3	3	3	1	1	3	3	3	2	3	3
53	3	2	2	3	1	1	0	0	0	3	1	1	3	2	2	3	2	2	3	2	2	3	1	1
54	0	0	0	3	1	1	0	0	0	0	0	0	3	2	2	3	1	1	3	2	3	3	1	1
55	0	0	0	0	0	0	0	0	0	0	0	0	3	1	1	0	0	0	2	1	1	0	0	0
56	3	1	1	3	1	1	0	0	0	0	0	0	3	1	2	3	1	2	0	0	0	3	1	2
57	3	2	2	3	1	1	3	1	2	3	1	2	3	1	2	3	2	3	3	1	2	3	1	1
58	3	1	1	3	1	1	3	1	1	3	1	1	3	2	2	3	3	2	3	3	3	3	2	3
59	3	2	3	3	2	3	0	0	0	3	2	2	3	2	2	3	2	3	0	0	0	3	1	1
60	3	2	1	3	2	2	3	1	0	3	1	2	3	2	2	3	2	3	0	0	0	3	2	2
61	3	1	1	0	0	0	3	1	2	3	1	2	3	1	2	3	1	1	3	3	3	3	1	1
62	3	2	3	2	2	2	3	1	1	3	1	2	3	2	3	3	2	3	3	3	3	3	2	2
63	3	2	3	3	1	2	0	0	0	3	1	2	3	2	3	3	3	3	3	3	3	3	2	2
64	3	2	2	3	1	2	3	1	1	3	1	2	3	2	3	3	3	3	3	2	3	3	1	1
65	3	2	3	3	1	2	0	0	0	3	1	2	3	2	2	3	2	3	3	1	2	3	2	2
66	3	1	1	0	0	0	0	0	0	3	1	1	3	1	2	3	1	2	3	2	2	3	1	1
67	3	1	1	3	1	1	3	1	2	3	1	2	3	2	3	3	2	2	3	1	1	3	1	2
68	3	2	2	3	1	1	3	2	3	3	1	2	3	2	3	3	1	2	3	2	2	3	1	1
69	3	1	1	3	1	2	3	2	3	3	1	2	3	1	2	3	2	3	3	3	3	3	2	2
70	3	1	2	3	0	0	3	1	1	3	1	1	3	1	2	3	2	2	0	0	0	3	1	1

Site #	1. Agriculture			2. Residential & comm. develop.			3. Energy Production & mining			4. Trans. & service corridors			5. Over-ex-ploitation, persecution & control			6. Human intrusions & disturbance			7. Natural systems mod.			8. Pollution		
#	Tm	Sc	Sv	Tm	Sc	Sv	Tm	Sc	Sv	Tm	Sc	Sv	Tm	Sc	Sv	Tm	Sc	Sv	Tm	Sc	Sv	Tm	Sc	Sv
71	3	1	1	3	1	2	3	1	1	3	1	1	3	2	3	3	2	3	0	0	0	3	1	1
72	3	2	2	1	1	0	0	0	0	3	1	1	3	2	2	3	2	2	3	3	3	3	1	1
73	2	1	1	3	1	2	3	2	3	2	1	1	3	2	2	3	2	3	3	2	3	3	2	1
74	3	1	1	0	0	0	3	2	2	1	0	0	3	1	2	3	1	2	3	1	1	3	1	1
75	3	1	1	3	2	1	3	2	1	2	2	1	3	3	3	3	3	3	3	3	3	3	2	2
76	3	1	1	3	1	2	3	2	3	3	1	1	3	3	3	3	1	2	3	2	3	3	1	2
77	3	1	1	3	1	2	3	2	3	3	1	1	3	3	3	3	3	3	3	2	3	3	1	2
78	0	0	0	0	0	0	0	0	0	3	1	2	3	2	3	3	1	2	0	0	0	0	0	0
79	3	1	2	0	0	0	3	2	2	3	1	1	3	1	2	3	2	2	0	0	0	3	1	1
80	3	2	3	3	2	2	3	2	3	3	1	1	3	2	2	3	2	3	0	0	0	3	2	1
81	3	1	1	3	1	1	3	1	2	3	2	2	3	1	2	3	2	3	3	1	2	3	2	3
82	3	1	1	3	2	3	3	1	2	3	1	2	3	1	2	3	1	2	3	2	3	3	2	2

INDEX OF SITES

#	Site Name	عربي	كردي	Site Code	Pg. No.
32	Parazan	بارازان	پارەزان	IQ034	155
33	Penjween	بنجوين	پێنجوین	IQ032	152
34	Peramagroon Mountain	جبل بيرامكرون	چیای پیرەمەگروون	IQ027	142
35	Qara Dagh	قرداغ	قەرەداغ	IQ039	163
36	Rawanduz Gorge	وادي رواندوز النهري	رەواندوز	IQ015	120
37	Sakran Mountain	جبل سكران	چیای سەکران	IQ019	126
38	Sangaw Area	منطقة سنكاو	سەنگاو	IQ038	161
39	Sargalu	سركلو	سەرگەڵوو	IQ028	145
40	Sari Hassan Bag Mountain	سري حسن بك	سەری حەسەن بەگ	IQ016	122
41	Amedi & Sulav	العمادية و سولاف	ئامێدی و سۆلاف	IQ008	108
42	Sharbazher Area	شاربجير	شارباژێر	IQ031	150
43	Taq Taq	طق طق	تەق تەق	IQ025	138
44	Zalm	زلم	زەلم	IQ041	168
45	Zawita	زاويته	زاویته	IQ011	112
46	Ajeel Himreen Hills	تلال عجيل حمرين		IQ048	181
47	Ga'ara	كعره		IQ055	193
48	Habbaniyah Lake	بحيرة الحبانية		IQ056	195
49	Haditha Wetlands & Baghdadi	اهوار حديثة والبغدادي		IQ050	184
50	Himreen Lake	بحيرة حمرين		IQ053	190
51	Jabal Makhool	جبل مكحول		IQ046	177
52	Mahzam and Al-Alam	محزم والعلم		IQ047	179
53	Mandli	مندلي		IQ054	192
54	Qadissiya Lake	بحيرة القادسية		IQ049	182
55	Samarra Wetlands	اهوار سامراء		IQ052	188
56	Tharthar Lake & Dhebaeji Field	بحيرة الثرثار وحقل الدبيجي		IQ051	186
57	Auda Marshes	هور عوده		IQ072	229
58	Central Marshes	الاهوار الوسطى		IQ075	236
59	Dalmaj Marsh	هور الدلمج		IQ064	211
60	Fao	الفاو		IQ082	251
61	Gharraf River	نهر الغراف		IQ065	214
62	Hammar, East	الحمار الشرقي		IQ077	242
63	Hammer, West	الحمار الغربي		IQ076	239
64	Hawizeh Marshes	هور الحويزه		IQ073	231
65	Hindiya Barrage	سدة الهندية		IQ060	203
66	Hoshiya and Saaroot	حوشية و صاروط		IQ066	217
67	Ibn Najm	ابن نجم		IQ063	209
68	Ibn Najm, North	ابن نجم الشمالي		IQ062	207
69	Jabal Senam	جبل سنام		IQ080	247
70	Jazman (Zurbatia)	جزمان (زرباطية)		IQ057	197

#	Site Name	عربي	كردي	Site Code	Pg. No.
71	Khor Az-Zubayr Marshes	خور الزبير		IQ081	249
72	Leheis	اللحيس		IQ079	246
73	Musayab	المسيب		IQ059	202
74	Razzaza Lake	بحيرة الرزازة		IQ058	199
75	Salman	السلمان		IQ078	245
76	Sawa Lake and Area	بحيرة ساوه		IQ071	227
77	Shuweicha Marsh	هور الشويجة		IQ061	205
78	Sinnaaf Seasonal Marshes	هور السناف		IQ069	223
79	Suwaibaat (or Sleibaat)	الصويبات (صليبات)		IQ074	234
80	Teeb Oasis & Zubaidaat	واحة الطيب و زبيدات		IQ067	219
81	Teeb Seasonal Wetlands	الطيب		IQ068	221
82	Wadi Al-W'eir & Sh'eeb Abu-Talha	وادي الوعير و شعيب ابوطلحة		IQ070	225